WIRELESS MESH NETWORKS

Edited by **Nobuo Funabiki**

Wireless Mesh Networks
Edited by Nobuo Funabiki

Published by InTech
Janeza Trdine 9, 51000 Rijeka, Croatia

Publishing Process Manager Iva Lipovic

Technical Editor Teodora Smiljanic

Cover Designer InTech Design Team

Additional hard copies can be obtained from orders@intechopen.com

Wireless Mesh Networks, Edited by Nobuo Funabiki
p. cm.
ISBN 978-953-307-519-8

For EU product safety concerns:
IN TECH d.o.o., Prolaz Marije Krucifikse Kozulić 3, 51000 Rijeka, Croatia,
info@intechopen.com or visit our website at intechopen.com.

Meet the editor

Nobuo Funabiki received the B.S. and PhD degrees in mathematical engineering and information physics from the University of Tokyo, Japan, in 1984 and 1993 respectively. He received the M.S. degree in electrical engineering from the Case Western Reserve University, USA in 1991. From 1984 until 1994 he was at the System Engineering Division, Sumitomo Metal Industries, Ltd., Japan. In 1994 he joined the Department of Information and Computer Sciences at Osaka University, Japan as an assistant professor and became an associate professor in 1995. He stayed at the University of California, Santa Barbara from 2000 until 2001 as a visiting researcher. In 2001 he moved to the Department of Communication Network Engineering at Okayama University as a professor. His research interests include computer networks, optimization algorithms, educational technology and Web technology. He is a member of IEEE, IEICE, and IPSJ. He was a vice chairman at IEEE Hiroshima section in 2009 and 2010. He served as a member of technical program committee in more than 10 international conferences. He has published more than 300 papers and coauthored two books.

Contents

Preface

The rapid advancements of low-cost small-size devices for wireless communications with their international standards and broadband backbone networks using optical fibers accelerate the deployment of wireless networks around the world. Using wireless networks people can enjoy network connections without bothering with wire cables between their terminals and connection points to backbone networks. This freedom of wireless connections dramatically increases the number of users of the Internet. Currently, wireless network services have become available at many places and organizations including companies, governments, schools and homes. Actually, wireless network services have been provided even at public spaces such as airports, stations, libraries, hotels and cafes. Through wireless networks people can access various Internet services from any place at any time by using portable computing terminals such as laptop personal computers and smart cellular phones.

The wireless mesh network has emerged as the generalization of the conventional wireless network. In wireless network the connection point or access point is usually connected to the wired network directly, where each user terminal or host is connected to the access point through a wireless link. Thus, the conventional wireless network can provide wireless connection services only to a limited area that can be covered by radio signal from a single access point. On the other hand, wireless mesh network can provide wireless connection services to a wider area by allowing multiple access points to be connected through wireless links. By increasing the number of allocated access points the service area can be flexibly and inexpensively expanded in wireless mesh network. As a result, a number of studies for the progress of wireless mesh network has been reported in literature. Even commercial products of wireless mesh network have appeared.

However, wireless mesh network has several problems to be solved before being deployed as the fundamental network infrastructure for daily use. These problems mainly come from the disadvantages in wireless network when compared to the wired network. They include the short signal propagation range, the limited spectrum assigned for wireless network by the government regulation, the small link bandwidth and the unstable link connection that can be affected even by human movements and weather changes. In designing the architecture, protocols and configurations of wireless network, multiple solutions may exist to solve some of these problems, where the tradeoff such as the cost vs. the performance and the priority vs. the fairness, always happens. Therefore, further great efforts should be made for the advancement of wireless mesh network.

This book is edited to specify some problems that come from the above-mentioned disadvantages in wireless mesh network and give their solutions with challenges. The contents of this book consist of two parts. Part I covers fundamental technical issues in wireless mesh network, including the signal transmission power management scheme, the access point allocation algorithm, the MAC (media access control) protocol design for the location-independent end-to-end delay, the self-adaptive multi-channel MAC protocol, the three-layered routing protocol, and QoS (quality of service) and QoE (quality of experience) considerations in the routing protocol. Part II covers administrative technical issues in wireless mesh network, including the throughput capacity estimation for the scalable wireless mesh network, the performance analysis of wireless mesh network with link failures, the performance evaluation of VoIP (voice over IP) applications in wireless mesh network, the distributed host location management service, security issues with the secure routing protocol, and the wireless network service pricing using the game theory.

This book can be useful as a reference for researchers, engineers, students and educators who have some backgrounds in computer networks and have interest in wireless mesh network. The book is a collective work of excellent contributions by experts in wireless mesh network. I would like to acknowledge their great efforts and precious time spent to complete this book. I would like to express my special gratitude for the support, encouragement and patience of Ms. Iva Lipovic at InTech Open Access Publisher. Finally, I appreciate my family for their constant encouragement, patience and warm hearts to me throughout this work.

Nobuo Funabiki
Okayama University
Japan

Part 1

Fundamental Technical Issues in Wireless Mesh Networks

Optimal Control of Transmission Power Management in Wireless Backbone Mesh Networks

Thomas Otieno Olwal[1,2,3], Karim Djouani[1,2], Barend Jacobus Van Wyk[1],
Yskandar Hamam[1] and Patrick Siarry[2]
[1]*Tshwane University of Technology,*
[2]*University of Paris-Est,*
[3]*Meraka Institute, CSIR,*
[1,3]*South Africa*
[2]*France*

1. Introduction

The remarkable evolution of wireless networks into the next generation to provide ubiquitous and seamless broadband applications has recently triggered the emergence of Wireless Mesh Networks (WMNs). The WMNs comprise stationary Wireless Mesh Routers (WMRs) forming Wireless Backbone Mesh Networks (WBMNs) and mobile Wireless Mesh Clients (WMCs) forming the WMN access. While WMCs are limited in function and radio resources, the WMRs are expected to support heavy duty applications, that is, WMRs have gateway and bridge functions to integrate WMNs with other networks such as the Internet, cellular, IEEE 802.11, IEEE 802.15, IEEE 802.16, sensor networks, et cetera (Akyildiz & Wang, 2009). Consequently, WMRs are constructed from fast switching radios or multiple radio devices operating on multiple frequency channels. WMRs are expected to be self-organized, self-configured and constitute a reliable and robust WBMN which needs to sustain high traffic volumes and long online time. Such complex functional and structural aspects of the WBMNs yield additional challenges in terms of providing quality of services (QoS)(Li et al., 2009). Therefore, *the main objective of this investigation is to develop a decentralized transmission power management (TPM) solution maintained at the Link-Layer (LL) of the protocol stack for the purpose of maximizing the network capacity of WBMNs while minimizing energy consumption and maintaining fault-tolerant network connectivity.*

In order to maximize network capacity, this chapter proposes a scalable singularly-perturbed weakly-coupled TPM which is supported at the LL of the network protocol stack. Firstly, the WMN is divided into sets of unified channel graphs (UCGs). A UCG consists of multiple radios, interconnected to each other via a common wireless medium. A unique frequency channel is then assigned to each UCG. A multi-radio multi-channel (MRMC) node possesses network interface cards (NICs), each tuned to a single UCG during the network operation. Secondly, the TPM problems are modelled as a singular-perturbation of both energy and packet evolutions at the queue system as well as a weak-coupling problem, owing to the interference across adjacent multiple channels. Based on these models, an optimal control problem is formulated for each wireless connection. Thirdly, differential Nash strategies are invoked to solve such a formulation. The optimization operation is implemented by means of an energy-efficient power selection MRMC unification protocol (PMMUP) maintained at the LL. The LL handles packet synchronization, flow control and adaptive channel coding (Iqbal & Khayam, 2009). In addition to these roles, the LL protocol effectively preserves the modularity of cross-layers and provides desirable WMN scalability (Iqbal & Khayam, 2009). Scalable solutions managed by the LL ensure that the network capacity does not degrade with an increase in the number of hops or nodes between the traffic source and destination. This is because the LL is strategically located just right on top of the medium access control (MAC) and just below the network layer. Message interactions across layers do not incur excessive overheads. As a result, dynamic transmission power executions per packet basis are expected to yield optimal power signals. Furthermore, if each node is configured with multiple MACs and radios, then the LL may function as a *virtual* MAC that hides the complexity of multiple lower layers from unified upper layers (Adya et al., 2004).

Finally, analytical results indicate that the optimal TPM resolves WMN capacity problems. Several simulation results demonstrate the efficacy of the proposed solution compared to those of recently studied techniques (Olwal et al., 2010**b**). The work in (Olwal et al., 2010**b**), furnishes an extensive review of the TPM schemes. In this chapter, however, only key contributions related to the MRMC LL schemes are outlined.

2. Related work

In order to make such MRMC configurations work as a single wireless router, a *virtual* medium access control (MAC) protocol is needed on top of the legacy MAC (Akyildiz & Wang, 2009). The virtual MAC should coordinate (unify) the communication in all the radios over multiple non-overlapping channels (Maheshwari et al., 2006). The first Multi-radio unification protocol (MUP) was reported in (Adya et al., 2004). MUP discovers neighbours, selects the network interface card (NIC) with the best channel quality based on the round trip time (RTT) and sends data on a pre-assigned channel. MUP then switches channels after sending the data. However, MUP assumes power unconstrained mesh network scenarios (Li et al., 2009). That is, mesh nodes are plugged into an electric outlet. MUP utilizes only a single selected channel for data transmission and multiple channels for exchanging control packets at high power.

Instead of MUP, this chapter considers an energy-efficient power selection multi-radio multi-channel unification protocol (PMMUP) (Olwal et al., 2009**a**). PMMUP enhances the functionalities of the original MUP. Such enhancements include: an energy-aware efficient power selection capability and the utilization of parallel radios over power controlled non overlapping channels to send data traffic simultaneously. That is, PMMUP resolves the need for a single mesh point (MP) node or wireless mesh router (WMR) to access mesh client network and route the backbone traffic at the same time (Akyildiz & Wang, 2009). Like MUP, the PMMUP requires no additional hardware modification. Thus, the PMMUP complexity is comparable to that of the MUP. PMMUP mainly coordinates local power optimizations at the NICs, while NICs measure local channel conditions (Olwal et al., 2009**b**). Several research papers have demonstrated the significance of the multiple frequency channels in capacity enhancement of wireless networks (Maheshwari et al., 2006; Thomas et al., 2007; Wang et al., 2006; Olwal et al., 2010**b**). While introducing the TPM design in such networks, some solutions have guaranteed spectrum efficiency against multiple interference sources (Thomas et al., 2007; Wang et al., 2006; Muqattash & Krunz, 2005), while some offer topology control mechanisms (Zhu et al., 2008; Li et al., 2008). Indeed, still other solutions have tackled cross-layer resource allocation problems (Merlin et al., 2007; Olwal et al., 2009**a**; 2009**b**).

In the context of interference mitigation, Maheshwari et al. (2006) proposed the use of multiple frequency channels to ensure conflict-free transmissions in a physical neighbourhood so long as pairs of transmitters and receivers can tune to different non-conflicting channels. As a result, two protocols have been developed. The first is called extended receiver directed transmission (xRDT) while the second is termed the local coordination-based multi-channel (LCM) MAC protocol. While the xRDT uses one packet interface and one busy tone interface, the LCM MAC uses a single packet interface only. Through extensive simulations, these protocols yield superior performance relative to the control channel based protocols (Olwal et al., 2010**b**). However, issues of optimal TPM for packet and busy tone exchanges remained untackled. Thomas et al. (2007) have presented a cognitive network approach to achieve the objectives of power and spectrum management. These researchers classified the problem as a two phased non-cooperative game and made use of the properties of potential game theory to ensure the existence of, and convergence to, a desirable Nash Equilibrium. Although this is a multi-objective optimization and the spectrum problem is NP-hard, this selfish cognitive network constructs a topology that minimizes the maximum transmission power while simultaneously using, on average, less than 12% extra spectrum, as compared to the ideal solution.

In order to achieve a desirable capacity and energy-efficiency balance, Wang et al. (2006) considered the joint design of opportunistic spectrum access (i.e., channel assignment) and adaptive power management for MRMC wireless local area networks (WLANs). Their motivation has been the need to improve throughput, delay performance and energy efficiency (Park et al., 2009; Li et al., 2009). In order to meet their objective, Wang et al. (2006) have suggested a power-saving multi-channel MAC (PSM-MMAC) protocol which is capable of reducing the collision probability and the wake state of a node. The design of the PSM-MMAC relied on the estimation of the number of active links, queue lengths and channel conditions during the ad hoc traffic indication message (ATIM) window. In terms of a similar perspective, Muqattash and Krunz (2005) have proposed POWMAC: a single-channel power-control protocol for throughput enhancement. Instead of alternating between the transmission of control (i.e., RTS-CTS) and data packets, as done in the 802.11 scheme (Adya et al., 2004), POWMAC uses an access window (AW) to allow for a series of RTS-CTS exchanges to take place before several concurrent data packet transmissions can commence. The length of the AW is dynamically adjusted, based on localized information, to allow for multiple interference-limited concurrent transmissions to take place in the same vicinity of a receiving terminal. However, it is difficult to implement synchronization between nodes during the access window (AW). POWMAC does not solve the interference problem resulting from a series of RTS-CTS exchanges.

In order to address MRMC topology control issues, Zhu et al. (2008) proposed a distributed topology control (DTC) and the associated inter-layer interfacing architecture for efficient channel-interface resource allocation in the MRMC mesh networks. In DTC, channel and interfaces are allocated dynamically as opposed to the conventional TPMs (Olwal et al., 2010**b**). By dynamically assigning channels to the MRMC radios, the link connectivity, topology, and capacity are changed. The key attributes of the DTC include routing which is agnostic but traffic adaptive, an ability to multiplex channel over multiple interfaces and the fact that it is fairly PHY/MAC layer agnostic. Consequently, the DTC can be integrated with various mesh technologies in order to improve capacity and delay performance over that of single-radio and/or single-channel networks (Olwal et al., 2010**b**). A similar TPM mechanism that solves the strong minimum power topology control problem has been suggested by Li et al. (2008). This scheme adjusts the limited transmission power for each wireless node and finds a power assignment that reserves the strong connectivity and achieves minimum energy costs. In order to solve problems of congestion control, channel allocation and scheduling algorithm for MRMC multi-hop wireless networks, Merlin et al. (2007) formulated the joint problem as a maximization of a utility function of the injected traffic, while guaranteeing stability of queues. However, due to the inherent NP-hardness of the scheduling problem, a centralized heuristic was used to define a lower bound for the performance of the whole optimization algorithm. The drawback is, however, that there are overheads associated with centralized techniques unless a proper TPM scheme is put in place (Akyildiz & Wang, 2009).

In Olwal et al. (2009**a**), an autonomous adaptation of the transmission power for MRMC WMNs was proposed. In order to achieve this goal, a power selection MRMC unification protocol (PMMUP) that coordinates Interaction variables (IV) from different UCGs and Unification variables (UV) from higher layers was then proposed. The PMMUP coordinates autonomous power optimization by the NICs of a MRMC node. This coordination exploits the notion that the transmission power determines the quality of the received signal expressed in terms of signal-to-interference plus ratio (SINR) and the range of a transmission. The said range determines the amount of interference a user creates for others; hence the level of medium access contention. Interference both within a channel or between adjacent channels impacts on the link achievable bandwidth (Olwal et al., 2009**b**).

In conclusion, the TPM, by alternating the dormant state and transmission state of a transceiver, is an effective means to reduce the power consumption significantly. However, most previous studies have emphasized that wake-up and sleep schedule information are distributed across the network. The overhead costs associated with this have not yet been thoroughly investigated. Furthermore, transmission powers for active connections have not been optimally guaranteed. This chapter will consequently investigate the problem of energy-inefficient TPM whereby nodes whose queue loads and battery power levels are below predefined thresholds are allowed to doze or otherwise participate voluntarily in the network. In particular, a TPM scheme based on singular perturbation in which queues on different or same channels evolve at different time-scales compared to the speed of transmission energy depletions at the multiple radios, is proposed (Olwal et al., 2010**a**). The new TPM scheme is also adaptive to the non orthogonal multi-channel problems caused by the diverse wireless channel fading. As a result, this paper provides an optimal control to the TPM problems in backbone MRMC wireless mesh networks (WMNs).

The rest of this chapter is organised as follows: The system model is presented in section 3. Section 4 describes the TPM scheme. In section 5, simulation tests and results are discussed. Section 6 concludes the chapter and furnishes the perspectives of this research.

3. System model

3.1 Unified channel graph model

Consider a wireless MRMC multi-hop WBMN assumed operating under dynamic channel conditions (El-Azouzi & Altman, 2003). Let us assume that the entire WBMN is virtually divided into $\|L\|$ UCGs, each with a unique non-overlapping frequency channel as depicted in Fig. 1. Further, let each UCG comprise $\|V\|=N_V$, NICs or radio devices that connect to each other, possibly via multiple hops (Olwal et al., 2009**a**). These transmit and receive NIC pairs are termed as network users within a UCG. It should further be noted that successful communication is only possible within a common UCG; otherwise inter-channel communication is not feasible. Thus, each multi-radio MP node or WMR is a member of at least one UCG. In practice, the number of NICs at any node, say node A denoted as $\|T_A\|$, is less than the number of UCGs denoted as $\|L_A\|$, associated with that node, i.e., $\|T_A\| < \|L_A\|$. If each UCG set is represented as $l \; \forall l \in L$, then the entire WBMN is viewed by the higher layers of the protocol stack as unions of all UCG sets, that is, $l_1 \cup l_2 \cup l_3 \cup \ldots \cup l_{\|L\|}$. Utilizing the UCG model, transmission power optimization can then be locally performed within each UCG while managed by the Link-Layer (LL). The multi-channel Link state information (LSI) estimates that define the TPM problem are coordinated by the LL (Olwal et al., 2009**b**). Through higher level coordination, independent users are fairly allocated shared memory, central processor and energy resources (Adya et al., 2004).

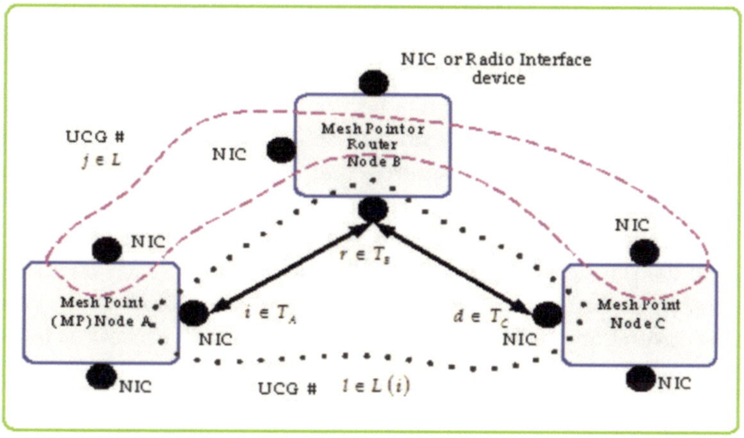

Fig. 1. MRMC multi-hop WBMN

Based on the UCG model depicted in Fig. 1, there exists an established logical topology, where some devices belonging to a certain UCG are *sources* of transmission, say $i \in T_A$ while certain devices act as 'voluntary' *relays*, say $r \in T_B$ to *destinations*, say $d \in T_C$. A sequence of connected *logical links* forms a *route* originating from source i. It should be noted that each asymmetrical physical link may be regarded as a multiple logical link due to the existence of multiple channels. Adjacent channels, actively transmitting packets simultaneously, cause adjacent channel interference (ACI) owing to their close proximity. The ACI can partly be reduced by dynamic channel assignment if implemented without run time overhead costs (Maheshwari et al., 2006). In this chapter, static channel assignment is assumed for every transmission time slot. Such an assumption is reasonable since the transmission power optimization is performed only by actively transmitting radios, to which channels have been assigned by the higher layers of the network protocol stack. It is pointless setting the time-scales for channel assignments to be greater than, or matching that, of power executions since the WMRs are assumed to be stationary. Furthermore, modern WMRs are built on multiple cheap radio devices to simultaneously perform multi-point to multi-point (M2M) communication. Indeed, network accessing and backbone routing functionalities are effective while using separate radios. Each actively transmitting user acquires rights to the medium through a carrier sensed multiple access with collision avoidance (CSMA/CA) mechanism (Muqattash and Krunz, 2005). Such users divide their access time into a transmission power optimization mini-slot time and a data packet transmission mini-slot time interval. For analytical convenience, time slots will be normalized to integer units $t \in \{0,1,2,...\}$ in the rest of the chapter.

3.2 Singularly-perturbed queue system

Suppose that N wireless links, each on a separate channel, emanate from a particular wireless MRMC node. Such links are assumed to contain N queues and consume N times energy associated with that node as illustrated by Fig. 2. It is noted that at the sender (and, respectively, the receiver), packets from a virtual MAC protocol layer termed as the PMMUP (respectively, multiple queues) are striped (respectively, resequenced) into multiple queues (respectively, PMMUP queues) (Olwal et al., 2009**b**; 2010**a**). Queues can be assumed to control the rates of the input packets to the finite-sized buffers. Such admission control mechanisms are activated if the energy residing in the node and the information from the upper layers are known a *priori*. Suppose that during a given time-slot, the application generates packets according to a Bernoulli process. Packets independently arrive at the multiple MAC and PHY queues with probability ϕ , where $\phi > 0$. Buffers' sizes of B packets are assumed. It should be considered that queues are initially nonempty and that new arriving packets are dropped when the queue is full; otherwise packets join the tail of the queue. The speed difference between the queue service rate and the energy level variations in the queue leads to the physical phenomenon called *perturbation*. Based on such perturbations, optimal transmission power is selected to send a serviced packet. It is noted that such a perturbation can conveniently be modelled by the Markov Chain process as follows:

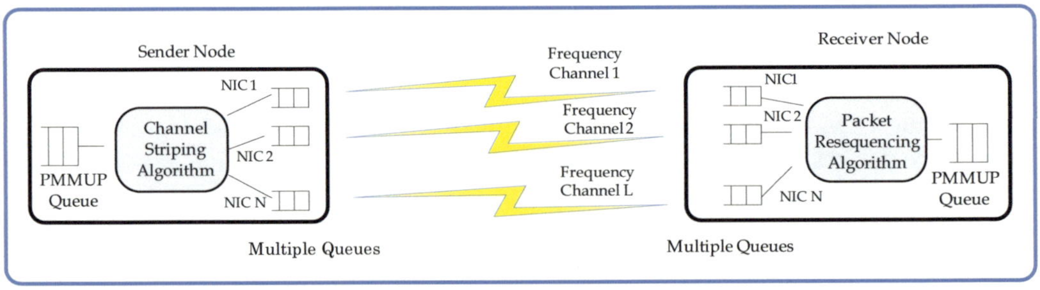

Fig. 2. Multiple queue system for a MRMC router-pair

Denote $i \in \mathrm{E}$, where $\mathrm{E} = \{1,2,...,i,...E\}$, as the energy level available for transmitting a packet over wireless medium by each NIC- pair (user). Denote φ_i , where $\varphi_i \in [0,1]$, as the probability of transmitting a packet with energy level i . The transition probability from energy state $X_n = i$ to state $X_{n+1} = j$ during the time transition $[n,n+1)$ is yielded by $\lambda_{ij} = \Pr(X_{n+1} = j \mid X_n = i)$. Let Λ, be the energy level transition matrix, where $\sum_{j=1}^{E} \lambda_{ij} = 1$ with the probability distribution denoted by $\vartheta = [\vartheta_1, \vartheta_2, ..., \vartheta_E]$ (El-Azouzi & Altman, 2003).

$$\Lambda = \begin{bmatrix} \lambda_{11} & \lambda_{12} & .. & \lambda_{1E} \\ \lambda_{21} & \lambda_{22} & .. & \lambda_{2E} \\ .. & .. & .. & .. \\ \lambda_{E1} & \lambda_{E2} & .. & \lambda_{EE} \end{bmatrix}, \quad (1)$$

It should be recalled that the power optimization phase requires information about the queue load and energy level dynamics. Denote $X(n)=\{X_n(i(n),j(n))\}$ as a two dimensional Markov chain sequence, where $i(n)$ and $j(n)$ are respectively the energy level available for packet transmission and the number of packets in the buffer at the nth time step. Let the packet arrival and the energy-charging/discharging process at each interface in time step $n+1$ be independent of the chain $X(n)$. Arrivals are assumed to occur at the end of the time step so that new arrivals cannot depart in the same time step that they arrive (Olwal et al., 2010**a**). Figure 3 depicts the two dimensional Markov chain evolution diagram with the transition probability matrix, $P_T(n)$, whose elements are $\lambda_{n,n+1}(i,j)$ for all $i=1,2,\ldots,E$ and $j=0,1,2,\ldots,B$. The notation, $\lambda_{n,n+1}(i,j)$ represents the transition probability of the ith energy level and the jth buffer level from state at n to state at $n+1$. In general, similar Markov chain representations can be assumed for other queues in a multi-queue system.

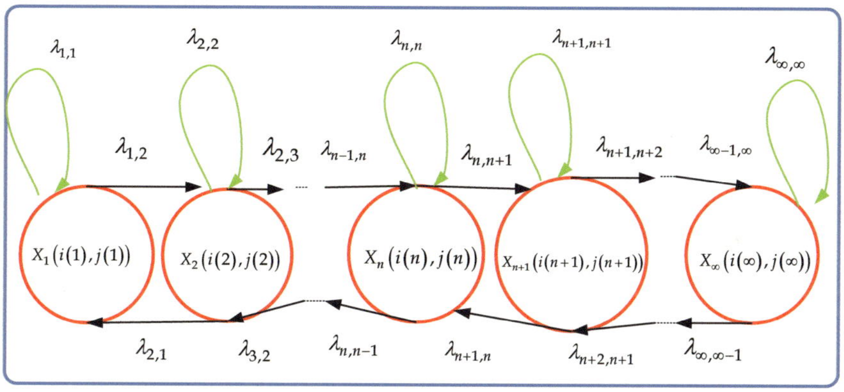

Fig. 3. Markov chain diagram

The transition probability $E(B+1)\times E(B+1)$ matrix of the Markov chain $X(n)$ is yielded by

$$P_T(n) = \begin{pmatrix} \mathbf{B}_0 & \mathbf{B}_1 & 0 & \cdots & \cdots & \cdots \\ \mathbf{A}_2 & \mathbf{A}_1 & \mathbf{A}_0 & 0 & \cdots & \cdots \\ 0 & \mathbf{A}_2 & \mathbf{A}_1 & \mathbf{A}_0 & 0 & \cdots \\ \vdots & \ddots & \ddots & \ddots & \ddots & \vdots \\ \vdots & \ddots & \ddots & \ddots & \mathbf{A}_1 & \mathbf{A}_0 \\ 0 & \cdots & \cdots & 0 & \mathbf{A}_2 & \mathbf{F}_1 \end{pmatrix} \begin{matrix} 0 \\ 1 \\ . \\ . \\ . \\ B \end{matrix}, \quad (2)$$

where $P_T(n)$ consists of $B+1$ block rows and $B+1$ block columns each of size $E\times E$. The matrices $\mathbf{B}_0$, $\mathbf{B}_1$, $\mathbf{A}_0$, $\mathbf{A}_1$, $\mathbf{A}_2$ and $\mathbf{F}_1$ are all $E\times E$ non-negative matrices denoted as $\mathbf{B}_0=\bar{\phi}\Lambda$, $\mathbf{B}_1=\phi\Lambda$, $\mathbf{A}_0=\mathbf{diag}(\phi\bar{\varphi}_i,\ i=1,\ldots,E)\Lambda$, $\mathbf{A}_1=\mathbf{diag}(\phi\varphi_i+\bar{\phi}\bar{\varphi}_i,\ i=1,\ldots,E)\Lambda$,

$\mathbf{A}_2 = \mathbf{diag}\left(\bar{\phi}\varphi_i\ i=1,...,E\right)\Lambda$ and $\mathbf{F}_1 = \mathbf{diag}\left(\phi\varphi_i + \bar{\varphi}_i,\ i=1,...,E\right)\Lambda$. Here $\bar{\phi}=1-\phi$ and $\bar{\varphi}_i = 1-\varphi_i$ respectively denote the probability that no packet arrives in the queue and no packet is transmitted into the channel when the available energy level is i. If one assumes that the energy level transition matrix Λ is irreducible and aperiodic[1] and that $\phi > 0$, then the Markov chain $X(n)$ is aperiodic and contains a single ergodic class[2]. A unique row vector of steady state (or stationary) probability distribution can then be defined as $\pi(i,j) = \lim_{n\to\infty} P_T\left(l(n)=i, b(n)=j\right),\ i=1,2,...,E,\ j=0,1,...,B$ and $\pi(i,j)\in\Re^{1\times i(j+1)} \geq 0$. Let $\pi(i,j,\varepsilon_s),\ i=1,...,i,...,E,\ j=0,1,...j...,B$ be the probability distribution of the state of the available energy and the number of packets in the system in a steady state. Such a probability distribution $\pi(i,j,\varepsilon_s)$ can uniquely be determined by the following system

$$\pi(\varepsilon_s)P_T(\varepsilon_s) = \pi(\varepsilon_s),\ \pi(\varepsilon_s)\mathbf{1} = 1,\ \pi(\varepsilon_s) \geq 0, \tag{3}$$

where ε_s denotes the *singular perturbation* factor depicting the speed ratio between energy and queue state evolutions. The first order Taylor series approximation of the perturbed Markov chain $X(n)$ transition matrix can be represented as $P_T(\varepsilon_s) = Q_0 + \varepsilon_s Q_1$, where Q_0 is the probability transition matrix of the unperturbed Markov chain corresponding to strong interactions while Q_1 is the generator corresponding to the weak interaction (El-Azouzi & Altman, 2003); that is,

$$Q_0 = \begin{pmatrix} \bar{\phi}I & \phi I & 0 & \cdots & \cdots & \cdots \\ \bar{\mathbf{A}}_2 & \bar{\mathbf{A}}_1 & \bar{\mathbf{A}}_0 & 0 & \cdots & \cdots \\ 0 & \bar{\mathbf{A}}_2 & \bar{\mathbf{A}}_1 & \bar{\mathbf{A}}_0 & 0 & \cdots \\ \vdots & \ddots & \ddots & \ddots & \ddots & \vdots \\ \vdots & \ddots & \ddots & \ddots & \ddots & \bar{\mathbf{A}}_0 \\ 0 & \cdots & \cdots & 0 & \bar{\mathbf{A}}_2 & \bar{\mathbf{F}}_1 \end{pmatrix},\ Q_1 = \begin{pmatrix} \tilde{\mathbf{B}}_0 & \tilde{\mathbf{B}}_1 & 0 & \cdots & \cdots & \cdots \\ \tilde{\mathbf{A}}_2 & \tilde{\mathbf{A}}_1 & \tilde{\mathbf{A}}_0 & 0 & \cdots & \cdots \\ 0 & \tilde{\mathbf{A}}_2 & \tilde{\mathbf{A}}_1 & \tilde{\mathbf{A}}_0 & 0 & \vdots \\ \vdots & \ddots & \ddots & \ddots & \ddots & \vdots \\ \vdots & \ddots & \ddots & \ddots & \ddots & \tilde{\mathbf{A}}_0 \\ 0 & \cdots & \cdots & 0 & \tilde{\mathbf{A}}_2 & \tilde{\mathbf{F}}_1 \end{pmatrix}, \tag{4}$$

where

$\bar{\mathbf{A}}_2 = \mathbf{diag}\left(\bar{\phi}\varphi_i, i=1,...,E\right)$, $\bar{\mathbf{A}}_1 = \mathbf{diag}\left(\phi\varphi_i + \bar{\phi}\bar{\varphi}_i, i=1,...,E\right)$, $\bar{\mathbf{A}}_0 = \mathbf{diag}\left(\phi\bar{\varphi}_i,\ i=1,...,E\right)$, $\bar{\mathbf{F}}_1 = \mathbf{diag}\left(\phi\varphi_i + \bar{\varphi}_i,\ i=1,...,E\right)$, $\tilde{\mathbf{B}}_0 = \bar{\phi}(\Lambda_1)$, $\tilde{\mathbf{B}}_1 = \phi(\Lambda_1)$, $\tilde{\mathbf{A}}_2 = \mathbf{diag}\left(\bar{\phi}\varphi_i,\ i=1,...E\right)\Lambda_1$, $\tilde{\mathbf{A}}_1 = \mathbf{diag}\left(\phi\varphi_i\bar{\phi}\bar{\varphi}_i,\ i=1,...,E\right)\Lambda_1$, $\tilde{\mathbf{A}}_0 = \phi\ \mathbf{diag}\left(\phi\bar{\varphi}_i, i=1,...,E\right)\Lambda_1$ and $\tilde{\mathbf{F}}_1 = \mathbf{diag}\left(\phi\varphi_i + \bar{\varphi}_i,\ i=1,...,E\right)\Lambda_1$.

Here,

$$\Lambda(\varepsilon_s) = I + \varepsilon_s\Lambda_1 \tag{5}$$

where Λ_1 is the generator matrix, representing an aggregated Markov chain $X(n)$.

The model in (2) to (5) leaves us with the perturbation problem under the assumption that an ergodic class exists (i.e., has exactly one closed communicating set of states), and Q_0

---

[1] A state evidences *aperiodic* behaviour if any return (returns) to the same state can occur at irregular multiple time steps.

[2] A Markov chain is called *ergodic* or *irreducible* if it is possible to go from every state to every other state.

contains E sub-chains (E ergodic class). The stationary probability $\pi(i,j,\varepsilon_s)$ from (3) of the perturbed Markov chain, therefore, takes a Taylor series expansion

$$\pi(i,j,\varepsilon_s) = \sum_{n=0}^{\infty} \pi^{(n)}(i,j)\varepsilon_s^n , \tag{6}$$

where ε_s^n is the nth order singularly-perturbed parameter. Denote the aggregate Markov chain probability distribution as $\bar{\vartheta} = \left[\bar{\vartheta}_1, \bar{\vartheta}_2, \ldots, \bar{\vartheta}_E\right]$. The unperturbed stationary probability is then yielded by $\pi^{(0)}(i,j) = \bar{\vartheta}_i \nu_{\zeta_i}(j)$ where ν_{ζ_i} is the probability distribution of the recurrent class ζ_i, i.e., $\sum_{j=0}^{B} \zeta_i(j) = 1$.

3.3 Weakly-coupled multi-channel system

Theoretically, simultaneous transmitting links on different orthogonal channels are expected not to conflict with each other. However, wireless links emanating from the same node of a multi-radio system do conflict with each other owing to their close vicinity. The radiated power coupling across multiple channels results in the following: loss in signal strength owing to inter-channel interference; hence packet losses over multi-channel wireless links. Such losses lead to packet retransmissions and hence queue instabilities along a link(s). Retransmissions also cause high energy consumption in the network. Highly energy-depleted networks result in poor network connectivity. Therefore, one can model the wireless cross-channel interference (interaction) as a weakly-coupled system (Olwal et al., 2010**a**). Each transmitter-receiver pair (user) operating on a particular channel (i.e., UCG) adjusts its transmission power dynamically, based on a sufficiently small positive parameter denoted as ε_w.

As an illustration, let us consider a two-dimensional node placement consisting of two co-located orthogonal wireless channels labelled i and j with simultaneous radial transmissions as depicted in Fig. 4. The coupled region is denoted by surface area A_ε. Since power coupling is considered, the weak coupling factor can be derived as a function of the region or surface A_ε, i.e., $O(d_{ij}^2)$, where d_{ij} is the distance between point i and j. From the geometry of Fig. 4, it is easy to demonstrate that the weak coupling parameter yields,

$$\varepsilon_{ij} = \frac{A_{\varepsilon i}}{A_\varepsilon} = \frac{d_i^2\left[\theta_i - \frac{\sin\theta_i}{\sqrt{2}}\right]}{d_i^2\left[\theta_i - \frac{\sin\theta_i}{\sqrt{2}}\right] + d_j^2\left[\theta_j - \frac{\sin\theta_j}{\sqrt{2}}\right]}, \quad \varepsilon_{ji} = \frac{A_{\varepsilon j}}{A_\varepsilon} = \frac{d_j^2\left[\theta_j - \frac{\sin\theta_j}{\sqrt{2}}\right]}{d_i^2\left[\theta_i - \frac{\sin\theta_i}{\sqrt{2}}\right] + d_j^2\left[\theta_j - \frac{\sin\theta_j}{\sqrt{2}}\right]} . \tag{7}$$

Thus, the weakly-coupled scalar is generally a function of the square of the transmission radii (d_i and d_j) and the coupling-sector angles (θ_i and θ_j). The weak coupling parameter is bounded by $0 < \varepsilon_{ij} = \varepsilon_w < 1$. The sectored angle has a bound, $0 \le \theta \le 2\pi$ in radians.

It should be noted that both the singular perturbation and weak coupling models at the multiple MACs and radio interfaces are coordinated by the virtual MAC protocol at the Link Layer. The motivation is to conceal the complexity of multiple lower layers from the higher layers of the protocol stack, without additional hardware modifications.

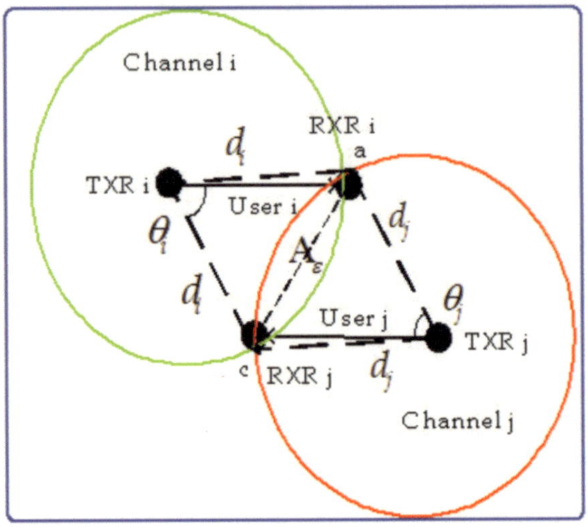

Fig. 4. A weakly-coupled wireless channel dual system of two simultaneously co-located transmitting users i and j described by infinitesimally small radiating points TXR i and RXR i pair, and TXR j and RXR j pair, respectively.

3.3 Optimal problem formulation

For N users at each WMR, the SPWC large-scale linear dynamic system is written as (Gajic & Shen, 1993; Mukaidani, 2009; Sagara et al., 2008),

$$\mathbf{x}_i(t+1) = \mathbf{A}_{ii}(\varepsilon)\mathbf{x}_i(t) + \mathbf{B}_{ii}(\varepsilon)\mathbf{u}_i(t) + \mathbf{W}_{ii}(\varepsilon)\mathbf{w}_i(t) + \sum_{\substack{j=1\\ j\neq i}}^{N} \varepsilon_{ij}\mathbf{A}_{ij}\mathbf{x}_j(t) + \sum_{\substack{j=1\\ j\neq i}}^{N} \varepsilon_{ij}\mathbf{B}_{ij}\mathbf{u}_j(t)$$

$$+ \sum_{\substack{j=1\\ j\neq i}}^{N} \varepsilon_{ij}\mathbf{W}_{ij}\mathbf{w}_j(t),$$

$$\mathbf{y}_i(t) = \mathbf{C}_{ii}(\varepsilon)\mathbf{x}_i(t) + \sum_{\substack{j=1\\ j\neq i}}^{N} \varepsilon_{ij}\mathbf{C}_{ij}\mathbf{x}_j(t) + \mathbf{v}_i(t),\ \mathbf{x}_i(0) = \mathbf{x}_i^0,\ i = 1,\dots,N, \tag{8}$$

where $\mathbf{x}_i \in \Re^{n_i}$ represents the state vector of the ith user, $\mathbf{u}_i \in \Re^{m_i}$ is the control input of the ith user, $\mathbf{w}_i \in \Re^{q_i}$ represents the Gaussian distributed zero mean disturbance noise vector to the ith user, $\mathbf{y}_i \in \Re^{l_i}$ represents the observed output and $\mathbf{v}_i \in \Re^{l_i}$ are the Gaussian distributed zero mean measurement noise vectors. The white noise processes $\mathbf{w}_i \in \Re^{q_i}$ and $\mathbf{v}_i \in \Re^{l_i}$ are independent and mutually uncorrelated with intensities $\Theta_{\mathbf{w}} > 0$ and $\Theta_{\mathbf{v}} > 0$, respectively. The system matrices **A**, **B**, **C** and **W** are defined in the same way as discussed in our recent invesitigation (Olwal et al., 2009**b**).

Let the partitioned matrices for the wireless MRMC node pair with the weak-coupling to the singular-perturbation ratio $0 < \varepsilon = \dfrac{\varepsilon_w}{\varepsilon_s} < \infty$, be defined as follows:

$$\mathbf{A}_{\varepsilon}=\begin{bmatrix}\mathbf{A}_{11}(\varepsilon) & \varepsilon_{12}\mathbf{A}_{12} & \dots & \varepsilon_{1N}\mathbf{A}_{1N}\\ \varepsilon_{21}\mathbf{A}_{21} & \mathbf{A}_{22}(\varepsilon) & \dots & \varepsilon_{2N}\mathbf{A}_{2N}\\ . & . & . & .\\ \varepsilon_{N1}\mathbf{A}_{N1} & \varepsilon_{N2}\mathbf{A}_{N2} & \dots & \mathbf{A}_{NN}(\varepsilon)\end{bmatrix},\ \mathbf{B}_{i\varepsilon}=\begin{bmatrix}\varepsilon^{1-\delta_{1i}}\mathbf{B}_{1i}\\ \varepsilon^{1-\delta_{2i}}\mathbf{B}_{2i}\\ .\\ \varepsilon^{1-\delta_{Ni}}\mathbf{B}_{Ni}\end{bmatrix},\ \delta_{ij}=\begin{cases}0\ (i\neq j)\\ 1\ (i=j)\end{cases},$$

$$\mathbf{W}_{\varepsilon}=\begin{bmatrix}\mathbf{W}_{11}(\varepsilon) & \varepsilon_{12}\mathbf{W}_{12} & \dots & \varepsilon_{1N}\mathbf{W}_{1N}\\ \varepsilon_{21}\mathbf{W}_{21} & \mathbf{W}_{22}(\varepsilon) & \dots & \varepsilon_{2N}\mathbf{W}_{2N}\\ . & . & . & .\\ \varepsilon_{N1}\mathbf{W}_{N1} & \varepsilon_{N2}\mathbf{W}_{N2} & \dots & \mathbf{W}_{NN}(\varepsilon)\end{bmatrix},\ \mathbf{C}_{\varepsilon}=\begin{bmatrix}\mathbf{C}_{11}(\varepsilon) & \varepsilon_{12}\mathbf{C}_{12} & \dots & \varepsilon_{1N}\mathbf{C}_{1N}\\ \varepsilon_{21}\mathbf{C}_{21} & \mathbf{C}_{22}(\varepsilon) & \dots & \varepsilon_{2N}\mathbf{C}_{2N}\\ . & . & . & .\\ \varepsilon_{N1}\mathbf{C}_{N1} & \varepsilon_{N2}\mathbf{C}_{N2} & \dots & \mathbf{C}_{NN}(\varepsilon)\end{bmatrix}. \quad (9)$$

Each strategy user is faced with the *minimization problem* along trajectories of a linear dynamic system in (8),

$$J_i(u_1,\dots,u_N,\mathbf{w},\mathbf{x}(0)) =\frac{1}{2}\mathrm{E}\left\{\lim_{t\to\infty}\frac{1}{t}\sum_{\tau=0}^{t-1}\left[\begin{array}{l}\mathbf{z}^T(\tau)\mathbf{z}(\tau)+\mathbf{u}_i^T(\tau)\mathbf{R}_{ii}\mathbf{u}_i(\tau)\\ +\sum_{\substack{j=1\\ j\neq i}}^{N}\varepsilon_{ij}\mathbf{u}_j^T(\tau)\mathbf{R}_{ij}\mathbf{u}_j(\tau)-\mathbf{w}^T(t)\Theta_{\mathbf{w}i\varepsilon}\mathbf{w}(t)\end{array}\right]\right\}, \quad (10)$$

where $\mathbf{z}\in\Re^{s}$ is the controlled output with dimension equal to s, given by (Gajic & Shen, 1993),

$$\mathbf{z}_i(t)=\mathbf{D}_{ii}(\varepsilon)\mathbf{x}_i(t)+\sum_{\substack{j=1\\ j\neq i}}^{N}\varepsilon_{ij}\mathbf{D}_{ij}\mathbf{x}_j(t), \quad (11)$$

with

$$\mathbf{D}_{\varepsilon}=\begin{bmatrix}\mathbf{D}_{11}(\varepsilon) & \varepsilon_{12}\mathbf{D}_{12} & \dots & \varepsilon_{1N}\mathbf{D}_{1N}\\ \varepsilon_{21}\mathbf{D}_{21} & \mathbf{D}_{22}(\varepsilon) & \dots & \varepsilon_{2N}\mathbf{D}_{2N}\\ . & . & . & .\\ \varepsilon_{N1}\mathbf{D}_{N1} & \varepsilon_{N2}\mathbf{D}_{N2} & \dots & \mathbf{D}_{NN}(\varepsilon)\end{bmatrix}\in\Re^{\bar{n}\times\bar{n}},\ \mathbf{R}_{ii}=\mathbf{R}_{ii}^T>0\in\Re^{m_i\times m_i},\ \mathbf{R}_{ij}=\mathbf{R}_{ij}^T\geq 0\in\Re^{m_j\times m_j},$$

$$\Theta_{\mathbf{w}i\varepsilon}=\textbf{block diag}\left(\varepsilon_{i1}^{-(1-\delta_{i1})}\Theta_{\mathbf{w}i1}\dots\varepsilon_{iN}^{-(1-\delta_{iN})}\Theta_{\mathbf{w}iN}\right)\geq 0\in\Re^{\bar{q}\times\bar{q}},\ i,\ j=1,\dots,N\ .$$

4. Transmission power management scheme

In order to manage SPWC optimal control problems at the complex MAC and PHY layers, a *singularly-perturbed weakly-coupled power selection multi-radio multi-channel unification protocol* (SPWC-PMMUP) is suggested. The SPWC-PMMUP firmware architecture is depicted in Fig. 5. The design rationale of the firmware is to perform an energy-efficient transmission power management (TPM) in a multi-radio system with minimal change to the existing standard compliant wireless technologies. Such TPM schemes may adapt even to a heterogeneous multi-radio system (i.e., each node has a different number of radios) experiencing singular perturbations. The SPWC-PMMUP coordinates the optimal TPM executions in UCGs. The key attributes are that the SPWC-PMMUP scheme minimizes the impacts of: (i) queue perturbations, arising between energy and packet service variations, and (ii) cross-channel interference problems owing to the violation of orthogonality of multiple channels by wireless fading. The proposed TPM scheme is discussed in the following sections: 4.1 and 4.2.

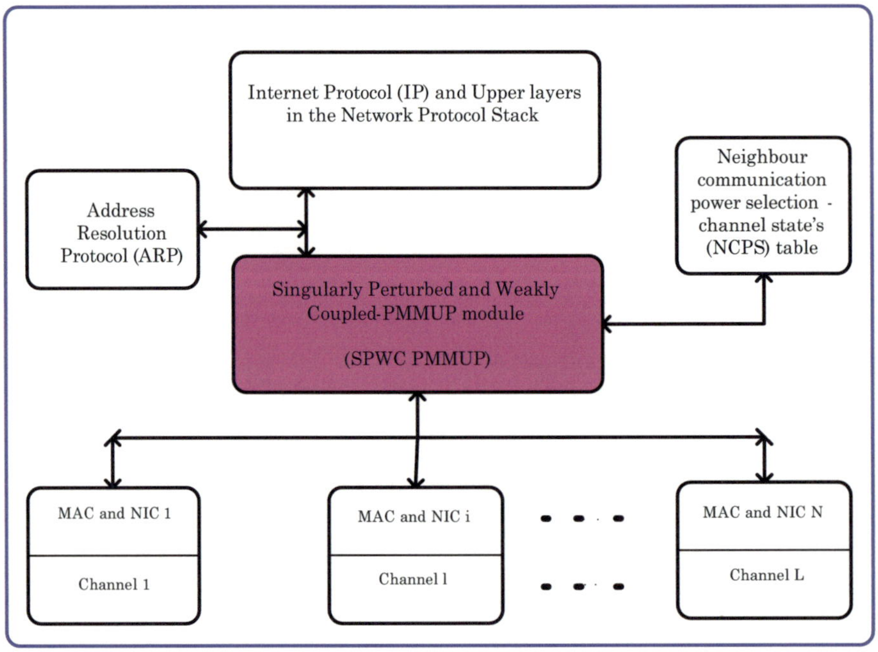

Fig. 5. Singularly-perturbed weakly-coupled PMMUP architecture

4.1 Timing phase structure

The SPWC-PMMUP contains L parallel channel sets with the *virtual* timing structure shown in Fig. 6. Channel access times are divided into identical time-slots. There are three phases in each time-slot after slot synchronization. Phase I serves as the channel probing or Link State Information (LSI) estimation phase. Phase II serves as the *Ad Hoc traffic indication message* (ATIM) window which is on when power optimization occurs. Nodes stay awake and exchange an ATIM (indicating such nodes' intention to send the queue data traffic) message with their neighbours (Wang et al., 2006). Based on the exchanged ATIM, each user performs an optimal transmission power selection (adaptation) for eventual data exchange. Phase III serves as the data exchange phase over power controlled multiple channels.

Phase I: In order for each user to estimate the number of active links in the same UCG, Phase I is divided into M mini-slots. Each mini-slot lasts a duration of channel probing time T_{cp}, which is set to be large enough for judging whether the channel is busy or not. If a link has traffic in the current time-slot, it may randomly select one probe mini-slot and transmit a busy signal. By counting the busy mini-slots, all nodes can estimate how many links intend to advertise traffic at the end of Phase I. Additionally, the SPWC-PMMUP estimates: the inter channel interference (i.e., weak coupling powers), the intra-UCG interference (i.e., the strong coupling powers), the queue perturbation and the LSI addressed in (Olwal et al.,

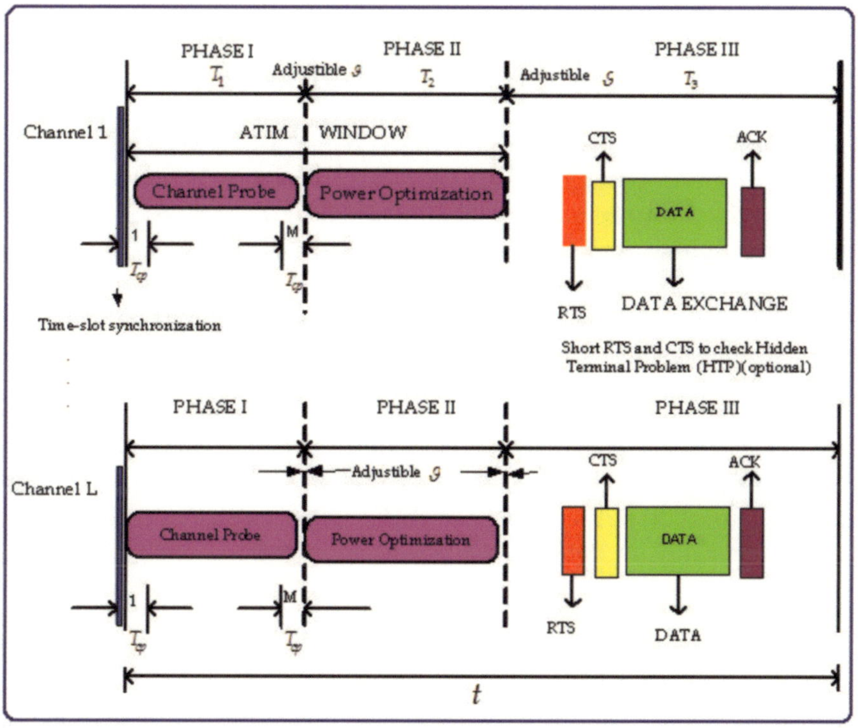

Fig. 6. The virtual SPWC-PMMUP timing structure

2009**a**). It should be noted that the number of links intending to advertise traffic, if not zero, could be greater than the observed number of busy mini-slots. This occurs because there might be at least one link intending to advertise traffic during the same busy mini-slot.

Denote the number of neighbouring links in the same UCG intending to advertise traffic at the end of Phase I as n. Given M and n, the probability that the number of observed busy mini-slots equals m, is calculated by

$$P_r(M,n,m) = \frac{\binom{M}{m}\binom{n-1}{m-1}}{\binom{n+M-1}{M-1}} \tag{12}$$

Let n remain the same for the duration of each time-slot t. Denote the estimate of the number of active links as $\hat{n}(t)$ and the probability mass function (PMF) that the number of busy mini-slots observed in the previous time-slot equals k as $f_k(t)$. Denote $m(t)$ as the number of the current busy mini-slots. The estimate $\hat{n}(t)$ is then derived from the estimation error process as,

$$\hat{n}(t) = \underset{n \geq m(t)}{\arg\min}\left\{\sum_{k=m(t)}^{n} P_r(M,n,k) - \sum_{k=m(t)}^{n} f_k(t)\right\}, \tag{13}$$

where $f_k(t)$ from one time-slot to other is updated as

$$f_k(t) = \begin{cases} (1-\alpha(t)) f_k(t-1), & k \neq m(t) \\ (1-\alpha(t)) f_m(t-1) + \alpha(t), & k = m(t) \end{cases} , \tag{14}$$

and $\alpha(t), (0 < \alpha(t) < 1)$ is the PMF update step size, which needs to be chosen appropriately to balance the convergence speed and the stability. Of course, selecting a large value of M when Phase I is adjusted to be narrower will imply short T_{cp} periods and negligible delay during the probing phase. Short channel probing phase time allows time for large actual data payload exchange, consequently improving network capacity.

Phase II: In this phase, the TPM problem and solution are implemented. Suppose the number of busy mini-slots is non-zero; then the SPWC-PMMUP module performs a power optimization following the $p-$ persistent algorithm or back-off algorithm (Wang et al., 2006). Otherwise, the transmission power optimization depends on the queue status only (i.e., the evaluation of the singular perturbation of the queue system). The time duration of the power optimization is denoted as T_2 and the minimal duration to complete power optimization as a function of the number of participating users in a p-persistent CSMA, is denoted as $T_{succ}(n, p^*)$. The transmission power optimization time allocation T_2 is then adjusted according to

$$T_2 = \min\left\{T_2^{\max}, \vartheta \sum_{n=1}^{\hat{n}} T_{succ}(n, p^*)\right\}, \tag{15}$$

where $T_2^{\max}$ is the power allocation upper bound time, ϑ is the power allocation time adjusting parameter and $\hat{n}$ is the estimated number of actively interfering neighbour links in the same UCG. The steady state medium access probability p in terms of the minimal average service time can be computed as (Wang et al., 2006),

$$p^* = \arg \min_{0<p<1} \{T_{succ}(n, p)\}. \tag{16}$$

It should be noted that due to energy conservation, T_1 and T_2 should be short enough and the optimal p^* can be obtained from a look up table rather than from online computation. The TPM solution is then furnished according to section 4.2.

Phase III: Data is exchanged by NICs over parallel multiple non-overlapping channels within a time period of T_3. The RTS/CTS are exchanged at the probe power level which is sufficient in order to resolve collisions due to hidden terminal nodes. Furthermore, the optimal medium access probability p^* resolves RTS/CTS collisions. After sending data traffic to the target receiver, each node may determine the achievable throughput according to,

$$Th_r(t) = \frac{\tilde{L}}{t}\left\{\sum_i \left\{P_i^{swp}\left(\vec{n}_{swp}, \vec{p}_{swp}, T_2\right) \times \sum_{l=1}^{L} \sum_j j\, P_j^{l,data}\left(\vec{n}_{l,data}, \vec{p}_{l,data}, T_3\right)\right\}\right\}. \tag{17}$$

Here, $\tilde{L}$ is the application/data packet length and t is the length of one virtual time-slot which equals $T_1 + T_2 + T_3$. Denote $P_i^{swp}(\vec{n}_{swp}, \vec{p}_{swp}, T_2)$ as the SPWC based probability that i actively interfering links successfully exchange ATIM in Phase II, given the number of links intending to advertise traffic as, $\vec{n}_{swp}$ and the medium access probability sequence as, $\vec{p}_{swp}$

during time T_2 period. Denote $P_i^{l,data}\left(\vec{n}_{l,data},\vec{p}_{l,data},T_3\right)$ as the probability that i data packets are successfully exchanged on channel l in Phase III, given the number sequence $\vec{n}_{l,data}$ and the medium access probability sequence as $\vec{p}_{l,data}$ during time T_3 period. The computations of such probabilities have been provided in (Li et al., 2009). If several transmissions are executed, then the average throughput performance can be evaluated. The energy efficiency in joules per successfully transmitted packets then becomes

$$E_{eff} = \frac{optimal\ transmission\ power\ per\ node\left(watts\right)}{average\ throughput\ per\ node\ \left(packets\ /\ \mathrm{s}\right)}\,. \tag{18}$$

It should be noted that a high throughput implies a low energy-efficiency for a given optimal power level, because of the high data payload needed to successfully reach the intended receiver within a given time slot. The use of an optimal power level is expected to yield a better spectrum efficiency and throughput measurement balance.

4.2 Nash strategies

The optimal solution to the given problem (10) with the conflict of interest and simultaneous decision making leads to the so called Nash strategies (Gajic & Shen, 1993) $\mathbf{u}_1^*,\ldots,\mathbf{u}_i^*,\ldots,\mathbf{u}_N^*$ satisfying

$$J_i\left(\mathbf{u}_1^*,\ldots,\mathbf{u}_i^*,\ldots,\mathbf{u}_N^*,\mathbf{x}(0)\right)$$

$$\leq J_i\left(\mathbf{u}_1^*,\ldots,\mathbf{u}_i,\ldots,\mathbf{u}_N^*,\mathbf{x}(0)\right),\mathbf{u}_i^*\neq\mathbf{u}_i,\ i=1,\ldots,N\,. \tag{19}$$

Assumption 1: *Each* ith *user has optimal closed-loop Nash strategies yielded by*

$$\mathbf{u}_i^*(t) = -\mathbf{F}_{i\varepsilon}^*\,\mathbf{x}(t),\quad i=1,\ldots,N\,. \tag{20}$$

Here, the decoupled $\mathbf{F}_{i\varepsilon}^*$ is the regulator feedback gain with singular-perturbation and weak-coupling components defined as

$$\mathbf{F}_{i\varepsilon} = \left[\varepsilon^{1-\delta_{1i}}\mathbf{F}_{1i}\ \varepsilon^{1-\delta_{2i}}\mathbf{F}_{2i}\ \ldots\ \varepsilon^{1-\delta_{Ni}}\mathbf{F}_{Ni}\right]\in\Re^{\bar{n}}, \tag{21}$$

with $\bar{n}=\sum_{i=1}^{N}n_i$, n_i is the size of the vector $\mathbf{x}_i$ and $\delta_{ij}=\begin{cases}0\ (i\neq j)\\ 1\ (i=j)\end{cases}$.

Define the N-tuple discrete in time Nash strategies by

$$\mathbf{u}_i^*(t) = -\mathbf{F}_{i\varepsilon}^*\,\mathbf{x}(t) = -\left(\mathbf{R}_{ii}+\mathbf{B}_{i\varepsilon}^T\mathbf{P}_{i\varepsilon}\mathbf{B}_{i\varepsilon}\right)^{-1}\mathbf{B}_{i\varepsilon}^T\mathbf{P}_{i\varepsilon}\mathbf{A}_{\varepsilon}\mathbf{x}(t),\ \ i=1,\ldots,N\,, \tag{22}$$

where $\left(\mathbf{F}_{1\varepsilon}^*,\ldots,\mathbf{F}_{N\varepsilon}^*\right)\in F_N$ and N-tuple $\mathbf{u}_i^*(t)$, form a soft constrained Nash Equilibrium represented as

$$J_i\left(\mathbf{F}_{1\varepsilon}^*\mathbf{x},\ldots,\mathbf{F}_{N\varepsilon}^*\mathbf{x},\,\mathbf{x}(0)\right) = \mathbf{x}(0)^T\,\mathbf{P}_{i\varepsilon}\mathbf{x}(0)\,. \tag{23}$$

Here, the decoupled $\mathbf{P}_{i\varepsilon}$ is a positive semi-definite stabilizing solution of the discrete-time algebraic regulator Riccati equation (DARRE) with the following structure:

$$\mathbf{P}_{i\varepsilon} = \mathbf{P}_{i\varepsilon}^{T} = \begin{bmatrix} \varepsilon_{i1}^{1-\delta_{i1}}\mathbf{P}_{i1} & \varepsilon_{i2}\mathbf{P}_{i12} & . & \varepsilon_{iN}\mathbf{P}_{i1N} \\ \varepsilon_{i2}\mathbf{P}_{i12}^{T} & \varepsilon_{i2}^{1-\delta_{i2}}\mathbf{P}_{i2} & . & \varepsilon_{iN}\mathbf{P}_{i2N} \\ . & . & . & . \\ \varepsilon_{iN}\mathbf{P}_{i1N}^{T} & \varepsilon_{iN}\mathbf{P}_{i2N}^{T} & . & \varepsilon^{1-\delta_{iN}}\mathbf{P}_{iN} \end{bmatrix}, \quad (24)$$
$$\in \Re^{\bar{n}\times\bar{n}}$$

where the DARRE is given by

$$\mathbf{P}_{\varepsilon} = \mathbf{D}_{\varepsilon}^{T}\mathbf{D}_{\varepsilon} + \mathbf{A}_{\varepsilon}^{T}\mathbf{P}_{\varepsilon}\mathbf{A}_{\varepsilon} - \mathbf{A}_{\varepsilon}^{T}\mathbf{P}_{\varepsilon}\mathbf{B}_{\varepsilon}\left(\mathbf{R}_{\varepsilon} + \mathbf{B}_{\varepsilon}^{T}\mathbf{P}_{\varepsilon}\mathbf{B}_{\varepsilon}\right)^{-1}\mathbf{B}_{\varepsilon}^{T}\mathbf{P}_{\varepsilon}\mathbf{A}_{\varepsilon}, \quad (25)$$

with

$$\mathbf{R} = \mathbf{diag}\left(\mathbf{R}_{1}. . . \mathbf{R}_{N}\right), \qquad \mathbf{D}_{\varepsilon} = \begin{bmatrix} \mathbf{D}_{11}(\varepsilon) & \varepsilon_{12}\mathbf{D}_{12} & \dots & \varepsilon_{1N}\mathbf{D}_{1N} \\ \varepsilon_{21}\mathbf{D}_{21} & \mathbf{D}_{22}(\varepsilon) & \dots & \varepsilon_{2N}\mathbf{D}_{2N} \\ . & . & . & . \\ \varepsilon_{N1}\mathbf{D}_{N1} & \varepsilon_{N2}\mathbf{D}_{N2} & \dots & \mathbf{D}_{NN}(\varepsilon) \end{bmatrix}.$$
$$\in \Re^{\bar{n}\times\bar{n}}$$

It should be noted that the inversion of the partitioned matrices $\mathbf{R}_{\varepsilon} + \mathbf{B}_{\varepsilon}^{T}\mathbf{P}_{\varepsilon}\mathbf{B}_{\varepsilon}$ in (25) will produce numerous terms and cause the DARRE approach to be computationally very involved, even though one is faced with the reduced-order numerical problem (Gajic & Shen, 1993). This problem is resolved by using bilinear transformation to transform the discrete-time Riccati equations (DARRE) into the continuous-time algebraic Riccati equation (CARRE) with equivalent co-relation.

The differential game Riccati matrices $\mathbf{P}_{i\varepsilon}$ satisfy the singularly-perturbed and weakly-coupled, continuous in time, algebraic Regulator Riccati equation (SWARREs) (Gajic & Shen, 1993; Sagara et al., 2008) which is given below,

$$\Omega_{i}\left(\mathbf{P}_{1\varepsilon}, . . . ,\mathbf{P}_{i\varepsilon}, . . . ,\mathbf{P}_{N\varepsilon}\right) = \mathbf{P}_{i\varepsilon}\left(\mathbf{A}_{\varepsilon} - \sum_{\substack{j=1 \\ j\neq i}}^{N}\mathbf{S}_{j\varepsilon}\mathbf{P}_{j\varepsilon}\right) + \left(\mathbf{A}_{\varepsilon} - \sum_{\substack{j=1 \\ j\neq i}}^{N}\mathbf{S}_{j\varepsilon}\mathbf{P}_{j\varepsilon}\right)^{T}\mathbf{P}_{i\varepsilon}$$

$$- \mathbf{P}_{i\varepsilon}\mathbf{S}_{i\varepsilon}\mathbf{P}_{i\varepsilon} + \sum_{\substack{j=1 \\ j\neq i}}^{N}\varepsilon_{ij}\mathbf{P}_{j\varepsilon}\mathbf{S}_{ij\varepsilon}\mathbf{P}_{j\varepsilon} + \mathbf{P}_{i\varepsilon}\mathrm{M}_{i\varepsilon}\mathbf{P}_{i\varepsilon} + \mathbf{D}^{T}{}_{i\varepsilon}\mathbf{D}_{i\varepsilon} = \mathbf{0}, \quad (26)$$

where

$$\mathbf{S}_{i\varepsilon} = \mathbf{B}_{i\varepsilon}\mathbf{R}_{ii}^{-1}\mathbf{B}_{i\varepsilon}^{T}, i = 1, . . ., N. \ \mathbf{S}_{ij} = \mathbf{B}_{j\varepsilon}\mathbf{R}_{jj}^{-1}\mathbf{R}_{ij}\mathbf{R}_{jj}^{-1}\mathbf{B}_{j\varepsilon}^{T}, \ i = 1, . . ., N.$$

$$\mathrm{M}_{i\varepsilon} = \mathbf{W}_{\varepsilon}\Theta_{\mathbf{w}i\varepsilon}^{-1}\mathbf{W}_{\varepsilon}, \ i = 1, . . ., N.$$

By substituting the partitioned matrices of $\mathbf{A}_{\varepsilon}$, $\mathbf{S}_{i\varepsilon}$, $\mathbf{S}_{ij\varepsilon}$, $\mathrm{M}_{i\varepsilon}$, $\mathbf{D}_{i\varepsilon}$, and $\mathbf{P}_{i\varepsilon}$ into SWARRE (26), and by letting $\varepsilon_w = 0$ and any $\varepsilon_s \neq 0$, then simplifying the SWARRE (26), the following reduced order (auxiliary) algebraic Riccati equation is obtained,

$$\mathbf{P}_{ii}\mathbf{A}_{ii} + \mathbf{A}_{ii}^T\mathbf{P}_{ii} - \mathbf{P}_{ii}\left(\mathbf{S}_{ii} - \mathrm{M}_{ii}\right)\mathbf{P}_{ii} + \mathbf{D}_{ii}^T\mathbf{D}_{ii} = 0, \tag{27}$$

where $\mathbf{S}_{ii} = \mathbf{B}_{ii}\mathbf{R}_{ii}^{-1}\mathbf{B}_{ii}^T$ and $\mathrm{M}_{ii} = \mathbf{W}_{ii}\Theta_{ii}^{-1}\mathbf{W}_{ii}^T$, and $\mathbf{P}_{ii}$, $i = 1,\ldots,N$ is the 0-order approximation of $\mathbf{P}_{i\varepsilon}$ when the weakly-coupled component is set to zero, i.e., $\varepsilon_w = 0$. It should be noted that a unique positive semi-definite optimal solution $\mathbf{P}_{i\varepsilon}^*$ exists if the following assumptions are taken into account (Mukaidani, 2009).

Assumption 2: *The triples* $\mathbf{A}_{ii}$, $\mathbf{B}_{ii}$ *and* $\mathbf{D}_{ii}$, $i = 1,\ldots,N$, *are stabilizable and detectable.*

Assumption 3: *The auxiliary (27) has a positive semidefinite stabilizing solution such that* $\tilde{\mathbf{A}} = \mathbf{A}_{ii} - \mathbf{S}_{ii}\mathbf{P}_{ii}$ *is stable.*

4.3 Analysis of SPWC-PMMUP optimality

Lemma 1: *Under assumption 3 there exists a small constant* ∂^* *such that for all* $\tilde{\varepsilon}(t) \in \left(0, \partial^*\right)$, *SWARRE admits a positive definite solution* $\mathbf{P}_{i\varepsilon}^*$ *represented as*

$$\mathbf{P}_{i\varepsilon} = \mathbf{P}_{i\varepsilon}^* = \mathbf{P}_i + O\left(\varepsilon(t)\right),\ i=1,\ldots,N \ \text{ and } \ \tilde{\varepsilon}(t) = \left|\sqrt{\varepsilon_w\varepsilon_s}\right|,$$

$$= \textbf{block diag}\left(0\ldots\mathbf{P}_{ii}\ldots 0\right) + O\left(\tilde{\varepsilon}(t)\right). \tag{28}$$

Proof: This can be achieved by demonstrating that the Jacobian of SWARRE is non-singular at $\tilde{\varepsilon}(t) = 0$ and its neighbourhood, i.e., $\varepsilon(t) \to +0$. Differentiating the function $\Omega_i\left(\tilde{\varepsilon}(t), \mathbf{P}_{1\varepsilon}, \ldots, \mathbf{P}_{N\varepsilon}\right)$ with respect to the decoupled matrix $\mathbf{P}_{i\varepsilon}$ produces,

$$\mathbf{J}_{ii} = \frac{\partial}{\partial\, vec\, \mathbf{P}_{i\varepsilon}} vec\, \Omega_i\left(\tilde{\varepsilon}(t), \mathbf{P}_{1\varepsilon}, \ldots, \mathbf{P}_{N\varepsilon}\right)^T = \Delta_{ii}^T \otimes I_{n_i} + I_{n_i} \otimes \Delta_{ii}^T.$$

$$\mathbf{J}_{ij} = \frac{\partial}{\partial\, vec\, P_{ij}} vec\, \Omega_i\left(\tilde{\varepsilon}(t), \mathbf{P}_{1\varepsilon}, \ldots, \mathbf{P}_{N\varepsilon}\right)^T,$$

$$= -\left(\mathbf{S}_{j\varepsilon}\mathbf{P}_{i\varepsilon} - \tilde{\varepsilon}_{ij}\mathbf{S}_{ij\varepsilon}\mathbf{P}_{j\varepsilon}\right)^T \otimes I_{n_i} - I_{n_i} \otimes \left(\mathbf{S}_{j\varepsilon}\mathbf{P}_{i\varepsilon} - \tilde{\varepsilon}_{ij}\mathbf{S}_{ij\varepsilon}\mathbf{P}_{j\varepsilon}\right)^T, \tag{29}$$

where $i \neq j$, $j = 1,\ldots,N$ and $\Delta = \mathbf{A}_{\varepsilon} - \sum_{\substack{j=1 \\ i \neq j}}^{N} \mathbf{S}_{j\varepsilon}\mathbf{P}_{j\varepsilon} + \mathrm{M}_{i\varepsilon}\mathbf{P}_{i\varepsilon}$.

Exploiting the fact that $\mathbf{S}_{j\varepsilon}\mathbf{P}_{i\varepsilon} = O(\tilde{\varepsilon}(t))$ for $i \neq j$, the Jacobian of SWARRE with $\tilde{\varepsilon}(t) \to +0$ can be verified as

$$\hat{\mathbf{J}} = \textbf{block diag}\left(\Delta_{11}\ldots\Delta_{NN}\right),\ \mathbf{J} = \textbf{block diag}\left(\hat{\mathbf{J}}\ldots\hat{\mathbf{J}}\right).$$

Since the determinant of $\Delta_{ii} = \mathbf{A}_{ii} - \mathbf{S}_{ii}\mathbf{P}_{ii} + \mathbf{M}_{ii}\mathbf{P}_{ii}$ with $\tilde{\varepsilon}(t) = 0$ is non-zero by following assumption 3 for all $i = 1, \ldots, N$, thus $\det \mathbf{J} \neq 0$ i.e., $\mathbf{J}$ is non-singular for $\tilde{\varepsilon}(t) = 0$. As a consequence of the implicit function theorem, $\mathbf{P}_{ii}$ is a positive definite matrix at $\tilde{\varepsilon}(t) = 0$ and for sufficiently small parameters $\tilde{\varepsilon}(t) \in \left(0, \partial^*\right)$, one can conclude that $\mathbf{P}_{i\varepsilon} = P_{ii} + O(\tilde{\varepsilon}(t))$ is also a positive definite solution.

Theorem 1: *Under assumptions 1-3, the use of a soft constrained Nash equilibrium* $\mathbf{u}_i^{(k)*}(t) = -\mathbf{F}_{i\varepsilon}^{(k)*}\mathbf{x}(t)$ *results in the following condition.*

$$J_i\left(\mathbf{u}_1^{(k)*}, \ldots, \mathbf{u}_N^{(k)*}, \mathbf{x}(0)\right) \approx J_i\left(\mathbf{u}_1^*, \ldots, \mathbf{u}_N^*, \mathbf{x}(0)\right) + O\left(\tilde{\varepsilon}^{2^k+1}\right). \tag{30}$$

Proof: Due to space constraints, we merely outline the proof. A detailed related analysis can be found in (Mukaidani, 2009; Sagara et al., 2008).

If the iterative strategy is $\mathbf{u}_i^{(k)*}(t) = -\mathbf{F}_{i\varepsilon}^{(k)*}\mathbf{x}(t)$ then the value of the cost function is given by

$$J_i\left(\mathbf{u}_1^{(k)*}, \ldots, \mathbf{u}_N^{(k)*}, \mathbf{x}(0)\right) = \mathbf{x}^T(0)\mathbf{Y}_{i\varepsilon}\mathbf{x}(0), \tag{31}$$

where $\mathbf{Y}_{i\varepsilon}$ is a *positive semi-definite solution* of the following algebraic Riccati equation

$$\mathbf{Y}_{i\varepsilon}\left(\mathbf{A}_\varepsilon - \sum_{\substack{j=1 \\ j\neq i}}^{N}\mathbf{S}_{j\varepsilon}\mathbf{P}_{j\varepsilon}^{(k)}\right) + \left(\mathbf{A}_\varepsilon - \sum_{\substack{j=1 \\ j\neq i}}^{N}\mathbf{S}_{j\varepsilon}\mathbf{P}_{j\varepsilon}^{(k)}\right)^T \mathbf{Y}_{i\varepsilon}$$

$$+ \mathbf{Y}_{i\varepsilon}\mathbf{M}_{i\varepsilon}\mathbf{Y}_{i\varepsilon} + \varepsilon\sum_{\substack{j=1 \\ j\neq i}}^{N}\mathbf{P}_{j\varepsilon}^{(k)}\mathbf{S}_{ij\varepsilon}\mathbf{P}_{j\varepsilon}^{(k)} - \mathbf{P}_{i\varepsilon}^{(k)}\mathbf{S}_{i\varepsilon}\mathbf{P}_{i\varepsilon}^{(k)} + \mathbf{D}_{i\varepsilon}^T\mathbf{D}_{i\varepsilon} = \mathbf{0}. \tag{32}$$

Let $\mathbf{Z}_{i\varepsilon} = \mathbf{Y}_{i\varepsilon} - \mathbf{P}_{i\varepsilon}$; then subtracting SWARRE (26) from (32) satisfies the following equation

$$\mathbf{Z}_{i\varepsilon}\bar{\mathbf{A}}_\varepsilon^{(k)} + \bar{\mathbf{A}}_\varepsilon^{(k)T}\mathbf{Z}_{i\varepsilon} + \sum_{\substack{j=1 \\ j\neq i}}^{N}\mathbf{P}_{i\varepsilon}\mathbf{S}_{j\varepsilon}\left(\mathbf{P}_{j\varepsilon} - \mathbf{P}_{j\varepsilon}^{(k)}\right)$$

$$+ \sum_{\substack{j=1 \\ j\neq i}}^{N}\left(\mathbf{P}_{j\varepsilon} - \mathbf{P}_{j\varepsilon}^{(k)}\right)\mathbf{S}_{j\varepsilon}\mathbf{P}_{i\varepsilon} + \varepsilon\left[\sum_{\substack{j=1 \\ j\neq i}}^{N}\left(\mathbf{P}_{j\varepsilon}^{(k)}\mathbf{S}_{ij\varepsilon}\mathbf{P}_{j\varepsilon}^{(k)} - \mathbf{P}_{j\varepsilon}\mathbf{S}_{ij\varepsilon}\mathbf{P}_{j\varepsilon}\right)\right] + \left(\mathbf{P}_{i\varepsilon} - \mathbf{P}_{i\varepsilon}^{(k)}\right)\mathbf{S}_{i\varepsilon}\left(\mathbf{P}_{i\varepsilon} - \mathbf{P}_{i\varepsilon}^{(k)}\right) = 0, \tag{33}$$

where $\bar{\mathbf{A}}_\varepsilon^{(k)} = \mathbf{A}_\varepsilon - \sum_{j=1}^{N}\mathbf{S}_{j\varepsilon}\mathbf{P}_{j\varepsilon}^{(k)} + \mathbf{M}_{i\varepsilon}\mathbf{P}_{i\varepsilon}^{(k)} + \mathbf{M}_{i\varepsilon}\left(\mathbf{P}_{i\varepsilon} - \mathbf{P}_{i\varepsilon}^{(k)}\right)$. Suppose $\left\|\mathbf{P}_{i\varepsilon} - \mathbf{P}_{i\varepsilon}^{(k)}\right\| \approx O\left(\tilde{\varepsilon}^{2^k}\right)$, (i.e., has a quadratic rate of convergence); then from the proof of theorem 1, one can have,

$$\theta(\mathbf{Z}_{i\varepsilon}) = \mathbf{Z}_{i\varepsilon}\left(\hat{\mathbf{J}} + O(\tilde{\varepsilon})\right) + \left(\hat{\mathbf{J}} + O(\tilde{\varepsilon})\right)^T\mathbf{Z}_{i\varepsilon} + \mathbf{Z}_{i\varepsilon}\mathbf{M}_{i\varepsilon}\mathbf{Z}_{i\varepsilon} + O\left(\tilde{\varepsilon}^{2^k+1}\right) = 0, \tag{34}$$

where $\theta(\mathbf{0}) = O\left(\tilde{\varepsilon}^{2^k+1}\right)$ and $\hat{\mathbf{J}} = \textbf{block diag}\left(\Delta_{11} \; . \; . \; . \; \Delta_{NN}\right)$ with $\Delta_{ii} = \mathbf{A}_{ii} - \left(\mathbf{S}_{ii} - \mathbf{M}_{ii}\right)\mathbf{P}_{ii}$ [258]. Thus, let $\left\|\mathbf{Z}_{i\varepsilon} - \mathbf{0}\right\| \leq O\left(\tilde{\varepsilon}^{2^k+1}\right)$ and from the cost function definition it is evident that:

$$\mathbf{x}^T(0)\mathbf{Z}_{i\varepsilon}\mathbf{x}(0) = \mathbf{x}^T(0)\mathbf{Y}_{i\varepsilon}\mathbf{x}(0) - \mathbf{x}^T(0)\mathbf{P}_{i\varepsilon}\mathbf{x}(0)$$

$$= J_i\left(\mathbf{u}_1^{(k)*},\ldots,\mathbf{u}_N^{(k)*},\mathbf{x}(0)\right) - J_i\left(\mathbf{u}_1^*,\ldots,\mathbf{u}_N^*,\mathbf{x}(0)\right)$$

$$\leq O\left(\tilde{\varepsilon}^{2^k+1}\right) \quad . \tag{35}$$

5. Simulation tests and results

5.1 Simulation tests

The efficiency of the proposed model and algorithm was studied by means of numerical examples. The MATLAB™ tool was used to evaluate the design optimization parameters, because of its efficiency in numerical computations. The wireless MRMC network being considered was modelled as a large scale interconnected control system. Upto 50 wireless nodes were randomly placed in a 1200 m by 1200 m region. The random topology depicts a non-uniform distribution of the nodes. Each node was assumed to have at most four NICs or radios, each tuned to a separate non-overlapping UCG as shown in Fig. 7. Although 4 radios are situated at each node, it should be noted that such a dimension merely simplifies the simulation. The higher dimension of radios per node may be used without loss of generality. The MRMC configurations depict the weak coupling to each other among different non-overlapping channels. In other words, those radios of the same node operating on separate frequency channels (or UCGs) do not communicate with each other. However, due to their close vicinity such radios significantly interfere with each other and affect the process of optimal power control. The ISM carrier frequency band of 2.427 GHz-2.472 GHz was assumed for simulation purposes only. Figure 7 illustrates the typical wireless network scenario with 4 nodes, each with 4 radio-pairs or users able to operate simultaneously. The rationale is to stripe application traffic over power controlled multiple channels and/or to access the WMCs as well as backhaul network cooperation (Olwal et al., 2009**a**).

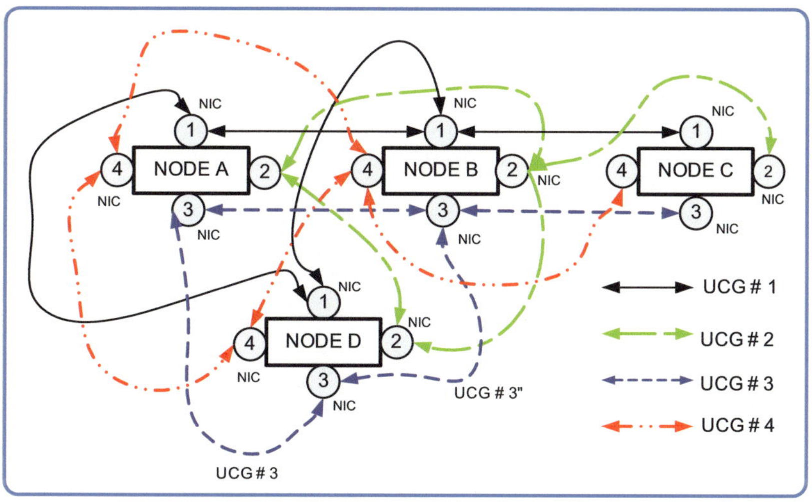

Fig. 7. MRMC wireless network

5.2 Performance evaluation

In order to evaluate the performance of the singularly-perturbed weakly-coupled dynamic transmission power management (TPM) scheme in terms of power and throughput, our simulation parameters, additional to those in section 5.1, were outlined as follows: The Distributed Inter Frame Space (DIFS) time = 50 μs, Short Inter Frame Space (SIFS) time = 10 μs and Back-off slot time = 20 μs. The number of mini-slots in the probe phase, M = 20, duration of probe mini-slot, T_{pc} = 40 μs and ATIM and power selection window adjustment parameter, ϑ = 1.2-1.5 as well as a virtual time-slot duration consisting of probe, power optimization and data packet transmission times, t = 100 ms.

An arrival rate of λ packets/sec of packets at each queue was assumed. For each arriving packet at the sending queue, a receiver was randomly selected from its immediate neighbours. Each simulation run was performed long enough for the output statistics to stabilize (i.e., sixty seconds simulation time). Each datum point in the plots represents an average of four runs where each run exploits a different randomly generated network topology. Saturated transmission power consumption and throughput gain performance were evaluated. Saturation conditions mean that packets are always assumed to be in the queue for transmission; otherwise, the concerned transmitting radio goes to doze/sleep mode to conserve energy (i.e., back-off amount of time).

The following parameters were varied in the simulation: the number of active links (transmit-receive radio-pairs) interfering (i.e., co-channel and cross-channel), from 2 to 50 links, the channels' availability, from 1 to 4 and the traffic load, from 12.8 packets/s to 128 packets/s. The maximum possible power consumed by a radio in the transmit state, the receive state, the idle state and the doze state was assumed as 0.5 Watt, 0.25 Watt, 0.15 Watt and 0.005 Watt, respectively. A user being in the transmitting state means that the radio at the head of the link is in the transmit state while the radio at the tail of the link is in the receive state. A user in the receive state, in the idle state, and in the doze state means that both the radio at the head of the link and the radio at the tail of the link are in the receive state, in the idle state, and in the doze state, respectively (Wang et al., 2006). In order to evaluate the transmission power consumption, packets must be assumed to be always available in all the sending queues of nodes. This is a condition of network saturation.

5.3 Results and disscussions

Figure 8 illustrates an average transmission power per node pair at steady state, versus the number of active radios relative to the total number of adjacent channels. During each time slot, each node evaluates steady state transmission powers in the ATIM phase. Average transmission power was measured as the number of active radio interfaces was increased at different values of the queue perturbations and the weak couplings of the MRMC systems. An increase in the number of active interfaces results in a linear increase in the transmission powers per node-pairs. At 80%, the number of radios relative to the number of adjacent channels with $\varepsilon=\sqrt{|\varepsilon_s\varepsilon_w|}=0.0001$ yields about 0.61%, 7.98%, 9.51% respectively, a greater power saving than with $\varepsilon=0.001$, 0.01 and 0.1. This is explained as follows. Stabilizing a highly perturbed queue system and strongly interfered disjoint wireless channels consumes more source energy. Packets are also re-transmitted frequently because of high packet drop rates. Retransmitting copies of previously dropped packets results in perturbations at the queue system owing to induced delays and energy-outages.

A number of previously studied MAC protocols for throughput enhancement were compared with the SPWC-PMMUP based power control scheme. The multi-radio unification protocol (MUP) was compared with the SPWC-PMMUP scheme because the latter is a direct extension of the former in terms of energy-efficiency. Both protocols are implemented at the LL and with the same purpose (i.e., to hide the complexity of the multiple PHY and MAC layers from the unified higher layers, and to improve throughput performance). However, the MUP scheme chooses only one channel with the best channel quality to exchange data and does not take power control into consideration. The power-saving multi-radio multi-channel medium access control (MAC) (PSM-MMAC) was compared with the SPWC-PMMUP scheme, because both protocols share the following characteristics: they are energy-efficient, and they select channels, radios and power states dynamically based on estimated queue lengths, channel conditions and the number of active links. The single-channel power-control medium access control (POWMAC) protocol was compared with the SPWC-PMMUP because both are power controlled MAC protocols suitable for wireless Ad Hoc networks (e.g., IEEE 802.11 schemes). Such protocols perform the carrier sensed multiple access with collision avoidance (CSMA/CA) schemes. Both protocols possess the capability to exchange several concurrent data packets after the completion of the operation of the power control mechanism. Both are distributed, asynchronous and adaptive to changes of channel conditions.

Figure 9 depicts the plots for energy-efficiency versus the number of active links per square kilometre of an area. Energy-efficiency is measured in terms of the steady state transmission power per time slot, divided by the amount of packets that successfully reach the target receiver. It is observed that low active network densities generally provide higher energy-efficiency gain than highly active network densities. This occurs because low active network densities possess better spatial re-use and proper multiple medium accesses. Except for low network densities, the SPWC-PMMUP scheme outperforms the POWMAC, the power saving multi-channel MAC (i.e., PSM-MMAC) and the MUP schemes. In low active network density, a single channel power controlled MAC (i.e., POWMAC) records a higher degree of freedom with spatial re-use. As a result, it indicates a low expenditure of transmission power. As the number of active users increases, packet collisions and retransmissions become significantly large. The POWMAC uses an adjustable access window to allow for a series of RTS/CTS exchanges to take place before several concurrent data packet transmissions can commence. Unlike its counterparts, the POWMAC does not make use of control packets (i.e., RTS/CTS) to silence neighbouring terminals. Instead, collision avoidance information is inserted in the control packets and is used in conjunction with the received signal strength of these packets to dynamically bound the transmission power of potentially interfering terminals in the vicinity of a receiving terminal. This allows an appropriate selection of transmission power that ensures multiple-concurrent transmissions in the vicinity of the receiving terminal. On the other hand, both SPWC-PMMUP and PSM-MMAC contain an adjustable ATIM window for traffic loads and the LL information. The ATIM window is maintained moderately narrow in order that less energy is wasted owing to its being idle. Statistically, the simulation results indicated that for between 4 and 16 users per deployment area, the POWMAC scheme was on average 50%, 87.50%, and 137.50% more energy-efficient than the SPWC-PMMUP, PSM-MMAC and MUP, respectively. However, between 32 and 50 users per deployment area, in the SPWC-PMMUP scheme, yielded on average 14.58%, 66.67%, and 145.83% more energy efficiency than the POWMAC, PSM-MMAC and MUP schemes, respectively.

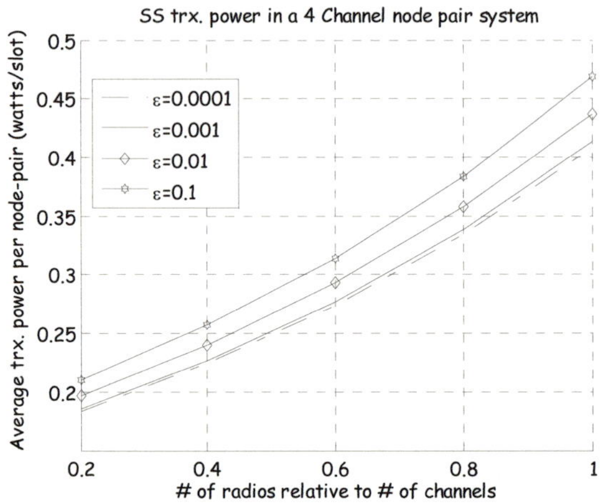

Fig. 8. Steady state transmission power versus relative number of radios per channel

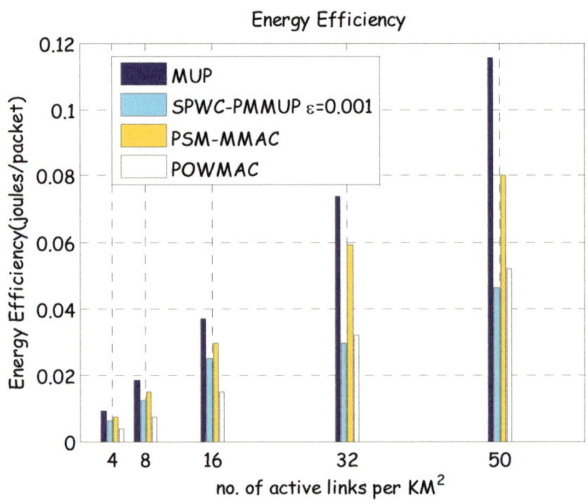

Fig. 9. Energy-efficiency versus density of active links

Figure 10 depicts the performance of the network lifetime observed for the duration of the simulation. The number of active links using steady state transmission power levels was initially assumed to be 36 links per square kilometre of area. Under the saturated traffic generated by the queue systems, different protocols were simulated and compared to the SPWC-PMMUP scheme. The links which were still alive were defined as those which were operating on certain stabilized transmission power levels and which remained connected at the end of the simulation time. The SPWC-PMMUP scheme evaluates the network lifetime based on the stable connectivity measure. That is, if a transmission power level, $p_{ij} = p_{ij}^*$ then the link (i,j) exists; otherwise if $p_{ij} < p_{ij}^*$, then there is no link between the transmitting interface i and the receiving interface j (i.e., the tail of the link). The notation, p_{ij}^* represents the minimum transmission power level needed to successfully send a packet to the target receiver at the immediate neighbours. After 50 units of simulation time, the SPWC-PMMUP scheme records, on average, 12.50%, 22.22% and 33.33%, of links still alive, more than the POWMAC, PSM-MMAC and MUP schemes, respectively. This is because

SPWC-PMMUP scheme uses a fractional power to perform the medium access control (i.e., RTS/CTS control packets are executed at a lower power than the maximum possible) while the conventional protocols employ maximum transmission powers to exchange control packets. The SPWC-PMMUP also transmits application or data packets using a transmission power level which is adaptive to queue perturbations, the intra and inter-channel interference, the receiver SINR, the wireless link rate and the connectivity range. The performance gains of the POWMAC scheme are explained as follows. The POWMAC uses a collision avoidance inserted in the control packets, and in conjunction with the received signal strength of these packets, to dynamically bound the transmission powers of potentially interfering terminals in the vicinity of a receiving terminal. This promotes mutual multiple transmissions of the application packets at a controlled power over a relatively long time. The PSM-MMAC scheme offers the desirable feature of being adaptive to energy, channel, queue and opportunistic access. However, its RTS/CTS packets are executed on maximum power. The MUP scheme does not perform any power control mechanism and hence records the worst lifetime performance.

Figure 11 illustrates an average throughput performance versus the offered traffic load at different singular-perturbation and weak-coupling conditions. Four simulation runs were performed at different randomly generated network topologies. The average throughput per send and receive node-pairs was measured when packets were transmitted using steady state transmission powers. Plots were obtained at confidence intervals of 95%, that is, with small error margins. In general, the average throughput monotonically increases with the amount of the traffic load subjected to the channels. The highly-perturbed and strongly-coupled multi-channel systems, that is, $\varepsilon=\sqrt{|\varepsilon_s\varepsilon_w|}=0.1$, degrade average per hop throughput performance compared to the lowly-perturbed and weakly-coupled system, that is, $\varepsilon = 0.0001$. On average, and at 100 packets/s of the traffic load, the system described by $\varepsilon = 0.0001$ can provide 4%, 16% and 28% more throughput performance gain over the system at $\varepsilon = 0.001$, $\varepsilon = 0.01$ and $\varepsilon = 0.1$, respectively. This may be explained as follows. In large queue system perturbations (i.e., $\varepsilon = 0.1$) the SPWC-PMMUP scheme wastes a large portion of the time slot in stabilizing the queue and in finding optimal transmission power

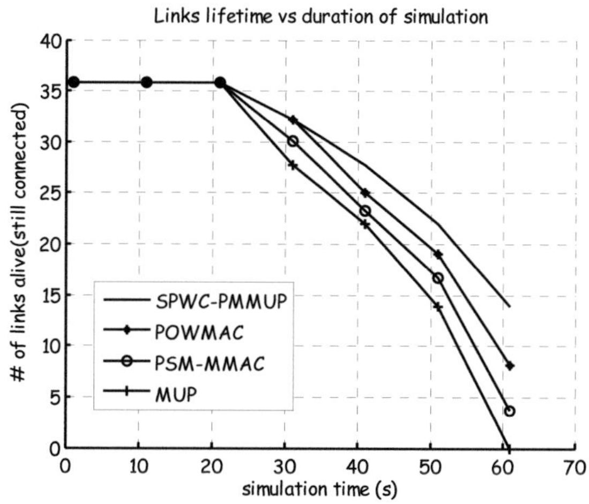

Fig. 10. Active links lifetime performance

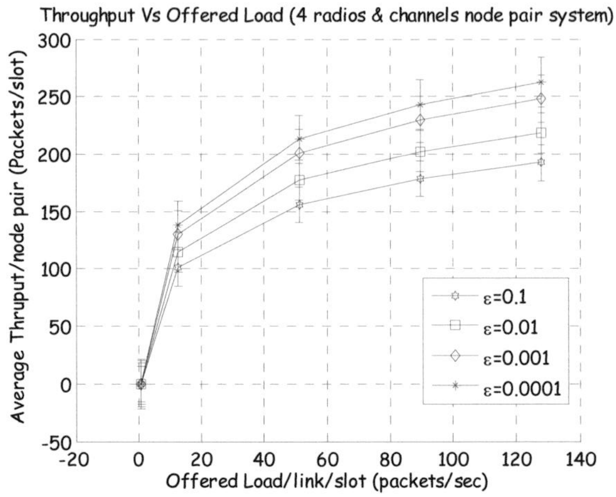

Fig. 11. Average Throughput versus offered Load

levels. This means that only lesser time intervals are allowed for actual application packet transmission. Furthermore, the inherently high inter-channel interference degrades the spatial re-use. Consequently, smaller volumes of application/data packets actually reach the receiving destination successfully (i.e., low throughput). Conversely, with a low perturbation and inter-channel interference (i.e., $\varepsilon = 0.001$), data packets have a larger time interval for transmission; the wireless medium is spectrally efficient and hence achieves an enhanced average throughput.

6. Conclusion

This chapter has furnished an optimal TPM scheme suitable for backbone wireless mesh networks employing MRMC configurations. As a result, an energy-efficient link-layer TPM called SPWC-PMMUP has been proposed for a wireless channel with packet losses. The optimal TPM demonstrated at least 14% energy-efficiency and 28% throughput performance gains over the conventional schemes, in less efficient channels (Figs. 9 & 11). However, a joint TPM, channel and routing optimization for MRMC wireless networks remains an open issue for future investigation.

7. References

Adya, A.; Bahl, P.; Wolman, A. & Zhou, L. (2004). A multi-radio unification protocol for IEEE 802.11 wireless networks, *Proceedings of international conference on broadband networks (Broadnets'04)*, pp. 344-354, ISBN: 0-7695-2221-1, San Jose, October 2004, IEEE, CA.

Akyildiz, I. F. & Wang, X. (2009). *Wireless mesh networks*, John Wiley & Sons Ltd, ISBN: 978-0-470-03256-5, UK.

El-Azouzi, R. & Altman, E. (2003). Queueing analysis of link-layer losses in wireless networks, *Proceedings of personal wireless communications*, pp. 1-24, ISBN: 3-540-20123-8, Venice, September 2003, Italy.

Gajic, Z. & Shen, X. (1993). *Parallel algorithms for optimal control of large scale linear systems*, Springer-Verlag, ISBN: 3-540-19825-3, New York.

Iqbal, A. & Khayam, S. A. (2009). An energy-efficient link-layer protocol for reliable transmission over wireless networks. *EURASIP journal on wireless communications and networking,* Vol. 2009, No. 10, July 2009, ISSN: 1687-1472.

Li, D.; Du, H.; Liu, L. & Huang, S. C.-H. (2008). Joint topology control and power conservation for wireless sensor networks using transmit power adjustment, In: *Computing and combinatorics,* X. Hu & J. Wang (Eds.), pp. 541-550, Springer Link, ISBN: 978-3-540-69732-9, Berlin.

Li, X.; Cao, F. & Wu, D. (2009). QoS-driven power allocation for multi-channel communication under delayed channel side information, *Proceedings of consumer communications & networking conference,* ISBN: 978-952-15-2152-2, Las Vegas, January 2009, Nevada, USA.

Maheshwari, R.; Gupta, H. & Das, S. R. (2006). Multi-channel MAC protocols for wireless networks, *Proceedings of IEEE SECON 2006,* ISBN: 978-1-580-53044-6, Reston, September 2006, IEEE, VA.

Merlin, S.; Vaidya, N. & Zorzi, M. (2007). *Resource allocation in multi-radio multi-channel multi-hop wireless networks,* Technical Report: 35131, July 2007, Padova University.

Mukaidani, H. (2009). Soft-constrained stochastic Nash games for weakly coupled large-scale systems. *Elsevier automatica,* Vol. 45, pp. 1272-1279, 2009, ISSN: 0005-1098.

Muqattash, A. & Krunz, M. (2005). POWMAC: A single-channel power control protocol for throughput enhancement in wireless ad hoc networks. *IEEE journal on selected areas in communications,* Vol. 23, pp. 1067-1084, 2005, ISSN: 0733-8716.

Olwal, T. O.; Van Wyk, B. J; Djouani, K.; Hamam, Y.; Siarry, P. & Ntlatlapa, N. (2009a). Autonomous transmission power adaptation for multi-radio multi-channel WMNs, In: *Ad hoc, mobile and wireless networks,* P. M. Ruiz & J. J. Garcia-Luna-Aceves (Eds.), pp. 284-297, Springer Link, ISBN: 978-3-642-04382-6, Berlin.

Olwal, T. O.; Van Wyk, B. J; Djouani, K.; Hamam, Y.; Siarry, P. & Ntlatlapa, N. (2009b). A multi-state based power control for multi-radio multi-channel WMNs. *International journal of computer science,* Vol. 4, No. 1, pp. 53-61, 2009, ISSN: 2070-3856.

Olwal, T. O.; Van Wyk, B. J; Djouani, K.; Hamam, Y. & Siarry, P. (2010a). Singularly-perturbed weakly-coupled based power control for multi-radio multi-channel wireless networks. *International journal of applied mathematics and computer sciences,* Vol. 6, No. 1, pp. 4-14, 2010, ISSN: 2070-3902.

Olwal, T. O.; Van Wyk, B. J; Djouani, K.; Hamam, Y.; Siarry, P. & Ntlatlapa, N. (2010b). Dynamic power control for wireless backbone mesh networks: a survey. *Network protocols and algorithms,* Vol. 2, No. 1, pp. 1-44, 2010, ISSN: 1943-3581.

Park, H.; Jee, J. & Park, C. (2009). Power management of multi-radio mobile nodes using HSDPA interface sensitive APs, *Proceedings of IEEE 11th international conference on advanced communication technology,* pp. 507-511, ISBN: 978-89-5519-139-4, Phoenix Park, South Korea, February 2009.

Sagara, M.; Mukaidani, H. & Yamamoto, T. (2008). Efficient numerical computations of soft-constrained Nash strategy for weakly-coupled large-scale systems. *Journal of computers,* Vol. 3, pp. 2-10, November 2008, ISSN: 1796-203X.

Thomas, R. W.; Komali, R. S.; MacKenzie, A. B. & DaSilva, L. A. (2007). Joint power and channel minimization in topology control: a cognitive network approach, *Proceedings of IEEE international conference on communication,* pp. 6538-6542, ISBN: 0-7695-2805-8, Glasgow, Scotland, 2007.

Wang, J.; Fang, Y. & Wu, D. (2006). A power-saving multi-radio multi-channel MAC protocol for wireless local area networks, *Proceedings of IEEE infocom 2006 conference*, pp. 1-13, ISBN: 1-4244-0222-0, Barcelona, Spain, 2006.

Zhou, H.; Lu, K. & Li, M. (2008). Distributed topology control in multi-channel multi-radio mesh networks, *Proceedings of IEEE international conference on communication,* ISBN: 0-7803-6283-7, New Orleans, USA, May 2008.

2

Access-Point Allocation Algorithms for Scalable Wireless Internet-Access Mesh Networks

Nobuo Funabiki
Graduate School of Natural Science and Technology, Okayama University
Japan

1. Introduction

With rapid developments of inexpensive small-sized communication devices and high-speed network technologies, the Internet has increasingly become the important medium for a lot of people in daily lives. People can access to a variety of information, data, and services that have been provided through the Internet, in addition to their personal communications. This progress of the Internet utilization leads to strong demands for high-speed, inexpensive, and flexible Internet access services in any place at anytime. Particularly, such ubiquitous communication demands have grown up among users using wireless communication devices. A common solution to them is the use of the *wireless local area network (WLAN)*. WLAN has been widely studied and deployed as the access network to the Internet. Currently, WLAN has been used in many Internet service spots around the world in both public and private spaces including offices, schools, homes, airports, stations, hotels, and shopping malls.

The *wireless mesh network* has emerged as a very attractive technology that can flexibly and inexpensively solve the problem of the limited wireless coverage area in a conventional WLAN using a single wireless router (Akyildiz et al., 2005). The wireless mesh network adopts multiple wireless routers that are distributed in the service area, so that any location in this area is covered by at least one router. Data communications between routers are offered by wireless communications, in addition to those between user hosts (clients) and routers. This cable-less advantage is very attractive to deploy the wireless mesh network in terms of the flexibility, the scalability, and the low installation cost.

Among several variations of under-studying wireless mesh networks, we have focused on the one targeting the Internet access service, using only access points (APs) as wireless routers, and providing communications between APs mainly on the MAC layer by the *wireless distribution system (WDS)*. From now, we call this *Wireless Internet-access Mesh NETwork* as *WIMNET* for convenience. At least one AP in WIMNET acts as a *gateway (GW)* to the Internet (Figure 1). To reduce radio interference among wireless links, the IEEE802.11a protocol at 5GHz can be adopted to links between neighbouring APs, and the 802.11b/g protocol at 2.4GHz can be to links between hosts and APs. Each protocol has several non-interfered frequency channels.

Here, we note that IEEE 802.11s is the standard to realize the wireless mesh network so that a variety of vendors and users can use this technology to communicate with each other without problems. On the other hand, WIMNET is considered as a general framework for the wireless

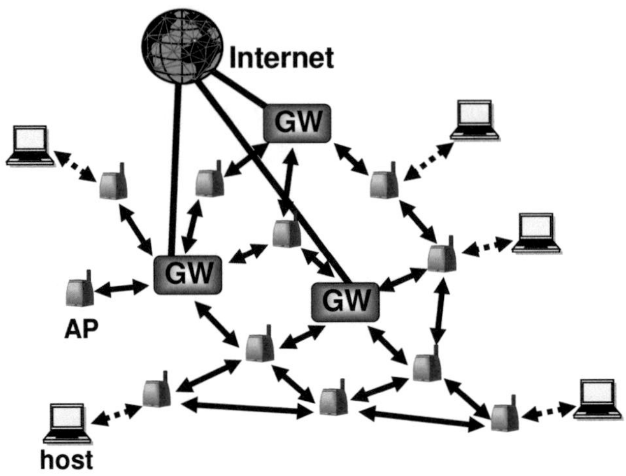

Fig. 1. An WIMNET topology.

Internet-access mesh network. Actually, WIMNET can be realized by adopting IEEE 802.11s on the MAC/link layers.

In WIMNET, all the packets to/from user hosts pass through one of the GWs to access the Internet. If a host is associated with an AP other than a GW, they must reach it through multihop wireless links between APs. In an indoor environment, where WIMNET is mainly deployed, the link quality can be degraded by obstacles such as walls, doors, and furniture. As a result, the APs must be allocated carefully in the network field, so that with the ensured communication quality, they can be connected to at least one GW directly or indirectly, and any host in the field can be covered by at least one AP. Besides, the performance of WIMNET should be maximized by reducing the maximum hop count (the number of links) between an AP and its GW (Li et al., 2000). At the same time, the number of APs and their transmission powers should be minimized to reduce the installation and operation costs of WIMNET. Thus, the efficient solution to this complex task in the AP allocation is very important for the optimal WIMNET design.

As the number of APs increases, WIMNET may frequently suffer from malfunctions of links and/or APs due to hardware faults in this large-scale system and/or to environmental changes in the network field. Even one link/AP fault can cause the disconnection of APs, which is crucial as the network infrastructure. Thus, the dependable AP allocation of enduring one link/AP fault becomes another important design issue for WIMNET. To realize this one link/AP fault dependability, redundant APs should be allocated properly, while the number of such APs should be minimized to sustain the cost increase.

In a large-scale WIMNET, the communication delay may inhibitory increase, because the traffic congestion around the GW exceeds the capacity for wireless communications, if a single GW is used. Besides, the propagation delay can inhibitory increase because of the large hop count between an AP and the GW. Then, the adoption of multiple GWs is a good solution to this problem, where the GW selection to each AP should be optimized at the AP allocation. Here, a set of the APs selecting the same GW is called the *GW cluster* for convenience. To reduce the delay by avoiding the bottleneck GW cluster, the maximum traffic load and hop count in one GW cluster should be minimized among the clusters as best as possible.

In this chapter, first, we present the AP allocation algorithm for WIMNET using the *path loss model* (Rappaport, 1996) to estimate the link quality in indoor environments. This algorithm is composed of the greedy initial stage and the iterative improvement stage. Then, we present the dependability extensions of this algorithm to find link/AP-fault dependable AP allocations tolerating one link/AP fault. Finally, we present the AP clustering algorithm for multiple GWs, which is composed of the greedy method and the variable depth search (VDS) method. The effectiveness of these algorithms is evaluated through network simulations using the WIMNET simulator (Yoshida et al., 2006). This chapter was written based on (Farag et al., 2009; Hassan et al., 2010; Tajima et al., 2010) that have been copyrighted by IEICE, where their reconstitutions in this chapter are admitted at 10RB0023, 10RB0024, and 10RB0025.

2. Access point allocation algorithm

2.1 Network model

A closed area such as one floor in an office/school building, a conference hall, or a library, is considered as the network field for WIMNET. Like (Lee et al., 2002; Li et al., 2007), we adopt the discrete formulation for the AP allocation problem. On this field, discrete points called *host points* are considered as locations where hosts and/or APs may exist. Every host point is associated with the number of possibly located hosts there. Besides, a subset of host points are given as *battery points* where the electricity can be supplied to operate APs. Thus, any AP location must be selected from battery points. Here, we note that some host points are allowed to be associated with zero hosts, so that some battery points can exist without any host association. A subset of battery points can be candidates for GWs to the Internet. This GW selection is also the important mission of the AP allocation problem.
In an indoor environment, the estimation of the signal strength received at a point is essential to determine the availability of the wireless link from its source node (host or AP) to this point, because it is strongly affected by obstacles between them. To estimate it properly, this chapter employs the following *log-distance path loss model* that has been used successfully for both indoor and outdoor environments (Rappaport, 1996; Faria, 2005; Kouhbor, Ugon, Rubinov, Kruger & Mammadov, 2006):

$$P_d = P_1 - 10 \cdot \alpha \cdot \log_{10} d - \sum_k n_k \cdot W_k + X_\sigma \tag{1}$$

where P_d represents the received signal strength (dBm) at a point with the distance d (m) from the source, P_1 does the received signal strength (dBm) at a point with 1 m distance from it when no obstacle exists, α does the path loss exponent, n_k does the number of type-k obstacles along the path between the source and the destination, W_k does the signal attenuation factor (dB) for the type-k obstacle, and X_σ does the Gaussian random variable with the zero mean and the standard deviation of σ (dB). Table 1 shows the signal attenuation factor associated with five types of obstacles often appearing in indoors (Kouhbor, Ugon, Mammadov, Rubinov & Kruger, 2006). Thus, the model determines the received signal strength not only by the distance between the source and the destination, but also by the effect from obstacles along the path between them.
The proper value for the parameter pair (α,σ) depends on the network environment. Measurements in literatures reported that α may exist in the range of 1.8 (lightly obstructed environment with corridors) to 5 (multi-floored buildings), and σ does in the range of 4 to 12 dB (Faria, 2005). After calculating the received signal strength at a point, we regard that the wireless link from its source can exist to this point if the strength is larger than the threshold.

concrete slab	13
block brick	8
plaster board	6
window	2
door	3

Table 1. Attenuation factors of five obstacle types.

2.2 AP allocation problem

2.2.1 Objectives of AP allocation

The proper AP allocation in WIMNET needs to consider several conflicting factors at the same time. First, the resulting WIMNET must be feasible as the Internet access network. That is, any AP must be connected to at least one GW to the Internet, and any host in the field be covered by at least one AP. Then, the performance of WIMNET should be maximized (de la Roche et al., 2006), while the AP installation/operation cost be minimized (Nagy & Farkas, 2000). The performance can be improved by reducing the maximum hop count (the number of hops) between an AP and the GW (Li et al., 2000). Besides, the maximum load limit for any AP should be satisfied to enforce the load balance between APs, where their proper load balance also improves the performance (Hsiao et al., 2001). Furthermore, the signal transmission power of an AP should be minimized to reduce the operation cost and the interference of links using the same radio channel. Hence, the objectives can be summarized as follows:

- to minimize the number of installed APs,
- to minimize the maximum hop count to reach a GW from any AP along the shortest path, and
- to minimize the transmission power of each AP.

2.2.2 Formulation of AP allocation problem

Now, we define the AP allocation problem for WIMNET.

- **Input:** A set of host points $HP = \{h_i\}$ with the number of possibly located hosts hn_i for the host point h_i, a set of battery points $BP = \{b_j\} \subseteq HP$ with the AP installation cost bc_j for the battery point b_j, a set of GW candidates $GC \subseteq BP$, the number of hosts that any AP can cover as the load limit L, and a set of discrete AP transmission powers TP for P_1.
- **Output:** A set of AP allocations S with the selected transmission power p_j for $b_j \in S$.
- **Constraint:** To satisfy the following six constraints:
    1) to cover every host point that has possibly located hosts by an AP,
    2) to connect every AP directly or indirectly,
    3) to allocate APs at battery points ($S \subseteq BP$),
    4) to include at least one GW ($S \cap GC \neq \phi$),
    5) to select one transmission power from TP for each AP, and
    6) to associate L or less hosts for any AP.
- **Objective:** To minimize the following cost function:

$$E = A \sum_{b_j \in S} bc_j + B \max_{b_j \in S} \left\{ \left| R_j \right| \right\} + C \sum_{b_j \in S} p_j \Big/ |S| \tag{2}$$

where A, B, and C are constant coefficients, $|R_j|$ is the hop count from the AP at the battery point b_j to the GW, and p_j is its transmission power. The A-term represents the total installation cost of APs, the B-term does the maximum hop count, and the C-term does the average transmission power.

2.3 Proof of NP-completeness

The *NP-completeness* of the decision version of the *AP allocation problem* in WIMNET is proved through reduction from the known *NP*-complete *connected dominating set problem for unit disk graphs* (Lichtenstein, 1982; Clark & Colbourn, 1990).

2.3.1 Decision version of AP allocation problem

The decision version of the AP allocation problem *AP-alloc* is defined as follows:

- **Instance:** The same inputs as the AP allocation problem and an additional constant E_0.
- **Question:** Is there an AP allocation to satisfy $E \leq E_0$?

2.3.2 Connected Dominating Set Problem for Unit Disk Graphs

The connected dominating set problem for unit disk graphs *CDS-UD* is defined as follows:

- **Instance:** a unit disk graph $G = (V, E)$ and a constant volume K, where a unit graph is an intersection graph of circles with unit radius in a plane.
- **Question:** Is there a connected subgraph $G_1 = (V_1, E_1)$ of G such that every vertex $v \in V$ is either in V_1 or adjacent to a member in V_1, and $|V_1| \leq K$?

2.3.3 Proof of NP-completeness

Clearly, *AP-alloc* belongs to the class NP. Then, an arbitrary instance of *CDS-UD* can be transformed into the following *AP-alloc* instance:

- **Input:** $HP = BP = GC = V$, $h_i = 1$, $bc_j = 1$, $L = \infty$, $TP = \{pw_0\}$, $A = 1$, $B = C = 0$, and $E_0 = K$.

pw_0 is the transmission power to generate a link between two APs whose distance corresponds to the unit radius in the unit disk graph. In this *AP-alloc* instance, the cost function E is equal to the number of vertices in *CDS-UD*, which proves the NP-completeness of *AP-alloc*.

2.4 AP allocation algorithm

In this subsection, we present a two-stage heuristic algorithm composed of the initial stage and the improvement stage for the AP allocation problem. Because the GW to the Internet is usually fixed due to the design constraint of the network field in practical situations, our algorithm finds a solution for the fixed GW. By selecting every point in *GC* as the GW and comparing the corresponding solutions, this algorithm can find an optimal solution for the AP allocation problem.

2.4.1 Initial stage

The initial stage consists of the *host coverage process* and the *load balance process* to allocate APs satisfying the constraints of the problem. Here, the maximum transmission power is always assigned to any AP in order to minimize the number of APs.

– **Host Coverage Process**
  The host coverage process repeats the sequential selection of one battery point that can cover the largest number of uncovered hosts without considering the load limit constraint, until every host point is covered by at least one AP.

  1. Initialize the AP allocation S by the given GW g ($S = \{g\}$).
  2. Assign the maximum transmission power in TP to the new AP, and calculate the path loss model in (1) to evaluate connectivity to APs, battery points, and host points.
  3. Terminate this process if every host point is covered by an AP in S.
  4. Select a battery point b_j satisfying the following four conditions:
     1) b_j is not included in S,
     2) b_j is connected with at least one AP in S,
     3) b_j can cover the largest number of uncovered hosts, and
     4) b_j has the largest number of incident links to selected APs in S (maximum degree) for tie-break, if two or more APs become candidates in 3).
  5. Go to 2.

– **Load Balance Process**
  The host coverage process usually does not satisfy the load limit constraint for host associations, where some APs may be associated with more than L hosts. If so, the following load balance process selects new battery points for APs to reduce their loads.

  1. Associate each host point to the AP such that the received power is maximum among the APs.
  2. Calculate the number of hosts associated with each AP.
  3. If every AP satisfies the load limit constraint, calculate the cost function E for the initial AP allocation and terminate this process.
  4. For each AP that does not satisfy this constraint, select one battery point closest from it into S.
  5. Go to 1.

2.4.2 Improvement stage

In the initial stage, the AP allocation can be far from the best one in terms of the cost function due to the greedy nature of this algorithm and to additional APs by the load balance process. Actually, the transmission power is not reduced at all. Thus, the improvement stage of our algorithm improves the location, the power transmission, and the host association jointly by using a local search method. At each iteration, the location is first modified by randomly selecting a new battery point for the AP, and removing any redundant AP due to this new AP. Then, associated APs to the host points around the effected APs are improved under the current AP allocation, and the transmission power is reduced if possible. During the iterative search process, the best solution in terms of the cost function E is always updated for the final output. In the improvement stage, the following procedure is repeated for a constant number of iterations T:

1. Randomly select a battery point $b_j \notin S$ that is connected to an AP in S, and add it into S with the maximum transmission power.

2. Apply the AP association refinement in 2.4.3.
3. Remove from S any AP that satisfies the following four conditions:
   1) it is different from b_j and the GW,
   2) all the host points can be covered by the remaining APs if removed,
   3) all the APs can be connected if removed, and
   4) the load limit constraint is satisfied if removed.
4. Change the transmission power of any possible AP to the smallest one in TP such that this AP can still cover any of the associated host and maintain the links necessary to connect all the APs.

2.4.3 AP association refinement

After locations of APs are modified, some host points may have better APs for associations in terms of the received power than their currently associating APs. To correct AP associations to such host points, the following procedure is applied:

1. Find the better AP for association to every host point in terms of the received power in (1).
2. Apply the following procedure for every host point that is associated with a different AP from the best:
   a) Change the association of this host point to the best AP, if its load is smaller than the load limit.
   b) Otherwise, swap the associated APs between such two host points, if this swapping becomes better.

2.5 Performance evaluation

We evaluate the AP allocation algorithm through simulations using the WIMNET simulator.

2.5.1 WIMNET simulator

The WIMNET simulator simulates least functions for wireless communications of hosts and APs that are required to calculate throughputs and delays, because it has been developed to evaluate a large-scale WIMNET with reasonable CPU time on a conventional PC. A sequence of functions such as host movements, communication request arrivals, and wireless link activations are synchronized by a single global clock called *a time slot*. Within an integral multiple of time slots, a host or an AP can complete the one-frame transmission and the acknowledgement reception.

From our past experiments (Kato et al., 2007) and some references (Proxim Co., 2003; Sharma et al., 2005), we set $30Mbps$ for the maximum transmission rate for IEEE 802.11a and $20Mbps$ for IEEE 802.11g. Note that this transmission rate can cover about 26 hosts (Gast, 2002; Bahri & Chamberland, 2005). Then, if the duration time of one time slot is set $0.2ms$ and each frame size is $1{,}500bytes$, two time slots can complete the $30Mbps$ link activation because $(1{,}500byte \times 8bit \times 10^{-6}M)/(0.2ms \times 2slot \times 10^{-3}s) = 30Mbps$, and three slots can complete the $20Mbps$ link activation because $(1{,}500byte \times 8bit \times 10^{-6}M)/(0.2ms \times 3slot \times 10^{-3}s) = 20Mbps$. We note that the different transmission rate can be set by manipulating the time slot length and the number of time slots for one link activation. When two or more links within their wireless ranges may be activated at the same time slot, randomly selected one link among them is

successfully activated, and the others are inserted into waiting queues to avoid collisions, supposing DCF and RTS/CTS functions.
In order to evaluate the throughput shortly, every host has 1,000 packets to be transmitted to the GW, and the GW has 125 packets to every host before starting a simulation. Then, when every packet reaches the destination or is lost, the simulation is finished. Here, no packet is actually lost by assuming the queue with the infinite size at any AP in our simulations. The packets for each request are transmitted along the shortest path that is calculated for the hop count by our algorithm. Only the connection-less communication is implemented this time, where the retransmission between end hosts is not considered.
The throughput comparison using this simple WIMNET simulator is actually sufficient to show the effectiveness of our algorithm, because it simulates the basic behaviors affecting the throughput of the wireless mesh network, such as the contention resolution among the interfered links and the packet relay action for the multihop communication. Note that our experimental results in a simple topology confirmed the throughput correspondence between the simulator and the measurement. The packet retransmission of the interfered link, if implemented, can worse the throughput by the poor AP allocation in comparisons, because it causes more interferences between links.

2.5.2 Algorithm parameter set

In our simulations, we select the following set of parameter values. For the path loss model in (1), we use $\alpha = 3.32$, and $P_1 = -20dBm$ as the maximum transmission power of an AP (Faria, 2005), with four additional choices with $10dBm$ interval ($TP = \{-20, -30, -40, -50, -60\}$). We set $X_\sigma = 0$ and consider only concrete slabs or walls with $W_k = 13$ as obstacles of the signal propagation in the field for simplicity. We select $-90dBm$ as the threshold of a link by referring the Cisco Aironet 340 card data sheet (Cisco Systems, Inc., 2003). For the cost function in (2), we use $A = B = 1$ and $C = 0.05$. For the improvement stage, we select $T = 10{,}000$ for iterations. Here, we note that in (Faria, 2005), $\alpha = 3.32$ is selected for the outdoor environment, whereas $\alpha = 4.02$ is for the indoor. However, $\alpha = 4.02$ represents the average attenuation in the environment with mixtures of walls, doors, windows, and other obstacles in a large room. On the other hand, $\alpha = 3.32$ represents the attenuation strongly affected by the wall, where the signal measured inside a building comes from the transmitter at the outside through one wall. In this chapter, we consider a floor in a building as the indoor environment, where walls separating rooms mainly cause the attenuation and their count along the propagation path is very important to estimate the received signal strength. Experimental results in our building (Kato et al., 2006) show that the wireless link between two APs is actually blocked if they are located in rooms separated by two walls without any glass window, and is active if separated by only one wall. In futures, we should use a proper value for α after measuring received signal strengths in the network field.
After the AP allocation with the routing (shortest path) is found by the proposed algorithm, the links in the routing are assigned channels by the algorithm in (Funabiki et al., 2007) for simulations, whose goal is to find the additional NIC assignment to congested APs for multiple channels and the channel assignment to the links. The first stage of this two-stage algorithm repeats one additional NIC assignment to the most congested AP until its given bound. Then, the second stage sequentially assigns one feasible channel to the link such that it can minimize the interference between links assigned the same channel. The link channel assignment is actually realized by assigning the same channel to the NICs at the both end APs of the link. If some NICs are not assigned any channel, they are moved to different APs and

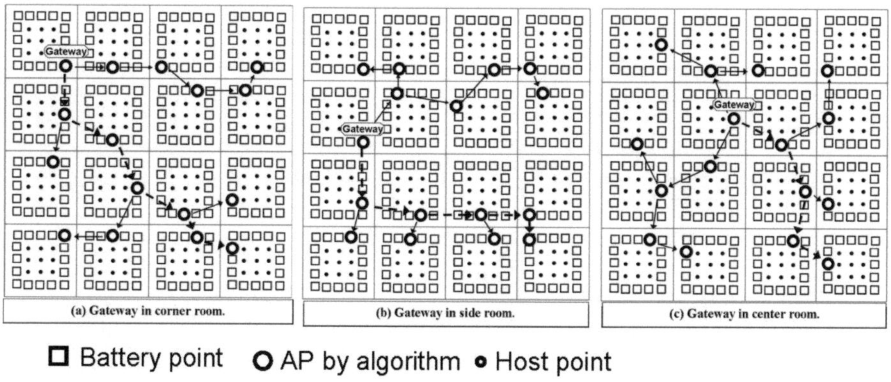

Fig. 2. AP allocations with three GW positions for network field 1.

the channel assignment is repeated from its first step.

2.5.3 Network field 1

To investigate the optimality of our algorithm in terms of the number of allocated APs and the maximum hop count, we adopt an artificial symmetric network field that is composed of 16 square rooms with the $60m$ side as the first field. In each room, $25(= 5 \times 5)$ host points are allocated with the $10m$ interval, and each host point is associated with one host. The 16 host points along the walls in a room are selected as battery points, because electrical outlets are usually installed on walls. The total of 400 host points are distributed regularly in the field. The maximum load constraint L is set 25, which indicates that the lower bound on the number of allocated APs becomes 16 to cover the host points from the calculation of $400(= \textit{total number of hosts})/25(= L)$. For this field, we examine the effect of the GW position for the AP allocation and the network performance. For this purpose, we prepare three GW positions as the input to our algorithm, namely in the *corner room*, in the *side room*, and in the *center room*. Figure 2 illustrates their AP allocations with routings found by our algorithm.

Table 2 summarizes the solution quality indices of our AP allocations for three GW positions in network field 1. The same single channel is used for every link in network simulations. The throughput is calculated by dividing the total amount of received packets by the simulation time. The average result among ten simulation runs using different random numbers for packet transmissions is used to avoid the bias of random number generations. Our algorithm finds lower bound solutions in terms of the number of APs (=16 APs) and the maximum hop count for any GW position. In this field, an AP in any room can communicate with an AP in its four-neighbor room at the maximum, due to the signal attenuation at the wall. Here, one side of the room is $60m$, and any host point is at least $10m$ away from the wall. The communication range of an AP is reduced to $52m$ when it passes through one wall, and to $21m$ when it passes through two walls. Thus, the minimum hop count to the farthest AP from the GW, which represents the lower bound on the maximum hop count, is six for the corner room GW, five for the side room GW, and four for the center room GW. The throughput comparison between three cases shows that the GW in the center room provides the best one with the smallest hop count.

GW position	corner	side	center
# of APs	16	16	16
its lower bound	16	16	16
max. hop count	6	5	4
its lower bound	6	5	4
throughput (Mbps)	12.02	12.08	13.48

Table 2. AP allocation results for network field 1

2.5.4 Simulation results for network field 2

Then, we adopt the second network field that simulates one floor in a building as a more practical case. This field is composed of two rows of the same rectangular rooms and one corridor between them. Each row has eight identical rooms with $5m \times 10m$ size. In each room, $15(= 3 \times 5)$ host points are allocated regularly with one associated host for each point, and the six host points along the horizontal walls (three along the external wall and three along the corridor wall) are selected as battery points. Besides, nine battery points are allocated in the corridor with no host association where the center one is selected as the GW. Thus, the total number of expected hosts is $240(= 15 \times 16)$. The maximum load limit L is set 25. As a result, the lower bound on the number of APs to satisfy the load constraint is $10(= \lceil \frac{240}{25} \rceil)$.

Figure 3 shows our AP allocation for this field using 10 APs that are represented by circles. Every AP other than the GW has a one hop distance from the GW. Thus, our algorithm found the lower bound solution. For the comparison, a manual allocation using 17 APs is also depicted there by triangles, where one AP is allocated to each room regularly. The maximum hop count of this manual allocation is two as shown by lines in the figure.

In network field 2, the effect of the multiple channels for throughputs is investigated by allocating two NICs (Network Interface Cards) to the GW (Raniwala et al., 2005), in addition to the single channel case. The channels of links are assigned by using the algorithm in (Funabiki et al., 2007). Table 3 compares throughputs between two allocations when 1 NIC or 2 NICs are assigned at the GW. Our allocation provides about 36% better throughput than the manual allocation for the practical case using the single NIC, by avoiding unnecessary link activations.

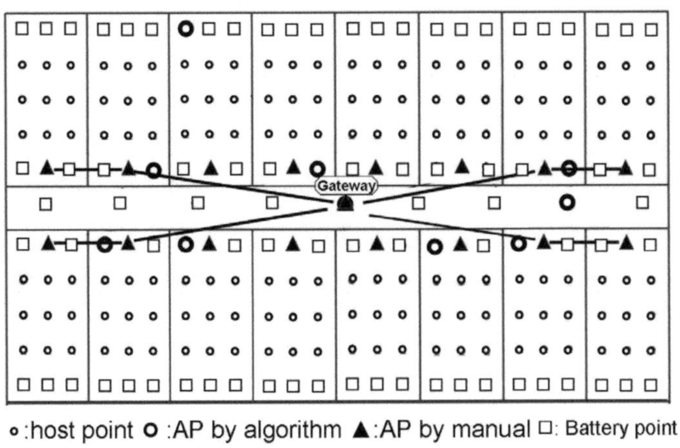

Fig. 3. AP allocations for network field 2.

method	algorithm	manual
# of APs	10	17
max. hop count	1	2
1-NIC throughput (Mbps)	30.96	22.79
2-NIC throughput (Mbps)	47.74	46.26

Table 3. AP allocation results for network field 2

However, for the 2-NIC case, the advantage becomes small by allowing the enough bandwidth for communications between APs.

2.5.5 Network field 3

Finally, we examine the third network field as the more practical and harder one similar to a building floor in our campus. This field is composed of two rows of different-sized rectangular rooms and one corridor. One row has 12 small square rooms with $5m \times 5m$ size with 4 host points, and another row has 5 large rectangular rooms with $10m \times 12.5m$ size with 20 host points. The host points along the walls parallel to the corridor are selected as battery points. Besides, 29 battery points are allocated with the same interval in the corridor with no host association. The battery point in front of the center of the fifth small room in the corridor is selected as the GW, so that in the manual allocation, each AP in the corridor can cover three small rooms regularly. The total number of expected hosts is $148 (= 4 \times 12 + 20 \times 5)$. The maximum load limit L is again set 25. Thus, the lower bound on the number of APs to satisfy the load constraint is $6 (= \lceil \frac{148}{25} \rceil)$.

Figure 4 shows our AP allocation using 6 APs for this field. Every AP other than the GW has one hop distance from the GW. Thus, our algorithm found the lower bound solution. For comparisons, a manual allocation using 9 APs is also depicted, where one AP is allocated to each large room and 4 APs are allocated in the corridor regularly. The maximum hop count of this manual allocation is two as shown by lines. Table 4 compares the throughputs between two allocations, where our allocation provides the better throughput than the manual one for both 1-NIC and 2-NIC cases.

2.5.6 Effect of estimation error of log-distance path loss model

The estimation error of the log-distance path loss model in (1) may have the considerable impact to the result of our algorithm. To estimate this impact briefly, we calculate the percentage of the received signal strength drop in the real world from the estimation that

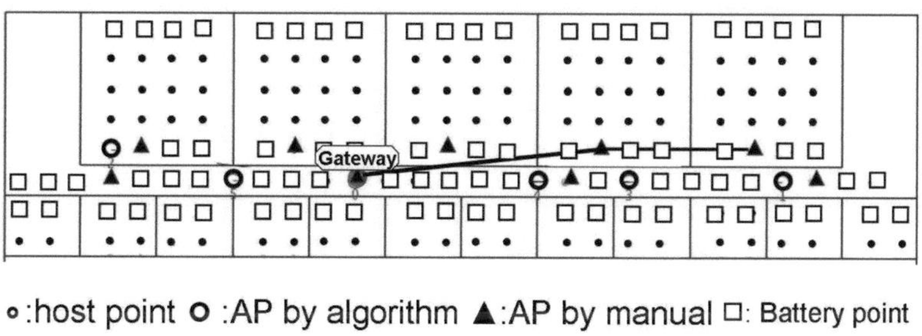

Fig. 4. AP allocations for network field 3.

method	algorithm	manual
# of APs	6	9
max. hop count	1	2
1-NIC throughput (Mbps)	33.19	27.16
2-NIC throughput (Mbps)	55.54	52.58

Table 4. AP allocation results for network field 3

causes the disconnection at the AP allocation. As shown in Table 5, this percentage is distributed from 3% in the network field 1 to 30% in the field 3. In our future works, we will improve our algorithm in terms of the robustness to the estimation error of the log-distance path loss model, such that the connectivity is maintained while the interference is curbed even if the model has the error.

2.6 Related works

Several papers have reported studies of AP placement algorithms for conventional WLANs. Within our knowledge, the same AP allocation problem in the wireless mesh network for the Internet access in indoor environments has not been reported before. In fact, most of the papers focus on the construction of WLAN without considering wireless connections between APs, or on the GW placement for the wireless mesh networks.

In (Lee et al., 2002), Lee et al. study simple ILP formulations for the AP placement and channel assignment problems in conventional WLANs, using discrete placement formulations. Their algorithm finds best AP associations of host points to minimize the maximum channel utilization among APs. In their WLANs, APs are connected with each other through wired connections, whereas our AP allocation problem must satisfy the connectivity among APs through wireless connections. This additional constraint makes the problem much harder, because it usually requires the more number of APs to provide wireless connections between them while the number of APs should be minimized to reduce the cost and the interference between links. Besides, their algorithm does not consider the minimization of APs and their transmission powers.

In (Kouhbor, Ugon, Rubinov, Kruger & Mammadov, 2006), Kouhbor et al. investigate the AP allocation problem in indoors for WLANs with a path loss model to calculate the coverage area of an AP. They present a continuous mathematical model of finding AP locations to cover every user while avoiding insecure locations, which is solved by their global optimization algorithm. The effectiveness is verified through simulating one real building floor. They observe that the dimension of the building, the number of users and their locations, the transmission power, and its received threshold have effects on the AP allocation. Unfortunately, they do no consider the wireless connection constraint, like (Lee et al., 2002).

In (Bahri & Chamberland, 2005), Bahri et al. study the problem of designing a conventional WLAN, and propose an optimization model for selecting the location, the power, and the channel for each AP. They propose a Tabu search heuristic algorithm to improve this solution.

network field 1			network field 2	network field 3
corner	side	center		
3%	3%	4%	12%	30%

Table 5. Percentage of received signal strength drops for AP allocation failure

The results are compared to lower bounds obtained by relaxing a subset of the constraints in their model, and show that this heuristic produces relatively good solutions rapidly. It is significant to develop the lower bound formulation in order to precisely evaluate the proposed heuristic, and to explore exact algorithms to solve small-size instances of the problem.

In (Chandra et al., 2004), Chandra et al. formulate the Internet transit access point placement problem under various wireless models. This problem aims to provide the Internet connectivity in multihop wireless networks. If we consider the Internet transit access point as a GW, their network model is the same as WIMNET where every AP becomes a GW.

In (Wu & Hsieh, 2007), Wu et al. investigate the impact of multiple wireless mesh networks that are overlapped in a service area. They formulate the resource sharing problem as an optimization problem, and present a general LP formulation. They consider the optimization of the number and the selection of bridge nodes. Simulation results show that if a proper interworking is provided between overlapped networks, significant performance gain can be obtained.

In (Li et al., 2007), Li et al. study the GW placement problem for the throughput optimization in wireless mesh networks, given the traffic demand for each node, the number of GWs, and the interference model. They present an LP formulation to find a periodic TDMA link scheduling to maximize the throughput for given GW locations. Then, by applying it with every possible combination of the grid points superimposed on the field for GW locations, they find the best GW layout.

In (Robinson et al., 2008), Robinson et al. study the GW placement problem for the wireless mesh network. They present a technique to efficiently compute the GW-limited fair capacity as a function of the contention at each GW, and two GW placement algorithms. The first *MinHopCount* adapts a local search algorithm for the capacitated facility location problem in (Pal et al., 2001) that is composed of *add*, *open*, and *close* operations. The second *MinContention* adopts a swap-based local search algorithm for the incapacitated k-median problem with a provable performance guarantee.

In (Naidoo & Sewsunker, 2007), Naidoo et al. discuss the use of Mesh technology as a strategy to extend coverage to provide rural telecommunication services. Their study investigates the range extension using a hybrid wireless local area network architecture running both infrastructure and client wireless mesh networks.

2.7 Conclusion

This section presented the two-stage AP allocation algorithm for WIMNET in indoor environments. The effectiveness was verified through simulations using the WIMNET simulator, where the significant performance improvement over manual allocation was observed. The future works may include the more precise consideration of indoor environments in the signal propagation model (Beuran et al., 2008), the algorithm improvement in terms of the robustness to the estimation error of the model, the adoption of the ILP formulation (Lee et al., 2002) and the global optimization algorithm (Kouhbor, Ugon, Rubinov, Kruger & Mammadov, 2006) to the AP allocation problem, and the application to the design of real wireless mesh networks.

3. Dependability extensions of AP allocation algorithm

3.1 Fault dependability in WIMNET

WIMNET may be disconnected by occurrence of even one link fault or one AP fault in the AP allocation found by the algorithm in the previous section. To improve the dependability of WIMNET, the AP allocation algorithm should be extended to find an AP allocation such that the APs can be connected even for one link fault or one AP fault occurrence. This dependability can be achieved by allocating redundant APs to provide backup routes (Ramamurthy et al., 2001). At the same time, the number of such APs and the maximum hop count should be minimized for the cost reduction and the performance improvement. Here, we summarize the design goal in dependability extensions of the AP allocation algorithm as follows:

1. to endure one link fault or one AP fault,
2. to minimize the number of additional APs, and
3. to minimize the maximum hop count.

3.2 Link-fault dependability extension

3.2.1 Constraint for link-fault dependability

First, we discuss the *link-fault dependability* extension of the AP allocation algorithm. To achieve the link-fault dependability, the network must be connected if any link is removed from there. Then, another constraint must be satisfied in the AP allocation in addition to the original six constraints in 2.2.2:

7) to provide the connectivity among the APs if any link is removed.

3.2.2 Algorithm extension for link-fault dependability

Then, we present the algorithm extension for the link-fault dependability. The idea here is that after maximizing the transmission power from any AP to increase the connectivity, we find any link whose removal disconnects the network, which is called the *bridge*. While bridges exist, we sequentially allocate an additional AP at the battery point that can resolve the maximum number of bridges until all of them are resolved. Then, we find the minimum-delay routing tree to this link-fault dependable AP allocation by applying the algorithm in (Funabiki et al., 2008). Finally, we minimize the transmission powers of APs such that the constraints of the problem are satisfied. The following procedure describes the link-fault dependability extension:

1. Input the AP allocation from the algorithm in (Farag et al., 2009).
2. Maximize the transmission power for any AP and find the links between two APs.
3. Find the set of bridges BR.
4. Apply the following procedure if $BR = \emptyset$:
    a. Apply the AP association refinement in 2.4.3.
    b. Apply the routing tree algorithm in (Funabiki et al., 2008).
    c. Minimize the transmission power of the APs such that all the constraints are satisfied.
    d. Terminate the procedure.
5. For every bridge in BR, find the set of battery points that can resolve this bridge if a new AP is allocated there. Let this set of the battery points found here be BS.
6. Calculate the number of bridges in BR for each battery point in BS that the AP allocated there can resolve.

7. Find the battery point in BS that can resolve the largest number of bridges in BR, and allocate an AP there.
8. Update BR.
9. Go to 4.

3.3 AP-fault dependability extension

3.3.1 Constraint for AP-fault dependability

Next, we discuss the *AP-fault dependability* extension of the AP allocation algorithm. To achieve the AP-fault dependability, the network must be connected, and every host must be covered by a remaining AP, if any AP is removed from there. Here, no GW is removed, assuming no fault at GW. Then, the following two constraints must be satisfied in the AP allocation in addition to the original six constraints in 2.2.2:

7) to cover any host by an existing AP if any AP is removed, and

8) to provide the connectivity among the APs if any AP is removed.

3.3.2 Algorithm extension for AP-fault dependability

We present the algorithm extension to the AP-fault dependability. For the AP-fault dependability, at least the link-fault dependability must be satisfied, because if one AP is removed from the network, its incident links are also removed. Thus, in this extension, we use the link-fault dependable AP allocation and maximize the transmission power of any AP as the initial state.

First, we find any host point that cannot be covered if one AP is removed from the network due to the fault, called the *critical point*, in the initial state. The critical point satisfies the following either condition:

1) only this fault AP covers it, or

2) all the backup APs reach association load limits, including the re-associated hosts by this AP fault.

While critical points exist, we sequentially allocate an additional AP to the battery point that can cover the maximum number of critical points until all of them are resolved. Then, we find any AP whose removal disconnects the network, called the *cut AP*. While cut APs exist, we sequentially allocate an additional AP to the battery point that can cover the maximum number of cut APs until all of them are resolved.

After these procedures, we apply the improvement stage in 3.3.3 for finding the better AP allocation. Then, we apply the algorithm in (Funabiki et al., 2008) to find the routing tree to the AP-fault dependable allocation. Finally, we minimize the transmission powers such that the constraints are satisfied. The following procedure describes the AP-fault dependability extension:

1. Input the link-fault dependable AP allocation.
2. Maximize the transmission power for any AP and find the links between APs.
3. Find the set of critical host points CR.
4. Apply the following *critical host resolution* procedure until $CR = \emptyset$:
   a. For every host point in CR, find the set of battery points that can cover this critical point if a new AP is allocated there. Let this set of the battery points found here be CS.

b. Calculate the number of critical points in CR for each battery point in CS that the AP allocated there can cover.

c. Find the battery point in CS that can cover the largest number of critical points in CR, and allocate an AP there.

d. Update CR.

5. Find the set of cut APs CA.
6. Apply the following *cut AP resolution* procedure until $CA = \emptyset$:

   a. For every cut AP in CA, find the set of battery points that can cover this cut AP if a new AP is allocated there. Let this set of the battery points found here be CB.

   b. Calculate the number of cut APs in CA for each battery point in CB that the AP allocated there can cover.

   c. Find the battery point in CB that can cover the largest number of cut APs in CA, and allocate an AP there.

   d. Update CA.
7. Apply the improvement stage in 3.3.3.
8. Apply the AP association refinement in 2.4.3.
9. Apply the routing tree algorithm in (Funabiki et al., 2008).
10. Minimize the transmission power of the APs such that all the constraints are satisfied.
11. Terminate the procedure.

3.3.3 Improvement stage

The improvement stage for the AP-fault dependable extension has been slightly modified from the corresponding one in the original AP allocation algorithm, such that any AP must be connected with at least two APs in order to preserve the link/AP fault dependability. The following procedure is repeated for a given constant number of iterations AT, where the best solution in terms of the cost function F is always kept for the final solution during the iterative search process:

1. Randomly select a battery point $b_j \notin S$ that is connected to at least two APs in S, and add it into S with the maximum transmission power.
2. Apply the AP association refinement in 2.4.3.
3. Remove from S any AP that satisfies the following four conditions:

   1) it is different from b_j and GW,

   2) all the host points associated with the AP can be re-associated with the remaining APs, where for the new association of each host point, the load limit constraint is checked from the AP whose signal power is largest if two or more APs can be associated,

   3) no cut AP appears if removed, and

   4) no critical host point appears if removed.
4. If removed, re-associate all the host points associated with this AP to the APs found in 2).
5. Change the transmission power of any possible AP to the smallest one in TP such that this AP can still cover any associated host and maintain the links necessary to connect all the APs.

3.4 Simulation results for dependability extensions

3.4.1 Simulated instances

In this subsection, we show simulation results for the dependability extension using the WIMNET simulator. A network field composed of 16 square rooms with 400 host points, and a field similar to the first floor in the central library at Cairo university as a practical one, are considered for simulated instances. Like the previous instance, each host point is associated with one host, and the maximum load limit L is set 25. In the latter field, the total size is $64m \times 32m$, and 411 host points are allocated, where the host points along the walls are selected as battery points. Note that the size of the largest room at the top right, called *Taha Hussin Hall*, is $18m \times 12m$ with 74 host points. The lower bound on the number of APs to satisfy the load constraint is $17(= \left\lceil \frac{411}{25} \right\rceil)$.

Figures 5 and 6 illustrate the network field and the AP allocation result with the routing tree for each instance, respectively. The white circle represents an AP allocated by the original algorithm, the gray circle does an additional AP by the link-fault dependability extension, and the black circle does an additional AP by the AP-fault dependability extension.

3.4.2 AP allocation results

First, we discuss the solution quality in terms of the number of APs in AP allocation results for dependability extensions. Table 6 compares the numbers of APs in the original AP allocation algorithm, the link-fault extension, and the AP-fault extension. For the artificial network field of 16 square rooms (Square field), our dependability extensions can provide the link-fault dependability with additional three APs, and the AP-fault dependability with additional ten APs. The latter result is much better than the trivial solution for the AP-fault dependability using 15 additional APs where two APs are allocated in each room. For the practical field in the central library (Library field), no additional AP is necessary for the link-fault dependability and only three additional APs for the AP-fault dependability. Because most APs can communicate with GW in one hop, any link can easily be backed up by other

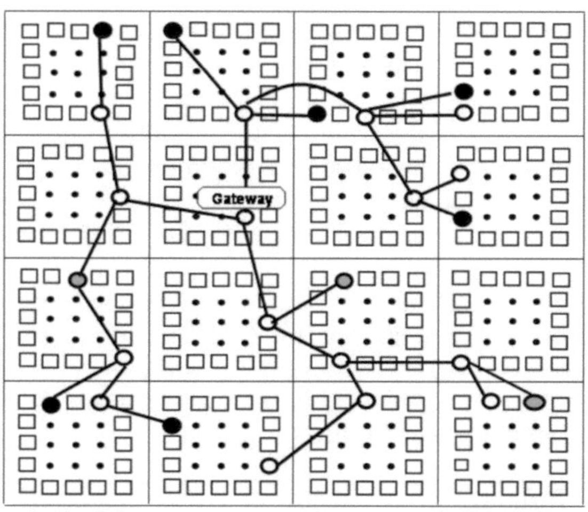

Fig. 5. AP allocation result for dependability extensions in 16 square-room field.

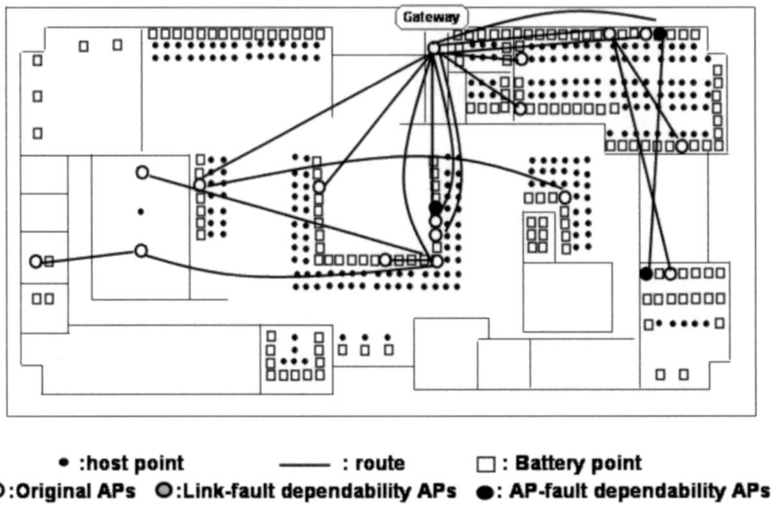

Fig. 6. AP allocation result for dependability extensions in central library field.

links. These results verify the effectiveness of our proposal for dependability extensions in WIMNET in terms of the AP allocation cost.

3.4.3 Throughput results

Then, we investigate throughput changes with or without link/AP faults among AP allocation results for dependability extensions. Table 7 compares total throughputs among AP allocations for the three cases when no link/AP has fault. The result indicates that the total throughput is slightly degraded as the number of APs increases for the fault dependability extensions due to the increase of the interference among wireless links between APs using the single channel.

Tables 8 and 9 show the average, maximum, and minimum throughputs in the link-fault dependable and AP-fault dependable allocations when one link or AP is removed from the network to assume the occurrence of a fault. By comparing these results, we conclude that our proposal can provide sufficient throughputs, even if one link fault or one AP fault occurs in WIMNET.

Here, we note that in the fault dependable AP allocation, some APs may become redundant. Thus, the routing without using such APs may be able to improve the performance by reducing the interference. Besides, if multiple NICs are used at APs for multiple channel communications, the results can be changed by reducing the interference. The performance evaluation in such cases will be in our future studies.

Instance	Original	Link-fault	AP-fault
Square field	16	19	26
Library field	17	17	20

Table 6. Numbers of allocated APs.

Instance	Original	Link	AP
Square field	13.0	12.9	12.6
Library field	23.9	23.9	23

Table 7. Total throughputs with no fault (Mbps).

Instance	Ave.	Max.	Min.
Square field	12.4	12.9	10.9
Library field	23.37	23.74	23

Table 8. Total throughputs for link-fault extension with one link fault (Mbps).

3.5 Related works

Several studies have been reported for the dependability in multihop wireless networks including wireless mesh networks. This subsection briefly introduces some of them.

In (Gupta & Younis, 2003), Gupta et al. presented efficient detection and recovery mechanisms of one failed GW or its link in a clustered wireless sensor network. The detection is based on the consensus of healthy GWs. The recovery reassociates the sensors that are managed by the failed GW to other clusters based on the range information. The effectiveness is verified through simulations.

In (Varshney & Malloy, 2006), Varshney et al. presented the multilevel fault tolerance design of wireless networks using adaptable building blocks (ABBs). The ABB has several levels of components such as base stations, base station controllers, databases, and links, similar to cellular networks, where the reliability such as MTBF/MTTR can differ significantly by using different number of components. The fault tolerance design is achieved at the three levels of the component and link, the building block, and the interconnection. If the computed dependability attributes are not acceptable, the process of adding the incremental redundancy at the three levels is repeated. They present an analytical model of measuring the dependability enhancement, and evaluate the network survivability and the network availability with different interconnection architectures, block-level redundancy, mobility, and fault tolerance at the three levels in ring, star, and SONET dual ring topologies.

In (Pan & Keshav, 2006), Pan et al. studied detection and repair methods of faulty APs for large-scale wireless networks. For the detection, they presented three algorithms. The first one is that if an AP gives reports to the network operation center, it is regarded as no fault. The second one modifies the first one such that the no-fault probability of an AP is exponentially decreased as the time interval of no report increases. The third one further improves it by considering the path of APs that the host is moving along, where if an AP along the path does not report, it can be regarded as a fault. For the repair, they presented the ellipse heuristic algorithm to find the best schedule of repairing faulty APs by minimizing the total moving length and the downtime of popular APs. They evaluate their proposal using the free data set available from Dartmouth College that includes log messages from client association, authentication, and others in their wireless networks for nearly four years.

Instance	Ave.	Max.	Min.
Square field	12.31	12.6	11.1
Library field	21.75	22.65	21

Table 9. Total throughputs for AP-fault extension with one AP fault (Mbps).

3.6 Conclusion

This section presented extensions of the AP allocation algorithm to find the link/AP-fault dependable AP allocations, to assure the connectivity and the host coverage in case of one link/AP fault by allocating redundant APs. The effectiveness was verified through simulations in regular and practical network fields using the WIMNET simulator. The future works may include the routing without using redundant APs, the evaluation of multiple channel communications, and the reduction of APs by considering backup APs in different GW clusters.

4. Access point clustering algorithm

4.1 AP clustering in WIMNET

As the number of APs increases in WIMNET with a single GW, the communication delay may inhibitory increase, because the links between APs around the GW become too crowded. Then, the adoption of multiple GWs is a reasonable solution to this problem, where the proper clustering of the APs into a set of disjoint GW clusters is important to maximize the performance of WIMNET.

The proper AP clustering is actually a hard task because it must consider several constraints and optimization indices simultaneously. The first constraint is that the number of APs in a cluster must not exceed the upper limit due to the WDS size constraint. The second one is that all APs in a cluster must be connected with each other. The third one is that one AP in a cluster must be selected as the GW that can deploy wired connections to the Internet. The fourth one is that the number of hosts associated with APs in a cluster must not exceed the limit, so that any cluster can ensure the communication bandwidth of hosts. As the optimizing indices, the number of GW clusters should be minimized to save installation and operation costs of the network. The communication delay between any AP and a GW in any cluster should be minimized to enhance the performance. As a result, the APs, the GW, and the communication routes between APs and the GW in every GW cluster must be found simultaneously.

4.2 AP clustering problem

4.2.1 Assumptions in AP clustering problem

In the AP clustering problem, we assume that the locations of the APs with battery supplies and the wireless links between APs in the network field have been given manually, or by using their corresponding algorithms during the design phase of WIMNET, as the inputs. The topology of the AP network is described by a graph $G = (V, E)$, where a vertex in V represents an AP and an edge in E represents a link. Each vertex is assigned the maximum number of hosts associated with the AP as the load, and each edge is assigned the transmission speed for the delay estimation, which are given as design parameters. A subset of V are designated as *GW candidates*, where wired connections are available for the Internet access. The number of GW clusters K greatly affects the installation and operation costs of WIMNET because the costly Internet GW is necessary in each cluster. Thus, the number of clusters K is given in the input, so that the network designer can decide it. Furthermore, the limit on the cluster size and the required bandwidth in one cluster are given to determine their constraints.

4.2.2 Formulation of AP clustering problem

Now, we formulate the AP clustering problem for WIMNET as a combinatorial optimization problem.

- **Input:** $G = (V, E)$: a network topology with N APs ($N = |V|$), h_i: the maximum number of hosts associated with the AP_i (the i-th AP) for $i = 1, \cdots, N$, s_{ij}: the transmission speed of the ij-th link ($link_{ij}$) from AP_i to AP_j in E, X ($\subseteq V$): a set of GW candidates, K: the number of GW clusters, H: the limit on the number of associated hosts in a GW cluster (bandwidth limit), and P: the limit on the number of APs in a GW cluster (cluster size limit).
- **Output:** $C = \{C_1, C_2, \cdots, C_K\}$: a set of GW clusters, g_k: the GW in C_k for $k = 1, \cdots, K$, and r_i: the communication route between AP_i and the GW.
- **Constraint:** to satisfy the following four constraints:
  - the number of APs in any GW cluster must be P or smaller: $|C_i| \leq P$ (cluster size constraint),
  - the number of associated hosts in any GW cluster must be H or smaller: $\sum_{j \in C_i} h_j \leq H$ (bandwidth constraint),
  - the APs must be connected with each other in any cluster (connection constraint),
  - one GW must be selected from GW candidates in X in any cluster (GW constraint).
- **Objective:** to minimize the following cost function F_c:

$$F_c = A \cdot \max\big(hop(AP_i)\big) + B \cdot \max\big(host(link_{ij}) + \sum_{kl \in intf(ij)} host(link_{kl})\big) \tag{3}$$

where A and B are constant coefficients, the function $\max(x)$ returns the maximum value of x, the function $hop(AP_i)$ returns the number of hops, or hop count, between AP_i and its GW, the function $host(link_{ij})$ returns the number of hosts using $link_{ij}$ in the shortest route to the GW to represent the link load, and the function $intf(ij)$ returns the link indices that may occur the primary conflict with $link_{ij}$. The A-term represents the maximum hop count, and the B-term does the maximum total load of a link and its primarily conflicting links. The minimization of the A- and B-terms intends the maximization of the network performance.

4.3 Proof of NP-completeness for AP clustering

The *NP-completeness* of the decision version of the AP clustering problem (*AP clustering*) is proved through reduction from the *NP-complete* bin packing problem (*Bin packing*) (Garey & Johnson, 1979).

4.3.1 Decision version of AP clustering problem

AP clustering is defined as follows:

- **Instance:** The same inputs as the AP clustering problem with an additional constant F_{c0}.
- **Question:** Is there an AP clustering with K clusters to satisfy $F_c \leq F_{c0}$?

4.3.2 Bin packing

Bin packing is defined as follows:

- **Instance:** $U = \{u_1, u_2, \cdots, u_{|U|}\}$: a set of items with various volumes, and L bins with a constant volume B.
- **Question:** Is there a way of partitioning all the items into the L bins ?

4.3.3 Proof of NP-completeness

Clearly, *AP clustering* belongs to the class *NP*. Then, an arbitrary instance of *Bin packing* can be transformed into the following instance of *AP clustering*. Thus, the NP-completeness of *AP clustering* is proved.

- **Input:** $G = (V, E) = K_N$: a complete graph with $N = |V| = |U|$, $s_{ij} = 1$, $h_i = u_i$ for $i = 1, \cdots, N$, $X = V$, $H = B$, $P = \infty$, $K = L$, and $F_{c0} = \infty$.
- **Output:** The set of GW clusters is equivalent to the bin packing, where any AP can be a GW and is one-hop away from the GW in each cluster.
- **Constraint:** to satisfy the following four constraints:
  - the number of APs in any cluster is not limited ($P = \infty$),
  - the number of associated hosts in any cluster must be $H = B$ or smaller,
  - the APs are connected with each other in any cluster ($G = K_N$), and
  - the GW is selected from GW candidates in any cluster ($X = V$).
- **Objective:** The condition $F_c \le F_{c0}$ is always satisfied with $F_{c0} = \infty$.

4.4 AP clustering algorithm

In this subsection, we present a two-stage heuristic algorithm for the AP clustering problem to avoid combinatorial explosions. As an efficient heuristic, our algorithm finds an initial solution by a greedy method, and improves it by the *Variable Depth Search (VDS)* method that can enhance the search ability of a local search method by expanding neighbor states flexibly (Yagiura et al., 1997). Our algorithm seeks the maximization of the network performance with the number of clusters K. If any feasible solution cannot be found with this number, our algorithm terminates after reporting the failure.

4.5 Check of number of clusters

First, the feasibility of the number of clusters K in the input is checked, because it has the trivial upper and lower limits that can be given by other inputs of the problem. The upper limit K_{max} is given by the number of GW candidates: $K_{max} = |X|$. The lower limit K_{min} is given by the following equation to satisfy the cluster size constraint and the bandwidth constraint:

$$K_{min} = \max\left\{ \lceil N/P \rceil, \left\lceil \sum_{i=1}^{N} h_i / H \right\rceil \right\} \tag{4}$$

where the ceiling function $\lceil x \rceil$ returns the smallest integer x or more. Then, if $K < K_{min}$ or $K > K_{max}$, our algorithm terminates after reporting the feasible range of K.

4.5.1 Initial GW selection

In our algorithm, K APs are randomly selected as initial GWs among GW candidates in X such that two selected APs are not adjacent to each other as best as possible. Starting from these selected APs, the initial GW clusters are constructed sequentially. Then, the clusters are iteratively improved by the VDS method. This AP clustering procedure is repeated by $\min(2N, {}_{|X|}C_K)$ times because initial GW APs are selected by different combinations, and the best solution in terms of the cost function is selected as the final solution.

4.5.2 Greedy construction

Our algorithm generates an initial AP clustering by repeating the following procedure:

1. Sort the APs adjacent to the clustered APs in descending order of its load h_i. If two or more such APs have the same load, resolve this tiebreak in ascending order of the number of incident links.
2. Apply the following procedure for each AP in step 1 from the top:
   a. Select the cluster of its adjacent AP as a cluster candidate for the AP, if the following two constraints are satisfied:
      - the number of APs in the cluster is smaller than P for the cluster size constraint, and
      - the total number of associated hosts in the cluster is H or smaller after the clustering for the bandwidth constraint.

   b. Cluster the AP as follows, if at least one cluster candidate is selected.
      - Select this cluster candidate if only one candidate exists, or otherwise
      - Select the cluster candidate that minimizes the cost function F_c.
3. Repeat steps 1–2 until every AP is clustered or no more AP can be clustered.

4.5.3 GW update

If the selected AP in the sequential AP clustering (let AP_k) is a GW candidate, the shortest path is calculated from every AP in the same cluster to this AP passing through only APs in this cluster, and the following GW cost function F_k is computed:

$$F_k = \max(host(link_{ij})). \tag{5}$$

If F_k becomes smaller, AP_k is selected as the new GW in the corresponding cluster.

4.5.4 Local search by VDS

Then, the initial AP clustering is improved iteratively by repeating the cluster changes of multiple APs at the same time using the VDS method. VDS is a generalization of a local search method, where the size of neighborhood is adaptively changed so that the algorithm can effectively traverse the large search space while keeping the amount of computational time reasonable. Actually, because each feasible state in this problem may have a different size of its neighborhood that satisfies the four constraints, VDS is suitable for this problem.

In our VDS for the AP clustering, a simple *move* operation is repeatedly tried until no further feasible operation is possible. Each *move* operation changes the cluster of an AP into a different feasible one such that the cost function F_c in (3) becomes minimum among the candidates. Then, only the subsequence of the *move* operations resulting into the smallest cost function are selected to be actually applied there, only if F_c after these operations becomes equal to or smaller than that of the previous state. If the cluster of any AP is not changed at one iteration or the cost function has not been improved during R iterations ($R = 10$ adopted in this chapter), the state is regarded as the local minimum. Then, the hill-climbing procedure is applied for the state to escape from it.

When the hill-climbing procedure is applied in T times ($T = 20$), the local search by VDS is terminated, and the best found solution is output as the final one. At this time, if an AP is not clustered at all, our algorithm regards that the K clustering of the APs is impossible and terminates after reporting the failure.

In summary, one iteration of this stage consists of three steps: 1) the cluster change trial, 2) the cluster change application, and 3) the hill-climbing. Here, we note that the unclustered APs in the initial AP clustering may be clustered in VDS.

Cluster Change Trial: The *cluster change trial* repeats the cluster change of the AP that satisfies the following three conditions until no more change is possible:

1. the AP has not been selected at this iteration,
2. the resulting clustering satisfies the constraints, and
3. the resulting clustering minimizes the cost function F_c among candidates.

Cluster Change Application: The cluster change trial always changes the AP cluster regardless of the increase of the cost function F_c as long as it satisfies the constraints. Thus, F_c may increase after some cluster changes. The *cluster change application* selects the subsequence of the cluster changes that minimizes F_c, and actually apply these cluster changes with the GW update procedure in 4.5.3 to the current solution, only if F_c becomes equal or smaller than that of the previous iteration. If the cluster changes are actually applied, another iteration is repeated from the cluster change trial.

Hill Climbing: The local search process using *move* operations in our VDS may be trapped into a local minimum where the solution cannot be improved without the hill-climbing step. In our algorithm, when either of the following two conditions is satisfied, the current state is regarded as a local minimum, and the hill-climbing procedure is applied to escape from it:

1. no cluster change is applied at one iteration, or
2. F_c has not been improved during R iterations ($R = 10$).

In the hill-climbing procedure, the following *random cluster change* operation is repeated until the clusters of S APs are actually changed, or no more APs can be changed ($S = 10$).

1. Enumerate any AP that satisfies the following three conditions for the random cluster change:
    a. it is not selected at this hill-climbing procedure,
    b. it is located on the boundary between different clusters, and
    c. its cluster change does not affect the connectivity of the other APs in the same cluster.
2. Randomly select one AP among them.
3. For this AP, find any cluster that can feasibly be changed into.
4. If such a cluster exists, change the cluster of this AP to a randomly selected cluster among them.
5. Otherwise, remove the cluster of this AP.

4.6 Performance evaluation by simulations

In this subsection, we discuss the performance evaluation of the AP clustering algorithm through network simulations using the WIMNET simulator. For this evaluation, the compared algorithm in 4.6.1 is also implemented. In each simulated instance, the minimum number of clusters such that each algorithm can find a feasible solution is given for the number of clusters K respectively, because we regard the minimization of K as the first priority task in the WIMNET design to reduce the installation and operation costs.

4.6.1 Compared algorithm

Within our knowledge, no algorithm has been reported for the same AP clustering problem in this chapter. Therefore, as the most analogous algorithm to our problem, the *Open/Close method* in (Prasad & Wu, 2006) has been implemented with some modifications for performance comparisons with our algorithm, where it does not consider the cluster size constraint and the distribution of associated hosts with APs. The procedure of this heuristic algorithm is described as follows.

Initial AP clustering

1. Generate the sorted list of the APs in descending order of the maximum number of associated hosts.
2. Select the first K APs in the list as GWs.
3. Assign the cluster to an unclustered AP that satisfies the following conditions:
   - the AP is adjacent to an AP clustered to this GW cluster,
   - the cluster size constraint is satisfied if added,
   - the bandwidth constraint is satisfied if added, and
   - the hop count (the number of hops between the AP and the GW) is minimized.
4. Repeat step 3 until no more AP can be assigned.
5. Calculate the sum of the hop count of every AP, if every AP is assigned a cluster, and save it.

AP clustering Improvement

The initial clustering is iteratively improved by repeating the following three operations:

1. *Close operation*
   a. Remove one GW randomly, and uncluster all the APs connected to this GW.
   b. Go to *Open operation*.
2. *Open operation*
   a. Select the first AP of the list in 4.6.1 as the GW that has not been selected.
   b. If no more AP is selected in step a, output the best-found solution if found, or output the error otherwise, and terminate the procedure.
   c. Assign the GW cluster to an unclustered AP that satisfies the four conditions in 4.6.1.
   d. Repeat step c until no more AP can be assigned.
   e. If every AP is assigned a cluster, calculate the sum of the hop count of every AP, and save it if the value is smaller than the best-found one. Return to *Close operation*.
   f. Otherwise, go to *Cluster adjustment*.
3. *Cluster adjustment*
   a. Assign the unassigned AP to one of the connectable GW clusters randomly.
   b. If the cluster size constraint or the bandwidth constraint is not satisfied as the result of the assignment in step a, APs in the cluster are unclustered one by one in ascending order of the hop count until the constraint is satisfied. If every AP in the cluster except the GW is unclustered but the constraint is not still satisfied, every unclustered AP is resumed and the cluster assignment in step a is discarded.

c. If every AP is assigned a cluster, calculate the sum of the hop count of every AP, and save it if the value is smaller than the best-found one. Return to *Close operation*.

d. If no feasible solution is obtained after repeating *Cluster adjustment* in 300 times, abort the procedure, and return to *Close operation*.

We note that the original Open/Close method assumes that each GW may have a different bandwidth for communications to/from wired networks to the Internet. In our implementation, we use the maximum number of associated hosts with an AP as this bandwidth.

4.6.2 Simulations for different traffic patterns

In our first simulations, the performance of our algorithm is evaluated through simple instances whose optimal solutions can be found easily, so that the optimality of our heuristic algorithm can be verified. For this purpose, we adopt the simple network topology of regularly allocated 24 (=6 × 4) APs, where each AP has wireless links with its four neighbor APs on the left, right, top, and bottom sides. This grid topology has been often used in wireless mesh network studies (Alicherry et al., 2006; Robinson & Knightly, 2007; Yan et al., 2008; Badia et al., 2008; Ye et al., 2007). To generate non-uniform traffics using simple loads, 8 APs among 24 are associated with 10 hosts, and the remaining 16 APs are with 1 host, which means the total of 96 hosts exist in the network. Then, by changing the locations of crowded APs in the field, we prepare 10 instances of different traffic patterns.

As the input parameters of the algorithm, the cluster size limit P is set 6 and the bandwidth size limit H is 24 where the lower limit on the number of clusters $K_{\min}$ is 4. Every link is assigned the same bandwidth $s_{ij} = 30\,\text{Mbps}$, and every AP becomes a GW candidate with $X = V$ for simplicity. The coefficients $A = B = 1$ are used for the cost function F_c, because our preliminary experiments using these instances observed no big difference in throughputs when A and B were changed from 1 to 3. To avoid the bias in random numbers, the average result among 10 runs using different random numbers is used in the evaluation for each instance. As example instances in our first simulations, Figure 7 illustrates traffic patterns and our clustering results with four clusters ($K = 4$) for four instances among them, where a black circle represents an AP associated with 10 hosts, and a white one represents an AP with 1 host. These results are actually optimum in these instances with the minimum number of clusters and cost functions.

Figure 8 compares the average number of clusters among 10 runs between two algorithms for each of 10 instances. Our algorithm (Proposal) always finds a feasible solution with the minimum number of clusters for any instance, whereas the compared one (Comparison) usually requires larger numbers. The reason may come from the fact that our algorithm seeks a feasible better solution with the fixed number of clusters, whereas the compared one does not explicitly minimize the number of clusters and may reduce it by chance through repeating open/close operations.

Then, to evaluate the AP clustering results in terms of the network performance, the WIMNET simulator is applied using the clustering results by both algorithms. Figure 9 compares the average total throughput for each instance between two algorithms, where our algorithm provides the larger throughput than the compared one by 24%–80% for any instance. Here, we analyze the reason why our algorithm achieves at least 150 Mbps. The total throughput of one GW cluster is determined by the summation of the GW throughput and the maximum communication throughput between APs in WIMNET. As shown in Fig. 8, the traffic load is evenly distributed among four clusters in our algorithm, which gives the same throughput for

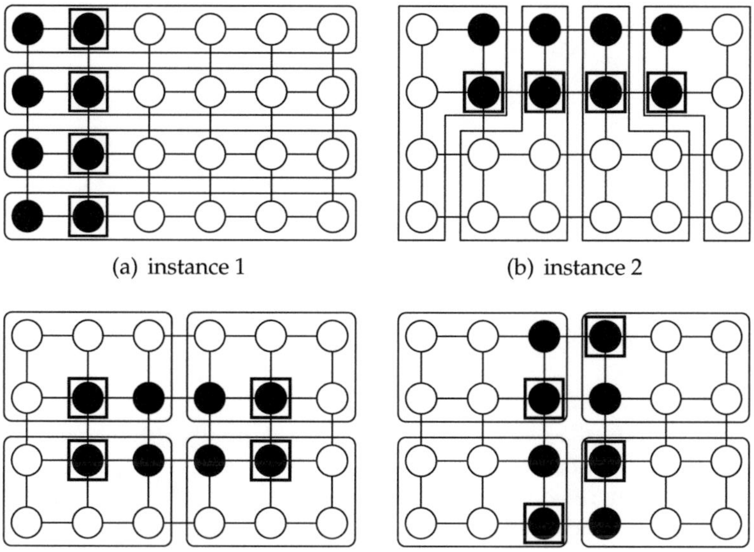

(a) instance 1 (b) instance 2

(c) instance 3 (d) instance 4

Fig. 7. Four traffic patterns and clustering results in first simulations.

every cluster. As a result, the total throughput of 150 Mbps or more comes from the formula of $((30 + \Delta) \times 4)$ Mbps where Δ represents the GW throughput by its associated hosts.

4.6.3 Simulations for verification of terms in cost function

The importance of each term in the cost function F_c is verified through simulations using the 10 instances in 4.6.2. Figure 10 compares the average throughput among the four different conditions for F_c, where AB represents the result using both terms, A does the result using the A-term only, B does the result using the B-term only, and *None* does the result without using F_c. This figure indicates that AB provides the best throughput in any simulated instance. Note that all of them find the solution with the least number of clusters. Thus, we conclude that the two terms in the cost function F_c are necessary for finding the high quality AP clustering.

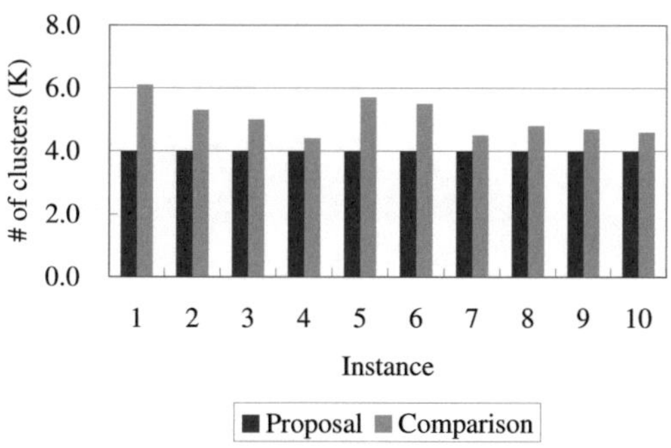

Fig. 8. Average number of clusters for different traffic patterns.

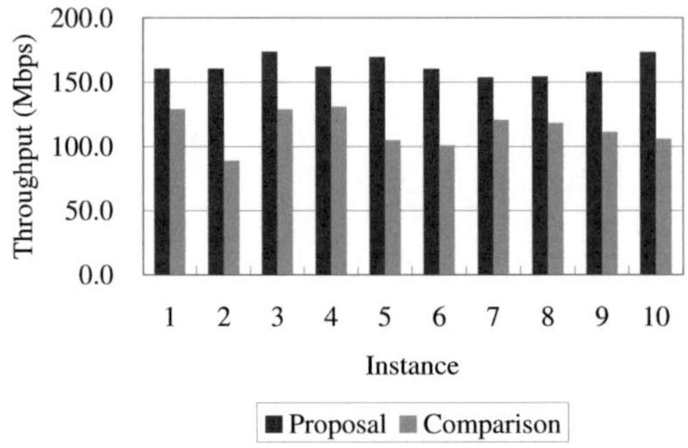

Fig. 9. Average throughputs for different traffic patterns.

4.6.4 Simulations for different bandwidth limits

In our second simulations, the performance for different bandwidth limits is investigated for instance 1 in Fig. 7. P is fixed with 8, and H is selected between 21 and 48, where $K_{\min}$ is 3, 4, or 5. Figures 11 and 12 compare the average number of clusters and the average total throughput, respectively. The number of clusters by our algorithm is always smaller than that by the compared one, and the throughput is larger by 10%–183%. Generally, as the bandwidth constraint becomes harder, both the number of clusters and the average throughput increase except for $H = 21$.

4.6.5 Simulations for different number of clusters

In our third simulations, the performance for different number of clusters is investigated using instance 4 in Fig. 7, where $P = 12$ and $H = 48$ are used, and the number of clusters K is changed from 2 to 24. Figure 13 shows changes of the throughputs by two algorithms and the cost function F_c in our algorithm. This result indicates that as K increases until certain values, F_c decreases and the throughput increases, and the throughput by our algorithm is always better than that by the compared one when it is not saturated. The results confirm the effectiveness of our algorithm for different number of clusters. Here, we note that the throughput are saturated at certain values of K because the communication bandwidth between an AP and a host (20 Mbps in simulations) becomes the bottleneck.

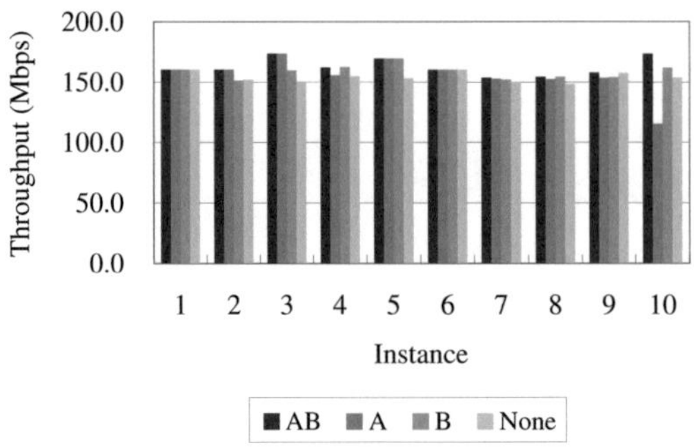

Fig. 10. Performance comparison of F_c with and without A or B-term.

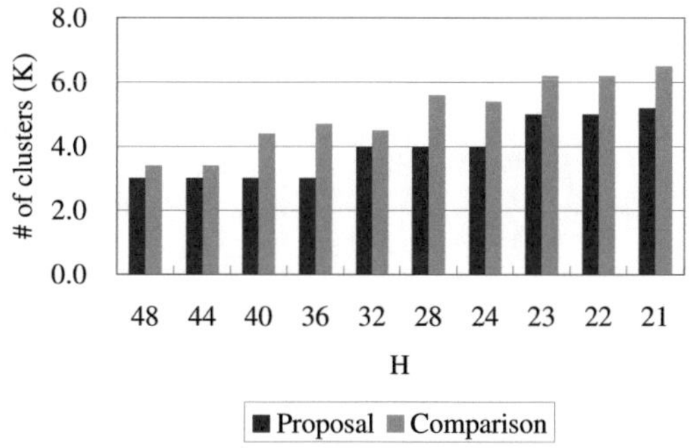

Fig. 11. Average number of clusters for different bandwidth limits.

4.6.6 Simulations for random networks

In our fourth simulations, the performance for random networks with 50 APs is investigated to evaluate our algorithm in more practical situations. The APs are randomly allocated on the network field (500 m × 500 m) such that the distance between any pair of APs is larger than the minimum one (50 m). Then, the wireless link is generated for any pair of APs within the distance of 110 m representing the wireless range in a free space. However, this wireless link can be blocked by obstacles such as walls and furniture in indoor environments as target fields for WIMNET. In order to consider the link failure stochastically, the following Waxman method is adopted to generate the link randomly, which has been often used in network studies (Waxman, 1988):

$$P(u,v) = \alpha e^{-d/(\beta D)} \tag{6}$$

where $P(u,v)$ is the probability of generating a link between AP_u and AP_v, α and β are constants satisfying $0 < \alpha, \beta \leq 1$ ($\alpha = 0.9$, $\beta = 0.8$), d is the distance between AP_u and AP_v, D is the largest distance between two APs in the network (on average, $D = 647.6$m). Then, the maximum number of hosts associated with each AP is randomly generated between 1 and 10 such that the total number of them becomes 200, in order to consider various network situations under the constant total load. As the constraints for GW clusters, $P = 6$ and $H = 25$ are used for $K_{min} = 9$.

By changing random numbers, 10 topologies are generated, and AP clusters are found by applying both algorithms to each topology in 10 times. Then, the WIMNET simulator is executed with each AP clustering in three times using different random numbers. As a result,

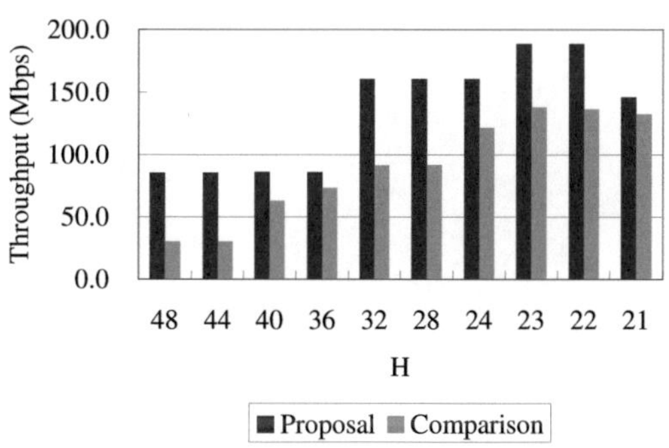

Fig. 12. Average throughputs for different bandwidth limits.

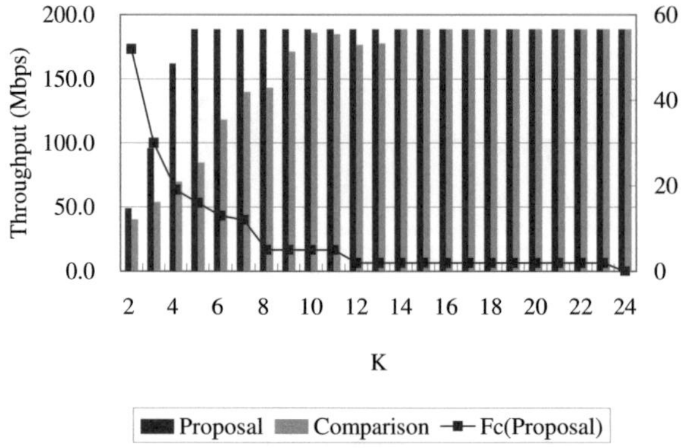

Fig. 13. Average throughputs and F_c for different number of clusters.

the average number of clusters and throughputs among the total of 30 trials for each of 10 topologies are evaluated in random network instances. Figure 14 illustrates two topologies with AP clusters and GWs found by our algorithm. Figure 15 and 16 compare the simulation results by both algorithms. The results show that our algorithm can find the AP clustering with the least number of clusters, which provides the better performance than the compared one for practical instances.

4.6.7 Simulations for load changes in random networks

In the AP clustering problem for WIMNET, the maximum number of associated hosts with each AP is given as the input. Normally, the number of associated hosts with an AP is frequently changing between 0 and this maximum number, because client hosts are often moving and are randomly connecting to the Internet through WIMNET.

In order to evaluate the performance of our algorithm in such normal situations, one random network instance is simulated when the number of associated hosts with each AP is changed randomly between the minimum and the given maximum. To vary the load, this minimum is changed from 1% of the maximum until reaching the maximum with the 1% interval. Figure 17 compares the throughputs between our algorithm and the compared one under 100 different loads. The result shows that the AP clustering by our algorithm provides the better throughput at any load than the compared one. Here, we note that if the maximum load for an AP is changed, the AP clustering should be redesigned by applying our algorithm.

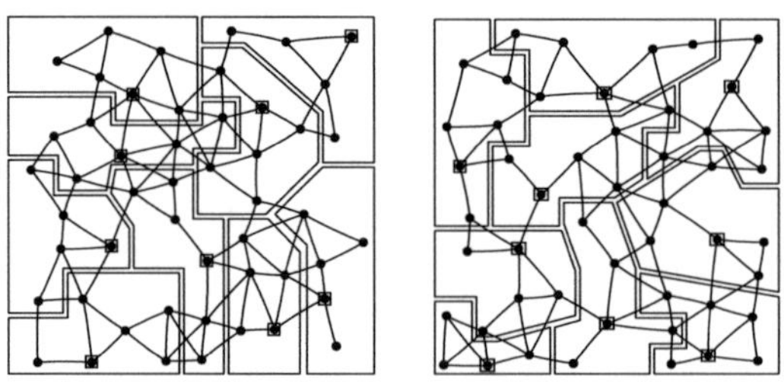

(a) instance 1 (b) instance 2

Fig. 14. Clustering results for two random networks.

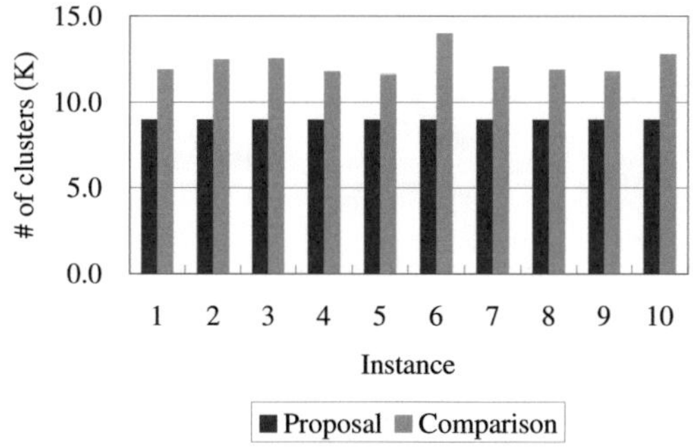

Fig. 15. Average number of clusters for random networks.

4.7 Related works

In this subsection, we introduce several related works to the AP clustering problem. Unfortunately, none of them deal with the four constraints in this problem including the GW cluster size constraint at the same time.

In (Aoun et al., 2006), Aoun et al. proposed a recursive dominating set algorithm based on (Chvatal, 1979) to find a clustering such that the maximum hop count, or radius, inside a cluster is smaller than the given limit. It first extracts a dominating set of the network, and generates a graph composed of this set and the edges connecting the two APs with two hops in the network. Then, it again extracts its dominating set, where any AP is connected with three hops to an AP in this set. This recursive procedure is repeated until the hop count surpasses the limit. This algorithm cannot generate clusters with an arbitrary hop count, and cannot always satisfy the constraints of the cluster size, the bandwidth, and the GW.

In (Lakshmanan et al., 2006), Lakshmanan et al. presented a multiple GW association model of allowing each host to be connected through more than one GWs to the Internet. They discuss its benefits in capacity, fairness, reliability, and security with its challenges. They presented the architecture using a super GW that controls the whole system, which can be a bottleneck, and the algorithms for the GW association and the packet transmission scheduling, which are just theoretical.

In (Li et al., 2007), Li et al. proposed a grid-based GW deployment method with a linear programming for a feasible interference-free TDMA link scheduling to maximize the throughput. By evaluating the throughput using the scheduling algorithm for every possible combination of K grid points in the field, the best locations of K GWs are found. Their

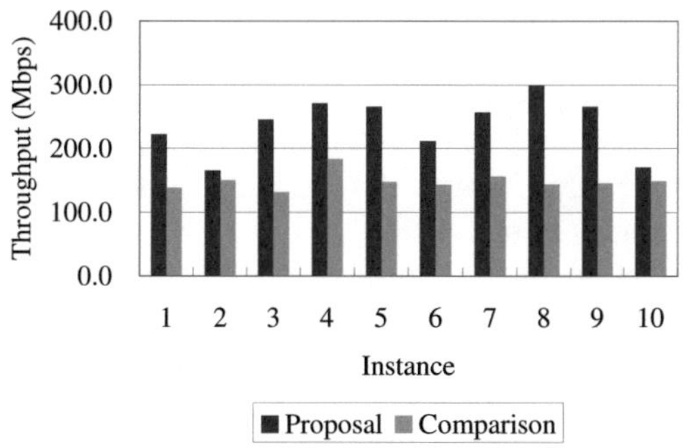

Fig. 16. Average throughputs for random networks.

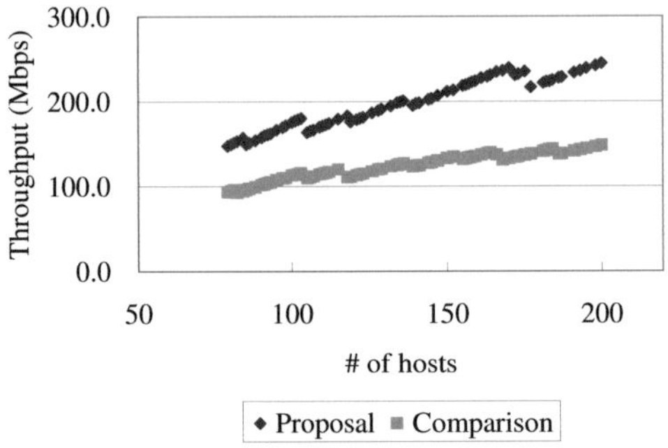

Fig. 17. Throughput changes for different numbers of hosts.

method can be extended to multi-channel and multi-radio networks. However, it assumes impractical TDMA operations for wireless mesh networks. Furthermore, it does not consider the constraints of the bandwidth, the cluster size, and the connection.

In (Park et al., 2006), Park et al. proposed a mesh router discovery scheme, and a QoS-driven mesh router selection mechanism for the dynamic GW selection by the traffic load. In (Nandiraju et al., 2006), Nandiraju et al. proposed a dynamic GW selection method for load balancing among multiple GWs. Unfortunately, it does not consider interference. These methods do not intend the allocation of GWs.

In (Hsiao & Kung, 2004), Hsiao et al. proposed a multiple network composition method with the same channel by using directional antennas. In their method, a lot of APs are necessary in the field so that each host can select its associated AP from multiple candidates for load balancing.

In (Huang et al., 2006), Huang et al. investigated AP deployments for intelligent transportation systems (ITS). They proposed an optimization algorithm of a mixed-integer nonlinear programming to determine the optimal number of APs in a cluster and the best cell radius for each AP. Because their proposal targets ITS, each cluster is composed of arrayed APs and the first AP becomes the GW.

In (Alicherry et al., 2006), Alicherry et al. formulated the joint problem of the channel assignment, the routing, and the scheduling for a special case of the wireless mesh network where every link activation was synchronously controlled by a single global clock, and presented its approximation algorithm that guarantees the order of approximation. Unfortunately, the realization of the synchronous wireless mesh network is very hard, and the superiority is actually not clear to the conventional asynchronous one. Furthermore, it assumes that every AP has the same number of associated hosts.

In (Denko, 2008), Denko studied the wireless mesh network with mobile Internet GWs using a multi-path routing scheme to increase the reliability and performance. However, the mobile GW is not practical because the wired connection to the Internet is static. Furthermore, the network may not work properly if the traffic of every router increases, because each router selects one route by the amount of its traffic.

In (Tokito et al., 2009), Tokito et al. proposed a routing method for multiple GWs in wireless mesh networks, called the GW load balanced routing (GLBR). GLBR reduces loads of congested GWs by changing the GW of a leaf node in the routing tree one by one, such that the new GW decreases the variance of loads at GWs and the length of the detouring path is shorter than the threshold. The initial routing tree is found by the shortest path algorithm. They show the advantage of their proposal over the shortest path routing in simulations.

However, because this algorithm can change the path for only one leaf node at one time, it can be easily trapped into a local minimum where simultaneous changes of multiple paths are often necessary to escape from.
In (Ito et al., 2009), Ito et al. studied a method of distributing traffics among multiple GWs on a session by session basis in wireless mesh networks. Their method first estimates the throughput for each GW from the traffic volume around there and the hop count, and then, selects the GW expecting the highest throughput. Through simulations using the network simulator *ns-2*, they show the effectiveness of their proposal by comparing the throughput and the fairness between the proposed session-distribution method and the packet-distribution method.

4.8 Conclusion

This section presented the AP clustering algorithm composed of the greedy method and the variable depth search method. The effectiveness was verified through network simulations using the WIMNET simulator, where the comparisons of the number of clusters and throughputs with an existing algorithm confirmed the superiority of our algorithm. The future works may include simulations with more realistic situations, the development of the distributed version of the AP clustering algorithm, and experiments using real networks.

5. References

Akyildiz, I. F., Wang, X. & Wang, W. (2005). Wireless mesh networks: a survey, *Comput. Network. ISDN Syst.* 47(4): 445–487.

Alicherry, M., Bhatia, R. & Li, L. (2006). Joint channel assignment and routing for throughput optimization in multiradio wireless mesh networks, *IEEE J. Select. Area. Commun.* 24(11): 1960–1971.

Aoun, B., Boutaba, R., Iraqi, Y. & G, K. (2006). Gateway placement optimization in wireless mesh networks with qos constraints, *IEEE J. Select. Area. Commun.* 24(11): 2127–2136.

Badia, L., Etra, A., Lenzini, L. & Zorzi, M. (2008). A general interference-aware framework for joint routing and link scheduling in wireless mesh networks, *IEEE J. Network* 22(1): 32–38.

Bahri, A. & Chamberland, S. (2005). On the wireless local area network design problem with performance guarantees, *Comput. Networks* 48: 856–866.

Beuran, R., Nakata, J., Okada, T., Nguyen, L. T., Tan, Y. & Shinoda, Y. (2008). A multi-purpose wireless network emulator: Qomet, *Proc. Int. Conf. Advanced Inform. Network. Applications (AINA2008).*

Chandra, R., Qiu, L., Jain, K. & Mahdian, M. (2004). Optimizing the placement of internet taps in wireless neighborhood networks, *Proc. Int. Conf. Network Protocols (ICNP)*, pp. 271–282.

Chvatal, V. (1979). A greedy heuristic for the set-covering problem, *Math. Oper. Res.* 4(3): 233–235.

Cisco Systems, Inc. (2003). Cisco aironet 1200 series access points,' product data sheet.

Clark, B. N. & Colbourn, C. J. (1990). Unit disk graphs, *Discrete Mathematics* 86: 165–177.

de la Roche, G., Rebeyrotte, R., JaffrRunser, K. & Gorce, J.-M. (2006). A qos-based fap criterion for indoor 802.11 wireless lan optimization, *Proc. IEEE Int. Conf. Commun. (ICC2006)*, pp. 5676–5681.

Denko, M. K. (2008). Using mobile internet gateways in wireless mesh networks, *Proc. Advanced Inform. Network. Applications (AINA)*, Vol. 1, pp. 1086–1092.

Farag, T., Funabiki, N. & Nakanishi, T. (2009). An access point allocation algorithm for indoor environments in wireless mesh networks, *IEICE Trans. Commun.* E92-B(3): 784–793.

Faria, D. B. (2005). Modeling signal attenuation in ieee 802.11 wireless lans - vol. 1, *Tech. Report TR-KP06-0118, Kiwi Project, Stanford Univ.* .

Funabiki, N., Nakanishi, T., Hassan, W. & Uemura, K. (2007). A channel configuration problem for access-point communications in wireless mesh networks, *Proc. IEEE Int. Conf. Networks (ICON)*.

Funabiki, N., Uemura, K., Nakanishi, T. & Hassan, W. (2008). A minimum-delay routing tree algorithm for access-point communications in wireless mesh networks, *Proc. Int. Conf. Research Innovation Vision for the Future (RIVF-2008)*, pp. 161–166.

Garey, M. R. & Johnson, D. S. (1979). Computers and intractability: A guide to the theory of np-completeness.

Gast, M. S. (2002). 802.11 wireless networks - the definitive guide.

Gupta, G. & Younis, M. (2003). Fault-tolerant clustering of wireless sensor networks, *Proc. IEEE Wireless Commun. Networking Conf. (WCNC)*, Vol. 3, pp. 1579–1584.

Hassan, W., Funabiki, N. & Nakanishi, T. (2010). Extensions of the access point allocation algorithm for wireless mesh networks, *IEICE Trans. Commun.* E93-B(6): 1555–1565.

Hsiao, P.-H., Hwang, A., Kung, H. T. & Vlah, D. (2001). Load-balancing routing for wireless access networks, *Proc. IEEE Infocom*, pp. 986–995.

Hsiao, P.-H. & Kung, H. T. (2004). Layout design for multiple collocated wireless mesh networks, *Proc. Vehicular Technology Conf. (VTC)*, Vol. 5, pp. 3085–3089.

Huang, J.-H., Wang, L.-C. & Chang, C.-J. (2006). Wireless mesh networks for intelligent transportation systems, *Proc. Systems, Man and Cybernetics (SMC)*, Vol. 1, pp. 625–630.

Ito, M., Shikama, T. & Watanabe, A. (2009). Proposal and evaluation of multiple gateways distribution method for wireless mesh network, *Proc. Int. Conf. Ubiquitous Inform. Manage. Commun. (ICUIMC)*, pp. 18–25.

Kato, H., Funabiki, N. & Nakanishi, T. (2007). Throughput measurements under various contention window size in wireless mesh networks, *IEICE Tech. Report, NS2007-115* pp. 55–60.

Kato, H., Nomura, Y., Funabiki, N. & Nakanishi, T. (2006). An experimental result of communication bands for a wireless mesh network, *IEICE Tech. Report, NS2006-139* pp. 5–8.

Kouhbor, S., Ugon, J., Mammadov, M., Rubinov, A. & Kruger, A. (2006). Nonsmooth optimization for the placement of access points to enhance security in wlan.

Kouhbor, S., Ugon, J., Rubinov, A., Kruger, A. & Mammadov, M. (2006). Coverage in wlan with minimum number of access points, *Proc. IEEE Vehi. Tech. Conf. (VTC 2006)*, pp. 1166–1170.

Lakshmanan, S., Sundaresan, K. & Sivakumar, R. (2006). On multi-gateway association in wireless mesh networks, *Proc. IEEE Workshop. Wireless Mesh Networks (WiMesh)*, pp. 64–73.

Lee, Y., Kim, K. & Choi, Y. (2002). Optimization of ap placement and channel assignment in wireless lans, *Proc. Work. Wireless Local Networks*.

Li, F., Wang, Y. & Li, X.-Y. (2007). Gateway placement for throughput optimization in wireless mesh networks, *Proc. IEEE Int. Conf. Commun. (ICC)*, pp. 4955–4960.

Li, J., Jannotti, J., Couto, D. S. J. D., Karger, D. R. & Morris, R. (2000). A scalable location service for geographic ad hoc routing, *Proc. Int. Conf. Mobile Comput. Network. (MobiCom)*, pp. 120–130.

Lichtenstein, D. (1982). Planar formulae and their uses, *Siam J. Comput* 11: 329–343.

Nagy, L. & Farkas, L. (2000). Indoor base station location optimization using genetic algorithms, *IEEE Int. Symp. Personal, Indoor, and Mobile Radio Commun. (PIMRC)*, Vol. 2, pp. 843–846.

Naidoo, K. & Sewsunker, R. (2007). 802.11 mesh mode provides rural coverage at low cost, *Proc. AFRICON 2007*.

Nandiraju, D., Santhanam, L., Nandiraju, N. & Agrawal, D. P. (2006). Achieving load balancing in wireless mesh networks through multiple gateways, *Proc. Mobile Adhoc and Sensor Systems (MASS)*, pp. 807–812.

Pal, M., Tardos, E. & Wexler, T. (2001). Facility location with nonuniform hard capacities, *Proc. IEEE Symp. Found. Comput. Science*, pp. 329–338.

Pan, H. J. & Keshav, S. (2006). Detection and repair of faulty access points, *Proc. IEEE Wireless Commun. Networking Conf. (WCNC)*, pp. 532–538.

Park, B.-N., Lee, W., Ahn, S. & Ahn, S. (2006). Qos-driven wireless broadband home networking based on multihop wireless mesh networks, *IEEE Trans. Consumer Electronics* 52(4): 1220–1228.

Prasad, R. & Wu, H. (2006). Gateway deployment optimization in cellular wi-fi mesh networks, *J. Networks* 1(3): 31–39.

Proxim Co. (2003). A detailed examination of the environmental and protocol parameters that affect 802.11g network performance.
URL: *http://www.proxim.com/learn/library/*

Ramamurthy, R., Bogdanowicz, Z., Samieian, S., Saha, D., Rajagopalan, B., Sengupta, S., Chauduri, S. & Bala, K. (2001). Capacity performance of dynamic provisioning in optical networks, *J. Lightwave Technol.* 19: 40–48.

Raniwala, A., Gppalan, K. & Chiueh, T. (2005). Architecture and algorithms for an ieee 802.11-based multi-channel wireless mesh networks, *Proc. IEEE Infocom*, Vol. 3, pp. 2223–2234.

Rappaport, T. S. (1996). Wireless communications - principles and practice.

Robinson, J. & Knightly, E. W. (2007). A performance study of deployment factors in wireless mesh networks, *Proc. Inform. Commun.(INFOCOM)*, pp. 2054–2062.

Robinson, J., Uysal, M., Swaminathan, R. & Knightly, E. (2008). Adding capacity points to a wireless mesh network using local search, *Proc. IEEE Infocom*.

Sharma, A., Raghavenda, R., Puttaswamy, K., Lundgren, H., Almeroth, K. & Belding-Ro, E. (2005). Experimental characterization of interference in a 802.11g wireless mesh network, *Tech. Paper, Univ. California Santa Barbara* .

Tajima, S., Funabiki, N. & Higashino, T. (2010). A wds clustering algorithm for wireless mesh networks, *IEICE Trans. Inform. Systems* E93-D(4): 800–810.

Tokito, H., Sasabe, M., Hasegawa, G. & Nakano, H. (2009). Routing method for gateway load balancing in wireless mesh networks, *Proc. Int. Conf. Networks (ICN)*, pp. 127–132.

Varshney, U. & Malloy, A. D. (2006). Multilevel fault tolerance in infrastructure-oriented wireless networks: framework and performance evaluation, *Int. J. Network Management* 16: 351–374.

Waxman, B. M. (1988). Routing of multipoint connections, *IEEE J. Select. Areas Commun.* 6(9): 1617–1622.

Wu, T.-W. & Hsieh, H.-Y. (2007). Interworking wireless mesh networks: performance characterization and perspectives, *Proc. IEEE Global Telecom. Conf. (GLOBECOM)*, pp. 4846–4851.

Yagiura, M., Yamaguchi, T. & Ibaraki, T. (1997). A variable depth search algorithm for the generalized assignment problem, *Proc. Int. Conf. Metaheuristic.*

Yan, Y., Cai, H. & Seo, S.-W. (2008). Performance analysis of ieee802.11 wireless mesh networks, *Proc. IEEE Int. Conf. Commun.(ICC)*, pp. 2547–2552.

Ye, F., Chen, Q. & Niu, Z. (2007). End-to-end throughput-aware channel assignment in multi-radio wireless mesh networks, *Proc. Global Telecommun.(GLOBECOM)*, pp. 1375–1379.

Yoshida, S., Funabiki, N. & Nakanishi, T. (2006). A development of wireless infrastructure mesh network simulator, *Proc. Ad-hoc Workshop*, pp. 1–9–1–12.

3

Performance Analysis of MAC Protocols for Location-Independent End-to-end Delay in Multi-hop Wireless Mesh Networks

Jin Soo Park[1], Yun Han Bae[2] and Bong Dae Choi[3]
[1]*USN Service Division, KT, Seoul*
[2,3]*Department of Mathematics and Telecommunication Mathematics Research Center, Korea University, Seoul*
France

1. Introduction

Backbone wireless mesh networks (WMNs) are emerging alternatives to conventional wired backbones for metropolitan and have attracted much attention from both academic and industrial world as an infrastructure network for realizing the ubiquitous computing environment. WMN is a generalization of Wireless Ad-Hoc Networks that considers the use of heterogeneous nodes (e.g., clients and routers) and both wired and wireless connections to exchange data between these devices. The basic architecture of a WMN consists of a backbone of mesh routers (MR) and the clients that access communication services through the use of this backbone. Therefore, this backbone serves as a last mile solution that is interconnected to provide direct communication between clients (i.e., without routing the interclient traffic through any other intermediate network). This characteristic of a WMN enables it to function as an isolated autonomous network or as a last mile solution depending on the telecommunication facilities available at the place where the WMN is deployed.

In a multi-hop WMN, communication between two nodes is basically carried out by forwarding packets through a number of intermediate nodes. In WMNs, nodes are comprised of mesh routers in fixed sites and mobile clients as shown in Fig.1. We call a mesh router (also called mesh node) with gateway functions a gateway node, which is equipped with wireline network interfaces to connect the internet backbone. In this chapter each mesh node operates not only as an access point (AP) for mobile clients in its own basic service set (BSS) but also as a router, forwarding packets on behalf of other nodes that may not be within direct wireless transmission range of their destinations (see Fig. 1). Mobile clients are attached to a node in their BSS. Data originating from mobile clients are relayed by intermediate relay nodes hop by hop and delivered to the gateway.

One of the important problems to be solved in WMNs is the unfair bandwidth sharing problem depending on the nodes' location. More specifically, the per node throughput may decrease and the end-to-end delay may dramatically increase with an increasing hop-count distance from the gateway. In particular, WMNs based on single radio, irrespective of its simplicity and high fault tolerance, face a significant limitation of limited network capacity. It has been shown (2) that the theoretical upper limit of the per node throughput is asymptotically limited by $O(1/\sqrt{n})$ where n is the number of nodes in the networks. Therefore, with increasing number of nodes in a WMN, the per node throughput becomes unacceptably low. It has also been found (3) through experiments using carrier sense multiple access with collision avoidance (CSMA/CA)-based MAC protocol such as IEEE 802.11 that on a string topology, the throughput degrades approximately to $1/n$ of the raw channel bandwidth.

To resolve the above problem in single radio WMNs, multi-radio WMNs are under intense research. Therefore, recent advances in WMNs are mainly based on a multi-radio approach. While multi-radio WMNs promise higher capacity compared with single radio WMNs, they also face several challenges. One of them is location-dependent problem (10; 11) as in single radio WMNs in the sense that the per-node throughput decreases and the end-to-end delay increases dramatically with an increasing hop count to the gateway node. In particular, delay-sensitive application such as VoIP is expected to be serviced in WMNs in the near future. Such a service requires to be delivered to the destination within a given delay requirement regardless of its generated location. Thus, to support delay-sensitive traffic in WMNs, we need a proper method to guarantee the location-independent end-to-end delay in WMNs. This chapter focuses in detail on these issues. As a preliminary, we survey recent studies which deal with location-dependent problem in WMNs or investigate the performance of WMNs in terms of throughput and end-to-end delay mostly based on the analytical modeling method.

1.0.1 Single radio WMNs

In recent years, there have been several studies focused on the unfairness problem of multi-hop wireless networks under single radio scenario. In (12; 32), queue management schemes for restoring the fairness in a WMN has been proposed, which in part share a common emphasis with this chapter which intends to devise packet management scheme in relay node for location-independent end-to-end delay in WMNs. Nandiraju et al.(32) showed that Queue management, at intermediate relay mesh nodes, plays an important role in limiting the performance of longer hop length flows. They (32) proposed a queue management algorithm for IEEE 802.11s based mesh networks that improves the performance of multihop flows by fairly sharing the available buffer at each mesh point among all the active source nodes whose flows are being forwarded. Gambiroza et al. (10) proposed a centralized scheme to solve the unfairness problem of IEEE 802.11 based multi-hop wireless networks. In this scheme, each mesh router collects information on the global topology including link capacities and offered traffic, and then calculates the optimal sending rate based on the information. Then, each node in the network limits its ingress rate according to the given optimal rate. They studied the critical relationship between fairness and aggregate throughput based on simulation. Above mentioned researches (10; 12; 32) were interested in throughput and were only based on the simulation method. Liu et al. (29) developed an analytic model to model throughput and end-to-end delay in wireless mesh networks with single radio and single channel. Based on their analytical model, they (29) proposed two network design strategies to provide fair resource sharing and minimize the end-to-end delay in wireless mesh networks. But, the study was carried out based on the simplified MAC protocol, not CSMA/CA protocol such as IEEE 802.11 DCF, which is characterized by only the parameter of successful transmission probability. Bisnik et al. (28) characterized the average end-to-end delay and capacity in random access MAC based WMNs with single radio. They (28) modeled residential area WMNs as open G/G/1 queuing networks. The analytical model takes into account the mesh client and router density, the random packet arrival process, the degree of locality of traffic and the collision avoidance mechanism of random access MAC. Even though the above mentioned studies (28; 29) developed analytical models for obtaining performance measures in WMNs such as end-to-end delay and throughput, the derivation are mainly based on the simplified MAC scheme apart from CSMA/CA protocol in IEEE 802.11. Without devising any queue management mechanism to give higher priority for the channel access to flows experiencing longer hops, they (28; 29) were interested in finding the achievable maximum throughput in WMNs while the end-to-end delay is guaranteed for a given value. Sarr et al.(30) developed an analytic model for evaluating average end-to-end delay in IEEE 802.11 multi-hop wireless networks with single radio.

1.0.2 Multi-radio WMNs

With a multi-radio functionality, the performance of WMN can be enhanced if relay nodes can transmit and receive simultaneously (6), (7) and nodes in different contention zones can transmit concurrently without any interference. For recent works on the performance of WMNs under multi-radio scenario, see (15; 19; 20). Raniwala et al. (15) aimed to expand WLAN into an enterprise scale backbone network technology by developing a multi-radio wireless mesh network architecture where each node equips with multiple transceivers and supports distributed channel assignment to increase the overall network throughput. The central design issues of multi-radio WMN architecture in (15) are channel assignment and routing. They (15) showed that even with just 2 radios on each relay node, it is possible to improve the network throughput by a factor of 6 to 7 when compared with the conventional single-channel ad hoc network architecture. Regarding the channel assignment issue in backbone WMNs as shown in Fig.1, we rely on the method in (15). Aoun et al. (33) showed the capacity of a WMN is constrained by the bottleneck collision domain; hence, placing an equal number of radios at all nodes is not necessary. They proposed that additional radios should be placed according to the distribution of traffic load in WMN. By giving the collision domains that need to support higher traffic load to more bandwidth by setting up additional radios, interfering wireless links would operate on different channels, avoiding interference and enabling multiple parallel transmissions. Duffy et al. (34) developed a tractable analytic model of throughput performance for 802.11 multi-hop networks where the relay node is equipped with multi-radio and each of them operates on different channel. They (34) tried to solve upstream/downstream unfairness problem induced by the 802.11 MAC at aggregation points in a relay node. With the use of the flexibility provided by the 802.11e standard (specifically, TXOP and CWmin adjustment), they proposed a scheme to restore fairness at relay aggregation points. But, their focus was not the unfairness problem depending on nodes' location in WMNs, but the well-known unfairness problem between access point (AP) and station in IEEE 802.11 one-hop network in the sense that AP and each station share the channel equally so that AP may be a bottleneck for the downstream. In (35), the authors mathematically modeled the channel and interface assignment problems by introducing link and node channel assignment binary vectors. They developed a formulation for cross-layer fair bandwidth sharing problem as a non-linear mixed-integer network utility maximization which takes into account the number of radios at each relay node, the number of channels, and the interference constraints. Lee et al.(20) proposed a fair throughput allocation scheme for nodes in 802.11-based WMN regardless of their hop distances to the gateway. To achieve the fair throughput, they differentiated the contention window size of 802.11 mesh routers according to their weights based on the number of active nodes attached to each router. This work (20) shares a common interest with this chapter from the view point of network topology, i.e., WMN with tree structure. But, the works (20; 35) did not deal with the end-to-end delay in WMNs. (19) proposed a multi-channel ring-based wireless mesh network. In (19), the WMNs are divided into several rings, which are allocated with different channels. In the proposed WMN, a simple ring-based frequency planning is used to effectively utilize the available multiple channels. They developed a cross-layer analytical framework to evaluate the end-to-end delay and throughput in the proposed WMN. Based on their analytical model, they provided a method to determine the optimal number of rings in a WMN and the associated ring widths to maximize the coverage for a WMN.

1.0.3 Location-independent end-to-end delay in WMNs

This chapter focuses on discussing the schemes for location-independent end-to-end delay in WMNs. More specifically, this chapter extends the result of most recent work (23) which has tried to guarantee location-independent end-to-end delay in WMNs where each relay node is equipped with the functionality of multi-channel and multi-radio. Furthermore, this chapter provides more details than the work by Bae et al. (23) by adding new results. They (23) proposed two packet management schemes, called the *differentiated CW policy* and the *strict priority policy*, which are employed by relay nodes to obtain almost equal end-to-end delay, independent of source nodes' locations. In (23), it is assumed that each node is equipped with multiple transceivers, each of which is tuned on a particular channel. With these employments, the WMN can be decomposed into disjoint zones such a way that each zone uses its own channel different from channels used in neighbor zones. At a relay node in each zone, relay packets are buffered in different queues according to their experienced hop count, we call the queue storing the packets passing by k hops as *priority queue of class k*. A queue in which packets passing by more hops are stored has a higher priority in the sense that it has a shorter CW_{min} at the *differentiated CW policy* and a higher priority for the service at the *strict priority policy*. For the differentiated CW policy, a relay node adopts IEEE 802.11e EDCA protocol where a higher priority queue has a shorter minimum contention window. For the strict priority policy, a relay node is regarded as a single queueing system where the service discipline among priority queues at the relay node follows strict priority. The relay node has shorter minimum contention window than that of end node.

In summary, i) a typical zone is modeled as a one-hop IEEE 802.11e EDCA network under non-saturation condition where nodes have different packet arrival rates and different minimum contention window sizes. The probability generating function (PGF) of the HoL-delay of packets priority queue of class-k at a relay node in a zone is derived. Eventually, the packet delay (the sum of the queueing delay and the HoL delay) in a zone is obtained, by modeling each queue as $M/G/1$ queue with the HoL-delay as a service time. ii) A method to determine the minimum contention window sizes of each priority queues satisfying almost same end-to-end delays of packets regardless of their source's location is presented.

The rest of this chapter is organized as follows. The network model is presented in Section 2. Section 3 describes the differentiated CW policy and presents its analytic model. The probability generating function (PGF) of the HoL-delay is derived and then the average end-to-end delay of packets is obtained. Section 4 deals with the strict priority policy.

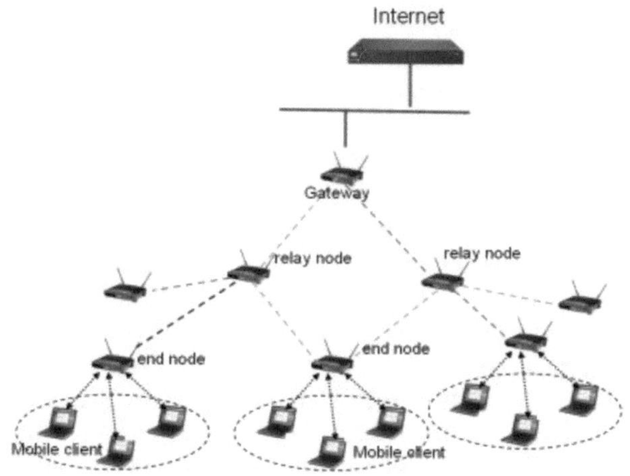

Fig. 1. Backbone Wireless Mesh Network

2. System model

2.1 Network model

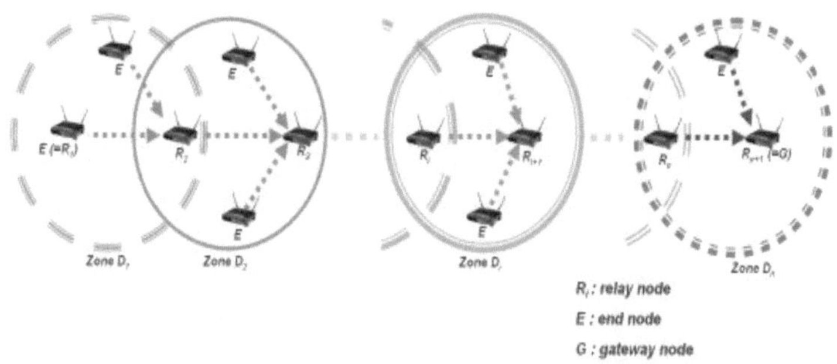

Fig. 2. A variant of n-hop linear network

In general, a backbone WMN can be viewed as a network with tree structure as shown in Fig.1, where the root node corresponds to the gateway. Backbone WMN consists of a gateway connected to the wired networks and multiple mesh nodes connected to the gateway through the multi-hop communications. In such a tree network, the traffic is skewed such that most packets flow toward or from the gateway node (33). Therefore, the nearer to the gateway a mesh node is, the more traffic the mesh node should have the capability to process. Thus it is required that an appropriate packet management scheme is employed in the mesh node to avoid the congestion and to deliver the traffic in time.

Since nowadays a gateway connected to wired networks are available in most of places, practically the coverage of WMN is not too wide. Due to this and also for the simplicity of analysis, we consider a variant of linear wireless mesh network as shown in Fig.2. This kind of topology is useful and quite general in the following aspects: Applying the static channel and interface assignment scheme (36; 37) to WMNs with tree structure, each channel and interface is permanently assigned to each relay node and the set of mesh nodes interfering with the

relay node forms a collision domain (20), which is not changed for a long period as long as a new mesh node is not deployed in the network or a node failure does not occur. In addition, if a static routing algorithm in which a routing path of a flow is not changed for a relatively long period (e.g, several minutes), is employed in WMNs, the set of relay nodes through which the flow passes from its source to the destination (gateway), and end nodes interfering with them, can be viewed as a variant of linear network as in Fig.2. Therefore, with the setting of multi-channel and multi-radio, and a static routing, WMNs with a tree structure can be decomposed into several linear networks, each of which forms a independent and separate sub-network without imposing any interference to each other. If there are enough channels available in the WMN, performance analysis of WMN with a tree structure can be obtained by the following similar method developed in Section 3 with complexity of expressions.
Uplink communication from nodes to the gateway is considered. In Fig.2, WMN consists of a gateway, n relay nodes and multiple end nodes where each end node is attached to a relay node. Mobile clients are attached to relay nodes and end nodes. Every *end node* E receives its local traffic from mobile clients in its BSS and sends the local traffic to an upstream node. Every *relay node* R_i forwards not only its local traffic from mobile clients in its BSS but also relay traffic from relay node R_{i-1} to relay node R_{i+1}. The n-hop linear network with $n-1$ relay nodes $R_2, \cdots, R_n$ is decomposed into n disjoint zones $D_1, D_2, \cdots, D_n$. It is assumed that there are enough channels available in WMN. A different channel is assigned to each zone: one for the transmission of relay traffic from R_i to R_{i+1} in zone D_i and the other for transmission of relay traffic from R_{i-1} to R_i in zone D_{i-1}, respectively. In other words, the WMN can be decomposed into disjoint zones as shown in Fig.2 so that nodes in a zone use one channel and those in neighbor zones use different channels in order to avoid the hidden node problem and the exposed node problem. As illustrated on Fig.2, zone D_i has a parent node R_{i+1} and child nodes consisting of one relay node R_i and several end nodes E. The parent node R_{i+1} plays a role as an AP in zone D_i, and child nodes consist of several end nodes E and a relay node R_i, (which plays a role as parent node in zone D_{i-1}). Relay node R_i has 3 transceivers operating on different channels: one for the uplink transmission with relay node R_{i+1} in zone D_i, another for communications with its child nodes in zone D_{i-1} and the other for communication with mobile clients in its BSS. By using multiple transceivers, the relay node can transmit and receive simultaneously. Each end node is equipped with 2 transceivers to communicate with its parent node and mobile clients. Assuming that every node in each zone is within the one-hop distance, collisions may occur only when two or more nodes within a zone transmit simultaneously. Since neighbor zones use different channels, there are no interferences between neighbor zones. Thus we may focus on the analysis of one zone D_i, and then the analytic results on one zone will be used in multi-hop WMN.

2.2 Modeling of a zone

As illustrated in dotted circle region in Fig. 4, we assume that the parent node R_{i+1} in zone D_i has total N_i child nodes: one relay node R_i and $N_i - 1$ end nodes E. (Note that in the case of a tree topology, the parent node may have multiple relay nodes as child nodes, the mathematical analysis in Section 3.2 can be extended to the tree topology with some tedious calculation.)
It is assumed that each end node E has one local uplink buffer where local packets transmitted from mobile clients in its BSS to the end node are stored before transmitting to relay node R_{i+1}. Packets' arrival at the local uplink buffer is assumed to follow a Poisson process with arrival rate $\lambda(/\text{sec})$.

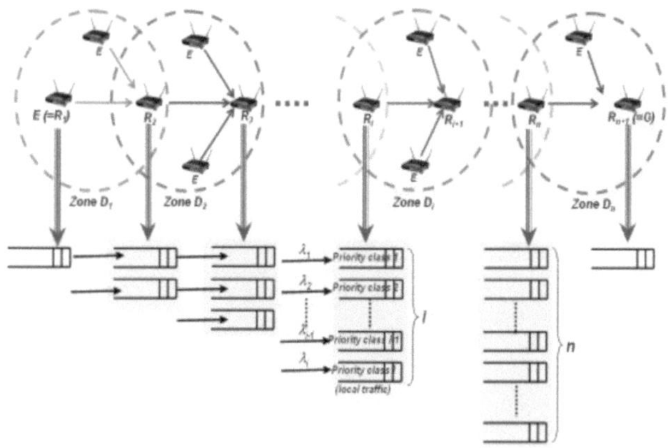

Fig. 3. Packet management scheme

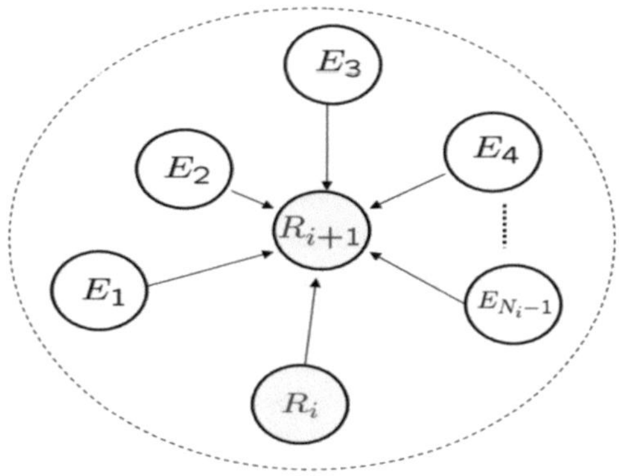

Fig. 4. Zone D_i

The relay node R_i has i uplink buffers, as shown in Fig.3: one is a local uplink buffer where local packets generated in its BSS are stored, and others are relay uplink buffers where relay packets forwarded from relay node R_{i-1} are stored before transmitting to relay node R_{i+1}. Relay packets in node R_i are stored in different uplink buffers according to their hop count passing by. To be precise, the kth uplink buffer ($k = 1, 2, \cdots, i$) in relay node R_i stores packets originated from all end nodes in the zone D_k and we call it *priority queue of class-k*, i.e., priority queue of class-1 is the highest priority class, which is for the relay packets originated from zone D_1 with the longest hop count to the relay node R_i. Priority queue of class-i is the lowest priority class for the local packets generated from the BSS of relay node R_i. Local packets' arrival at the local uplink buffer of each relay node R_i is assumed to follow a Poisson process with arrival rate λ just as the arrival at an end node.
In general, relay node R_i is heavily loaded compared to end nodes, and it relays both local traffic and relay traffic forwarded from zone D_{i-1}. Thus it is necessary to give more opportunities for transmission to the relay node R_i than end nodes E in order to shorten the end-to-end delay. Also, among all relay packets at the relay node, it is necessary to give a higher priority to relay packets experiencing more hops to reduce the end-to-end delay. The

differentiated CW policy and the *strict priority policy* for relay packets in relay node R_i in zone D_i are described in the next two sections.

3. Differentiated CW policy

3.1 Description of differentiated CW policy

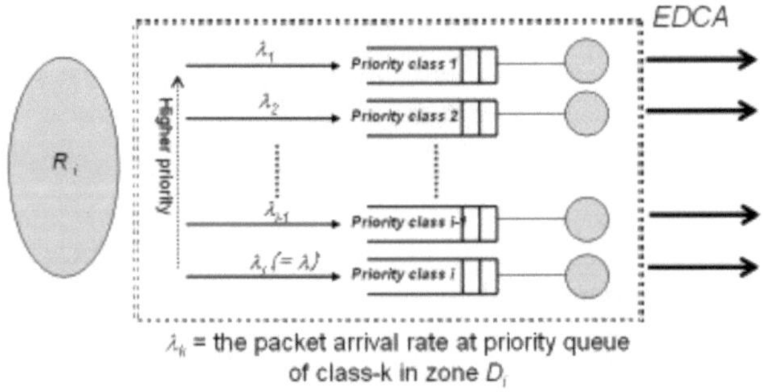

Fig. 5. Priority Queues at Relay Node R_i in zone D_i

First, *differentiated CW policy* (23) which adopts the functionality of IEEE 802.11e EDCA, is described. Fig.5 describes priority queues of class-k, $(k = 1, 2, \cdots, i)$, at relay node R_i in zone D_i. Each priority queue of class-k is regarded as a separate entity with EDCA. Here each priority queue of class-k uses different CW_{min}, more specifically, a priority queue of higher class has a shorter CW_{min} value than priority queues of lower class. Each priority queue of class-k competes with each other and also end nodes to transmit a packet to the next relay node R_{i+1}. By assigning the shorter CW_{min} value to the priority queue of higher class, it has more opportunities to access channel than priority queues of lower class and other end nodes, and thus the delay of the packet passing by more hops can be shortened.

3.2 HoL-delay of packet at priority queue in zone D_i

In order to obtain the end-to-end delay of packets in WMN under the differentiated CW policy, first of all, we focus on zone D_i and find the probability generating function (PGF) of the HoL-delay of packets at each priority queue in zone D_i, where the HoL-delay is defined as the duration from the instant when a packet arrives at the head of the queue to the instant when its successful transmission is completed.

For the differentiated CW policy, priority queue of class-k have different traffic arrival rates and different contention window sizes, $k = 1, 2, \cdots, i$. An end node is regarded as priority queue of class-i in zone D_i because the end node and priority queue of class-i have the same type of local packets. Thus there are one priority queue of class-k for each $k = 1, 2, \cdots, i-1$, and N_i priority queues of class-i in zone D_i. Thus there are $N_i + i - 1$ contending entities in zone D_i. The packet arrival process to priority queue of class-k is assumed to follow a Poisson process with rate λ_k(/sec), where λ_k is the total arrival rate of packets generated at end nodes in zone D_k, $k = 1, 2, \cdots, i$, (note that the arrival rate λ_i(/sec) of packets generated in D_i is λ(/sec) in Subsection 2.2). Thus zone D_i can be modeled as non-saturated IEEE 802.11e EDCA model where there are $N_i + i - 1$ stations and i different queues have different arrival rates and different CW_{min}s, respectively.

For mathematical simplicity, assume $i = 4$, that is, there are 4 priority queues in zone D_i, and so there are N_i priority queues of class-4 and one priority queue of class-k, $k = 1, 2, 3$.
The payload sizes of all packets are equal, and let T_p be the duration for one packet to transmit. Slots are distinguished by following types:

- idle slot with length σ when no nodes transmit.
- successful slot when only one node transmits; the slot duration is $T_s = T_p + SIFS + t_{ACK} + DIFS$. Let us denote $T_s^* = T_s/\sigma$
- collision slot when two or more nodes transmit simultaneously; the slot duration is $T_c = T_p + DIFS$. Let us denote $T_c^* = T_c/\sigma$

Let τ_k be the transmission probability of the priority queue of class-k in a generic slot, $k = 1, 2, 3, 4$, which will be given by Eq.(3). With each transmission attempt and regardless of the number of retransmissions, each packet of priority queue of class-k is assumed to collide with the constant probability p_k as in (31). p_k is a conditional collision probability, meaning that this is the probability of a collision seen by a packet at the time of its being transmitted on the channel. Then the probability p_k is given by

$$p_k = 1 - \prod_{1 \le j \le 3, j \ne k} (1 - \tau_j)(1 - \tau_4)^{N_i}, 1 \le k \le 3 \tag{1}$$

$$p_4 = 1 - \prod_{1 \le j \le 3} (1 - \tau_j)(1 - \tau_4)^{N_i - 1} \tag{2}$$

For the analysis, the following parameters and probabilities are defined:

- Let $W_0^{[k]}$ be the minimum contention window size CW_{min} of priority queue of class-k and $W_i^{[k]} = 2^i W_0^{[k]}$, $k = 1, 2, 3, 4$.
- The maximum backoff stage is set to m and the retry limit is infinite, i.e, no packet is discarded.
- Let S_k be the HoL-delay (measured in idle slot length σ) of priority queue of class-k.

In order to obtain packet delay of IEEE 802.11 DCF under non-saturation condition, first it is essential to find the transmission probability τ_k. Once τ_k is obtained, performance measures such as delay and throughput can be expressed in terms of τ_k. For calculating transmission probability τ_k, there have been two different approaches in the literatures: i) Markov chain-based approach initiated by Bianchi (31) and ii) a method developed by (24). In the former approach, transmission probability τ is determined by the steady state probability of Markov chain. In this chapter, the second approach is adopted to obtain the transmission probability τ_k. The paper (24) developed an analytic model for IEEE 802.11 DCF in the non-saturated and homogeneous condition in the sense that all stations use the same contention parameter and packet arrival process to each station is identical. The analytical model (24) is extended to modeling of IEEE 802.11 DCF in the non-saturated heterogeneous condition in the sense that the packet arrival rate of per-node is different and also the minimum contention window of per-node is different.
The transmission probability is calculated as follows: In the saturated condition, the average backoff window $\overline{W}^{[k]}$ of priority queue of class-k is given (31) by

$$\overline{W}^{[k]} = \left(\frac{1 - p_k - p_k(2p_k)^m}{1 - 2p_k} \right) \frac{W_0^{[k]}}{2}.$$

In the saturated case, the probability that priority queue of class-k transmits a packet in a randomly chosen time slot is equal to $\frac{1}{W^{[k]}}$ (31). For the non-saturated case, the conditional probability P[priority queue of class-k transmits|the queue is not empty] will be approximated by $\frac{1}{W^{[k]}}$ (24). Now let the traffic intensity of the priority queue of class-k be ρ_k, which is defined by $\rho_k = \lambda_k E[S_k]\sigma$ and we assume that $\rho_k < 1$, where HoL-delay S_k of the priority queue of class-k regarded as service time will be given by (7) below. Then the transmission probability τ_k of priority queue of class-k is given by

$$\tau_k = 0 \cdot (1 - \rho_k) + \frac{1}{W^{[k]}} \rho_k. \tag{3}$$

We define the probabilities representing the channel state during the backoff procedure of the priority queue of class-k. During the backoff procedure of the priority queue of class-k, the channel is in one of the following states; idle, collision transmission and successful transmission.

The probability P_{idle}^k of the channel being sensed idle during the backoff procedure of the priority queue of class-k, $(k = 1,2,3,4)$, is given by

$$P_{idle}^k = \prod_{1 \le j \le 3, j \ne k} (1 - \tau_j)(1 - \tau_4)^{N_i} \text{ for } k = 1,2,3, \tag{4}$$

$$P_{idle}^4 = \prod_{1 \le j \le 3} (1 - \tau_j)(1 - \tau_4)^{N_i - 1}. \tag{5}$$

The probability P_s^k of the channel being sensed busy due to successful transmission of other priority queues during the backoff procedure of the priority queue of class-k is given by

$$\begin{aligned}
P_s^1 &= \tau_2(1-\tau_3)(1-\tau_4)^{N_i} + (1-\tau_2)\tau_3(1-\tau_4)^{N_i} \\
&\quad + N_i(1-\tau_2)(1-\tau_3)\tau_4(1-\tau_4)^{N_i-1} \\
P_s^2 &= \tau_1(1-\tau_3)(1-\tau_4)^{N_i} + (1-\tau_1)\tau_3(1-\tau_4)^{N_i} \\
&\quad + N_i(1-\tau_1)(1-\tau_3)\tau_4(1-\tau_4)^{N_i-1} \\
P_s^3 &= \tau_1(1-\tau_2)(1-\tau_4)^{N_i} + (1-\tau_1)\tau_2(1-\tau_4)^{N_i} \\
&\quad + N_i(1-\tau_1)(1-\tau_2)\tau_4(1-\tau_4)^{N_i-1} \\
P_s^4 &= \tau_1(1-\tau_2)(1-\tau_3)(1-\tau_4)^{N_i-1} \\
&\quad + (1-\tau_1)\tau_2(1-\tau_3)(1-\tau_4)^{N_i-1} \\
&\quad + (1-\tau_1)(1-\tau_2)\tau_3(1-\tau_4)^{N_i-1} \\
&\quad + (N_i-1)(1-\tau_1)(1-\tau_2)(1-\tau_3)\tau_4(1-\tau_4)^{N_i-2}
\end{aligned}$$

The probability P_c^k of the channel being sensed busy due to collision transmission of other priority queues during the backoff procedure of the priority queue of class-k is given by

$$P_c^k = 1 - P_{idle}^k - P_s^k.$$

As for deriving the PGF of HoL-delay in the non-saturated and homogeneous conditions, refer to (22). To derive the distribution of HoL-delay S_k, the method (22) can be extended to the heterogeneous condition (23).

Let us consider a priority queue of class-k as the tagged station. Let X, Y and Z be the number of collision slots of other stations, successful transmission slots of the other stations and empty slots experienced until the backoff counter of tagged station becomes zero during a backoff stage, respectively. As a remainder, during the backoff process of the tagged station the length of the slot is empty slot time σ with probability P^k_{idle} or collision time T^*_c with probability P^k_c, or successful transmission time T^*_s with probability P^k_s. If the value of the backoff counter of the tagged station is chosen by a at a given backoff stage, then (X, Y, Z) has a trinomial distribution whose probability mass function is given by

$$P\{X = j, Y = h, Z = l\} = \frac{a!}{j!h!l!}(P^k_c)^j (P^k_s)^h (P^k_{idle})^l, \;\; j + h + l = a. \tag{6}$$

Denoting $T^k_i(z)$ by the PGF of the time duration that the tagged priority queue of class-k stays at the i-th backoff stage, $T^k_i(z)$ is obtained as follows:

$$\begin{aligned}
T^k_i(z) &= \sum_{a=0}^{W^{[k]}_i - 1} \frac{1}{W_i} \Big(\sum_{j=0}^{a} \sum_{h=0}^{a-j} P\{X = j, Y = h, Z = a - j - h\} z^{jT^*_c + hT^*_s + (a-j-h)} \Big) \\
&= \sum_{a=0}^{W^{[k]}_i - 1} \frac{1}{W^{[k]}_i} \Big(\sum_{j=0}^{a} \sum_{h=0}^{a-j} \frac{a!}{j!h!(n-j-h)!} P^j_c P^h_s P^{a-j-h}_i z^{jT^*_c + hT^*_s + (a-j-h)} \Big) \\
&= \sum_{a=0}^{W_i - 1} \frac{1}{W_i} (P^k_c z^{T^*_c} + P^k_s z^{T^*_s} + P^k_{idle} z)^a = \sum_{a=0}^{W^{[k]}_i - 1} \frac{1}{W^{[k]}_i} B^k(z)^a \\
&= \frac{B(z)^{W^{[k]}_i} - 1}{W^{[k]}_i (B^k(z) - 1)},
\end{aligned}$$

where $B^k(z) = P^k_c z^{T^*_c} + P^k_s z^{T^*_s} + P^k_i z$ and $B^k(z)$ is the PGF of the length of one slot.

By conditioning on the number of collisions experienced until the packet transmitted successfully, we obtain the PGF of S_k as follows:

$$\begin{aligned}
E[z^{S_k}] &= \sum_{n=0}^{\infty} E[z^{S_k} | N = n] P\{N = n\} \\
&= \sum_{n=0}^{\infty} \prod_{i=0}^{n} T^k_i(z) (z^{T^*_c})^n z^{T^*_s} (p_k)^n (1 - p_k) \\
&= \sum_{n=0}^{m} \prod_{i=0}^{n} T^k_i(z) (z^{T^*_c})^n z^{T^*_s} (p_k)^n (1 - p_k) \\
&+ \prod_{i=0}^{n} T^k_i(z) T^k_m(z) \frac{(p_k z^{T^*_c})^{m+1} (1 - p_k)}{1 - T^k_m(z) z^{T^*_c} p_k}
\end{aligned} \tag{7}$$

3.3 Extension to the tree structure

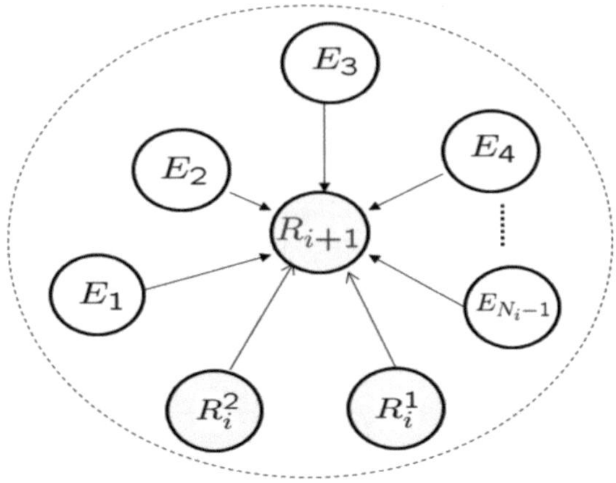

Fig. 6. Zone D_i in the case of tree structure

In the case of tree structure, if we decompose the WMN into disjoint zones, each zone D_i can contain relay nodes more than one as shown in Fig.6. In Fig..6, zone D_i contains two relay nodes R_i^1 and R_i^2 which compete each other and other end nodes to transmit their packets to the relay node R_{i+1}. Assuming that relay node R_i^1 should deliver packets generated at $x-1$ zones, R_i^1 has x uplink buffers for relay traffic. On the other hand, assuming that relay node R_i^2 should forward packets generated at $y-1$ zones, R_i^2 has y uplink buffers for relay traffic. Assume that $x \geq y$. Since the priority queue with packets experiencing the same hop-count uses the same CW_{min}, we note that there are x priority queues in zone D_i where each of them uses a different CW_{min} and has a different arrival rate. The number of contending entities is equal to $(N_i - 1) + x + y$. Thus, with the complexity of expression, it is straightforward to extend the analytical method presented in Section 3.2 to the case of tree structure.

3.4 Average end-to-end delay

The packet delay in a zone is obtained from $M/G/1$ queueing theory. The packet delay $A_{i,k}$ at priority queue of class-k in zone D_i is defined by the sum of queueing delay and service time (HoL-delay), and is given by

$$E[A_{i,k}] = \frac{\lambda_k E[S_k^2]}{2(1-\rho_k)} + E[S_k],\ k = 1,2,3,4 \tag{8}$$

where the first and second moments of S_k are obtained from (7).
Then, the end-to-end delay of a packet generated at an end-node until reaching the gateway node, is obtained. Thus the end-to-end delay $W_{end}^{(i)}$ of the local packet generated at the priority queue of class-i (i.e., end node) in zone D_i is given by

$$W_{end}^{(i)} = E[A_{i,i}] + E[A_{i+1,i}] + \cdots + E[A_{n,i}].$$

3.5 A method to determine CW_{min}s for location-independent end-to-end delay and numerical results

3.5.1 Analytic method to determine CW_{min}

channel bit rate	11Mbps
DIFS	50 μsec
slot size(σ)	20 μsec
SIFS	10 μsec
transmission time of PHY header	192 μsec
MAC header	34 byte
Payload length	1500 byte
ACK	14 byte + PHY header

Table 1. System parameters

The goal in (23) is to guarantee that the end-to-end delays of local packets generated at each end node are almost same by choosing the appropriate minimum contention window size of each priority queue of class-k at zone D_i. More specifically, it is a goal to find a natural number CW_{min} of each priority queue of class-k at zone D_i so that the end-to-end delays $W_{end}^{(i)}$ of the local packets generated at each zone D_i, $(i = 1, 2, \cdots, n)$ should satisfy the following equalities approximately:

$$W_{end}^{(1)} \approx W_{end}^{(2)} \approx \cdots \approx W_{end}^{(n)}. \tag{9}$$

Note that zone D_i has priority queue of class-k $(k = 1, 2, \cdots, i)$. Since each priority class uses different value of CW_{min}, there are i CW_{min}s to be determined in zone D_i. Thus the total number of CW_{min}s to be determined is $1 + 2 + \cdots + n = \frac{(n+1)n}{2}$. On the other hand, (9) provides ${}_nC_2 = \frac{(n-1)n}{2}$ equations. Thus we have $\frac{n(n+1)}{2}$ equations with $\frac{n(n+1)}{2}$ unknown variables. Therefore, if n CW_{min}s of highest priority queue in each zone are given initially, we have $\frac{n(n-1)}{2}$ non-linear equations with $\frac{n(n-1)}{2}$ unknown variables of CW_{min}. Thus we find one of the solutions CW_{min}s to satisfy Eq.(9) approximately. Thus we obtain one of solutions CW_{min}s satisfying (9) numerically by *trial and error* method.

For numerical example, system parameters for the numerical example are given by table 1 and the number of hops is set to $n = 3$. We display the end-to-end delays of packets as the arrival rate λ(/sec) of each end node increases, for the case that the number N_i of end nodes in zone D_i is set to 5 and 7 for all zones, respectively. We set CW_{min} of the highest priority queue in each zone as 32. One of solutions CW_{min}s satisfying criterion (9) approximately can be obtained in table 2 and 3 for $N_i = 5$ and $N_i = 7$. The simulation is performed using Matlab software under same environment as assumptions in our analytic models. Table 2 and 3 show that as arrival rate λ increases, the end-to-end delay increases. For a given packet arrival rate λ, we see that end-to-end delays of packets are almost equal regardless of source nodes' locations and the number of end nodes in each zone. Also, table 2 and 3 shows that analytical results match well with the simulation results.

3.5.2 Heuristic method to determine CW_{min}

Above mentioned method to determine CW_{min} is a centralized one which requires a coordinator to control overall WMNs. The central coordinator should know the network information such as the network size (maximum hop count), the number of mesh nodes and the packet generation rate of each zone. Based on those information, the controller

	$\lambda = 10/sec$			$\lambda = 11/sec$			$\lambda = 12/sec$		
	D_1	D_2	D_3	D_1	D_2	D_3	D_1	D_2	D_3
CW_1	32	32	32	32	32	32	32	32	32
CW_2		69	76		66	73		64	71
CW_3			192			180			171

ete-delay(ms)	$W^{(1)}_{end}$	$W^{(2)}_{end}$	$W^{(3)}_{end}$	$W^{(1)}_{end}$	$W^{(2)}_{end}$	$W^{(3)}_{end}$	$W^{(1)}_{end}$	$W^{(2)}_{end}$	$W^{(3)}_{end}$
analysis	7	7	7	7.5	7.2	7.2	7.6	7.5	7.5
simulation	7.1	6.2	6.2	7.1	6.3	6.4	7.4	6.6	6.5

	$\lambda = 13/sec$			$\lambda = 14/sec$			$\lambda = 15/sec$		
	D_1	D_2	D_3	D_1	D_2	D_3	D_1	D_2	D_3
CW_1	32	32	32	32	32	32	32	32	32
CW_2		61	67		59	65		57	63
CW_3			159			151			143

ete-delay(ms)	$W^{(1)}_{end}$	$W^{(2)}_{end}$	$W^{(3)}_{end}$	$W^{(1)}_{end}$	$W^{(2)}_{end}$	$W^{(3)}_{end}$	$W^{(1)}_{end}$	$W^{(2)}_{end}$	$W^{(3)}_{end}$
analysis	7.8	7.6	7.6	7.9	7.6	7.6	8	7.8	7.8
simulation	7.2	6.7	6.8	7.4	6.7	6.8	7.5	6.8	6.9

Table 2. $CW_k(= CW_{min})$ of the priority queue of class-k and end-to-end (ete) delay vs. arrival rate for $N_i = 5$

	$\lambda = 10/sec$			$\lambda = 11/sec$			$\lambda = 12/sec$		
	D_1	D_2	D_3	D_1	D_2	D_3	D_1	D_2	D_3
CW_1	32	32	32	32	32	32	32	32	32
CW_2		58	65		55	62		43	60
CW_3			149			138			131

ete-delay(ms)	$W^{(1)}_{end}$	$W^{(2)}_{end}$	$W^{(3)}_{end}$	$W^{(1)}_{end}$	$W^{(2)}_{end}$	$W^{(3)}_{end}$	$W^{(1)}_{end}$	$W^{(2)}_{end}$	$W^{(3)}_{end}$
analysis	7.9	7.5	7.5	8.1	7.8	7.8	8.4	8.1	8.3
simulation	7.5	6.5	6.2	7.4	6.6	6.4	7.7	6.8	6.9

	$\lambda = 13/sec$			$\lambda = 14/sec$			$\lambda = 15/sec$		
	D_1	D_2	D_3	D_1	D_2	D_3	D_1	D_2	D_3
CW_1	32	32	32	32	32	32	32	32	32
CW_2		51	58		49	55		48	56
CW_3			124			116			119

ete-delay(ms)	$W^{(1)}_{end}$	$W^{(2)}_{end}$	$W^{(3)}_{end}$	$W^{(1)}_{end}$	$W^{(2)}_{end}$	$W^{(3)}_{end}$	$W^{(1)}_{end}$	$W^{(2)}_{end}$	$W^{(3)}_{end}$
analysis	8.7	8.5	8.9	9.1	9.0	9.6	9.4	9.8	10.9
simulation	8.0	7.0	7.2	8.3	7.3	7.4	8.5	7.6	8.0

Table 3. $CW_k(= CW_{min})$ of the priority queue of class-k and end-to-end (ete) delay vs. arrival rate for $N_i = 7$

periodically (or if necessary) calculates the CW_{min} of each node according to the rule presented above and informs each node to use newly calculated contention window. As an alternative to centralized method, a method to distributively determine CW_{min} in each zone is considered. The principle of our proposed packet management scheme is that relay packets experiencing longer hops are buffered into the higher priority queue. By assigning a smaller value of CW_{min} to a higher priority queue of class, we intend that the packet in the higher priority queue with passing by longer hops is served faster than the packet in a lower priority queue. The question is how smaller value of CW_{min} is assigned to the higher priority queue of class.

Let $f(x)$ be a nondecreasing nonnegative function of x, where x denotes a hop count as a positive integer. Let μ_k denote the service rate of packets in priority queue of class-k, $(k = 1, 2, \cdots, i)$, in zone D_i, i.e, μ_k is the reciprocal of the mean HoL-delay derived in Subsection 3.2, which is given by $\mu_k = 1/E[S_k]\sigma$, and is a function of CW_{min}'s of priority queue of class-k. We introduce the following criterion to differentiate CW_{min}s betweens priority queues of classes:

$$\frac{\mu_j}{f(i-j)} = \frac{\mu_k}{f(i-k)}, \quad j \neq k. \tag{10}$$

In zone D_i, relay packets passing by $i-k$ hops are stored in priority queue of class-k. Since $f(x)$ is a nondecreasing function, according to the criterion (10), a higher priority queue of class occupies a larger portion of the serving capacity by having smaller value of CW_{min}. In each zone D_i, Eqs. (2), (3) and (10) can be solved using numerical technique to obtain the minimum contention window size for each priority queue of class. It is worth noting that, in the heuristic method, CW_{min} of nodes in each zone is independently determined without considering other zones.

For the numerical example, the considered topology is 4-hop linear WMN as shown in fig.2. We assume that there are five end nodes and one relay node in zone $D_i, i = 1, 2, 3, 4$. We set CW_{min} of the highest priority queue in each zone as 31. Lower priority classes including end nodes use the CW_{min} determined by the constraint (10), which is larger than 31. As a weight function for our proposed scheme, we set $f(k) = k$ for numerical examples, that is, $\mu_1 : \mu_2 : \mu_3 : \mu_4 = 4 : 3 : 2 : 1$. Table 4 depicts the CW_{min} of each priority queue in each zone determined by criterion (10) and the end-to-end delays of local packets generated at each zone versus the arrival rate λ. From table 4, we see that our heuristic method achieves almost equal end-to-end delays of packets regardless of their generated zones under moderate packet arrival rate. But as the packet arrival rate λ is high, there is a little, but not great, difference of end-to-end delays depending on the generated zone. This result is expected due to the weight function which we heuristically choose.

λ	15				16				17				18			
zone	D_1	D_2	D_3	D_4	D_1	D_2	D_3	D_4	D_1	D_2	D_3	D_4	D_1	D_2	D_3	D_4
CW_1	32	32	32	32	32	32	32	32	32	32	32	32	32	32	32	32
CW_2		41	50	49		39	50	49		38	49	48		38	48	47
CW_3			73	82			71	80			68	78			66	75
CW_4				159				152				144				135
ete-delay(ms)	11.9	11.9	12.3	12.6	12.5	12.6	13.4	13.9	13.2	13.6	15.3	15.6	14.2	15.1	19.6	18.2

Table 4. $CW_i(= CW_{min})$ of the priority class of i and end-to-end (ete) delay vs. arrival rate

4. Strict priority policy

4.1 Description of the strict priority policy

Next another packet management scheme at the relay node called the *strict priority policy* (23) is presented. As similar to the differentiated CW policy, in the strict priority policy, relay node R_i in zone D_i has i uplink buffers for relay packets as depicted in Fig.7, and the priority queue of class-k stores relay packets originated from zone D_k, which pass by $i-k$ hops,

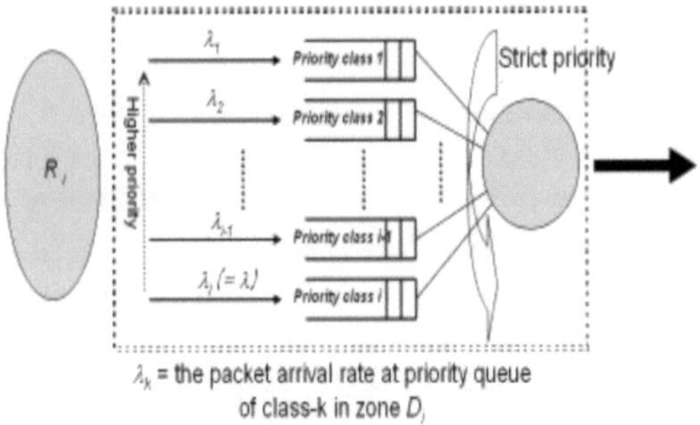

Fig. 7. Strict priority policy

$(k = 1, 2, \cdots, i)$. However, in the strict priority policy, there is only one contending entity of CSMA/CA protocol in relay node R_i. The relay node and end node use different CW_{min}s, respectively. Thus relay node R_i competes with end nodes in a zone D_i to access the channel. If relay node R_i has an opportunity to transmit a packet, the transmission occurs in the order of high priority classes among i uplink buffers. The service discipline among priority classes follows the order of strict priority

4.2 HoL-Delay of packet at priority queue in zone D_i

Under the strict priority policy, zone D_i can be modeled as a non-saturated IEEE 802.11e EDCA model with two different kinds of nodes, where CW_{min} values are differentiated between relay node and end node and also packet arrival rates are different from each other. We assume that the packet arrival processes to relay node R_i and an end node follow Poisson processs with rate Λ_i and λ, respectively, where Λ_i is given by $\Lambda_i = \sum_{k=1}^{i} \lambda_k$ and λ_k is the packet arrival rate to priority queue of class-k, which is the total arrival rate of local packets generated in zone D_k far away $i - k$ hops from relay node R_i as illustrated in Fig.3. Recall that there are N_i contending nodes in zone D_i; $N_i - 1$ end nodes and one relay node R_i. Thus we model zone D_i under the strict priority policy as the non-saturated IEEE 802.11e EDCA network and this model can be regarded as that with two classes under the differentiated CW policy. Therefore the PGF of the HoL-delay can be obtained directly from (7) by simply changing the parameters such as the arrival rate and the number of contending nodes in a zone D_i.

Similar to the argument in Section 3.3, even for the strict priority policy, the analytical method to derive the PGF of HoL-Delay can be extended to the case of tree structure with the complexity of expression.

4.3 Average end-to-end delay

As illustrated in Fig.7, relay node R_i in zone D_i has priority queue of class-k, $(k = 1, 2, \cdots, i)$. The priority queue of class-k stores the packets passing by $i - k$ hops and originated from zone D_k, as shown in Fig. 3. Then, relay node R_i can be modeled as $M/G/1$ queueing system with strict priority as shown in Fig.7.

Let $W_{i,k}, (k = 1, 2, \cdots, i)$, denote packet delay of priority queue of class-k at relay node R_i, where packet delay is defined as the sum of the queueing delay and the service time

	$\lambda = 10/\text{sec}$			$\lambda = 11/\text{sec}$			$\lambda = 12/\text{sec}$		
	D_1	D_2	D_3	D_1	D_2	D_3	D_1	D_2	D_3
CW_1	32	32	32	32	32	32	32	32	32
CW_2		144	195		137	182		130	170
ete-delay(ms)	$W^{(1)}_{end}$	$W^{(2)}_{end}$	$W^{(3)}_{end}$	$W^{(1)}_{end}$	$W^{(2)}_{end}$	$W^{(3)}_{end}$	$W^{(1)}_{end}$	$W^{(2)}_{end}$	$W^{(3)}_{end}$
analysis	7.2	7.4	7.8	7.2	7.6	8	7.4	7.5	8
simulation	6.9	7.2	7.4	7.1	7.2	7.5	7.3	7.4	7.6
	$\lambda = 13/\text{sec}$			$\lambda = 14/\text{sec}$			$\lambda = 15/\text{sec}$		
	D_1	D_2	D_3	D_1	D_2	D_3	D_1	D_2	D_3
CW_1	32	32	32	32	32	32	32	32	32
CW_2		124	160		118	151		113	142
ete-delay(ms)	$W^{(1)}_{end}$	$W^{(2)}_{end}$	$W^{(3)}_{end}$	$W^{(1)}_{end}$	$W^{(2)}_{end}$	$W^{(3)}_{end}$	$W^{(1)}_{end}$	$W^{(2)}_{end}$	$W^{(3)}_{end}$
analysis	7.6	7.8	8.1	7.6	7.8	8.2	7.8	8	8.2
simulation	7.3	7.5	7.6	7.4	7.5	7.6	7.6	7.6	7.7

Table 5. $CW_i(= CW_{min})$ of relay node and end node, respectively, and end-to-end (ete) delay vs. arrival rate for $N_i = 5$

	$\lambda = 10/\text{sec}$			$\lambda = 11/\text{sec}$			$\lambda = 12/\text{sec}$		
	D_1	D_2	D_3	D_1	D_2	D_3	D_1	D_2	D_3
CW_1	32	32	32	32	32	32	32	32	32
CW_2		148	147		139	136		131	126
ete-delay(ms)	$W^{(1)}_{end}$	$W^{(2)}_{end}$	$W^{(3)}_{end}$	$W^{(1)}_{end}$	$W^{(2)}_{end}$	$W^{(3)}_{end}$	$W^{(1)}_{end}$	$W^{(2)}_{end}$	$W^{(3)}_{end}$
analysis	7.5	8.5	8.0	7.6	8.7	8.2	7.8	8.9	8.5
simulation	7.2	7.3	7.0	7.4	7.6	7.5	7.5	7.8	7.6
	$\lambda = 13/\text{sec}$			$\lambda = 14/\text{sec}$			$\lambda = 15/\text{sec}$		
	D_1	D_2	D_3	D_1	D_2	D_3	D_1	D_2	D_3
CW_1	32	32	32	32	32	32	32	32	32
CW_2		123	117		116	109		110	101
ete-delay(ms)	$W^{(1)}_{end}$	$W^{(2)}_{end}$	$W^{(3)}_{end}$	$W^{(1)}_{end}$	$W^{(2)}_{end}$	$W^{(3)}_{end}$	$W^{(1)}_{end}$	$W^{(2)}_{end}$	$W^{(3)}_{end}$
analysis	8.0	9.2	8.7	8.2	9.5	9.1	8.5	10.0	9.5
simulation	7.6	8.0	7.8	7.8	8.5	7.7	8.1	8.9	8.1

Table 6. $CW_i(= CW_{min})$ of relay node and end node, respectively, and end-to-end (ete) delay vs. arrival rate for $N_i = 7$

(HoL-delay). The average of packet delay for priority class-k is given (25) by

$$E[W_{i,k}] = \frac{\sum_{j=1}^{i} \lambda_j E[S^2]}{2(1-\rho^+_{k-1})(1-\rho^+_k)} + E[S] \tag{11}$$

where

$$\lambda^+_k \triangleq \sum_{j=1}^{k} \lambda_j, \ \rho^+_k \triangleq \sum_{j=1}^{k} \rho_j. \tag{12}$$

and, $E[S]$ and $E[S^2]$ are obtained from Eq.(7).
In zone D_i, the local packet generated at the end node should traverse $n-i+1$ hops to reach the gateway node. Thus, the end-to-end delay $W^{(i)}_{end}$ of the local packet generated at the priority queue of class-i (i.e., end node) in zone D_i is given by

$$W^{(i)}_{end} = E[A_i] + E[W_{i,i}] + E[W_{i+1,i}] + \cdots + E[W_{n,i}],$$

where $E[A_i]$ is the average of packet delay of generated at an end node in zone D_i and is given by Eq.(8).

4.4 A method to determine CW_{min} for location-independent end-to-end delay and numerical Results

4.4.1 Analytic method to determine CW_{min}

The goal is to guarantee that the end-to-end delays of packets generated at each zone are almost same. We want to find natural number CW_{min}s of relay node and end node so that the end-to-end delay of the local packet generated in each zone should satisfy the following equalities approximately:

$$W^{(1)}_{end} \approx W^{(2)}_{end} \approx \cdots \approx W^{(n)}_{end}. \tag{13}$$

We should determine the minimum contention window sizes of relay node and end node in each zone, which satisfy (13) approximately. Each end-to-end delay in (13) is a function of CW_{min}s of relay node and end node. Under the strict priority policy, relay node and end nodes in each zone compete with each other via different values of CW_{min}s. Note that under the strict policy $E[A_{i,k}]$ involves 2 unknown CW_{min}s and therefore Eq.(13) have $2n$ unknown CW_{min}s. On the other hand, Eq.(13) provides $\frac{n(n+1)}{2}$ equations. Initially setting CW_{min}s of relay node in each zone as 32, respectively, then we can find one of the solutions CW_{min}s to satisfy Eq.(9) numerically by *trial and error* method.
For numerical example, system parameters for the numerical example are given by table 1 and the number of hops is set to $n=3$. Table 5 and 6 display the end-to-end delays of packets as the arrival rate $\lambda(/\text{sec})$ of each end node increases, for the case that the number N_i of end nodes in zone D_i is set to 5 and 7 for all zones, respectively. CW_{min} of the relay node in each zone is set to 32. One of solutions CW_{min}s satisfying criterion (13) approximately can be obtained in Table 5 and 6 for $N_i=5$ and $N_i=7$.
Table 5 and 6 show that as the arrival rate λ increases, the end-to-end delay of packet increases. As depicted in Table 5 and 6, we see that each end-to-end delays of packets are almost same regardless of source node's location and the number of end nodes in each zone.
In Table 2, 3, 5 and 6, we see that two packet management schemes achieve almost equal end-to-end delay of packets regardless of their generated location, respectively. Comparing end-to-end delays between two schemes, we see that there is almost no difference. Since (9) of differentiated CW policy involves more unknown CW_{min}s than (13) of strict priority policy, only computational complexity of differentiated CW policy is higher than that of strict priority policy.

4.4.2 A heuristic method to determine CW_{min}

Unlike the differentiated CW policy, in the *strict priority policy*, relay node R_i unifying all priority queues of classes contends with other end nodes and has a different contention

window size from that of an end node. Thus we cannot adopt the weight function depending on hop count as in Eq.(10) of the differentiated CW policy.
In general, relay node R_i is heavily loaded compared to end nodes since relay node R_i forwards its local traffic and relay traffics from zone D_{i-1}. Thus it is necessary to give more chances to access the channel to relay node R_i than end nodes E to avoid congestion. To assign the more chance to relay node R_i, we differentiate the CW_{min}s between relay node R_i and end nodes in zone D_i. The question is how smaller contention window size is assigned to the relay node compared to an end node. We define $g(\lambda)$ as the weight function of the arrival rate λ, which is a nonnegative nondecreasing function of arrival rate λ. Thus we introduce the following constraint to differentiate nodes:

$$\frac{\mu}{g(\lambda)} = \frac{\mu_i}{g(\Lambda_i)} \tag{14}$$

where Λ_i and λ are the arrival rates to relay node R_i and an end node E in zone D_i, respectively, and μ_i and μ are the service rates (reciprocal of the mean HoL-delay) of the relay node and the end node in zone D_i, respectively, which are functions of CW_{min}s. Eq.(14) says that the CW_{min} of the relay node and the end node are determined in such a way that their service rates are proportional to their arrival rates, respectively. By doing so, the highly loaded relay node may have more chances to access the channel, and so the relay node can avoid the congestion.
For the numerical example, the considered topology is 4-hop linear WMN as shown in fig.2. It is assumed that there are five end nodes and one relay node in zone $D_i, i = 1,2,3,4$. CW_{min} of the relay node in each zone is set as 31. An end node use the CW_{min} determined by the constraint (14). As a weight function, $g(x) = \sqrt{x}$, that is, $\mu : \mu_i = \sqrt{\lambda} : \sqrt{\Lambda_i}$. Table 7 illustrates the CW_{min} of relay node and end node in each zone determined by criterion (10) and the end-to-end delays of local packets generated at each zone versus the arrival rate λ. From table 4, we see that the heuristic method to determine CW_{min} achieves the almost equal end-to-end delay of packets regardless of their generated zones. Compared with table 4, we see that strict priority policy with constraint (14) achieves less end-to-end delay than the differentiated CW policy with constraint (10) and moreover end-to-end delays of packets are almost equal regardless of source nodes' locations.

λ	15				16				17				18			
zone	D_1	D_2	D_3	D_4	D_1	D_2	D_3	D_4	D_1	D_2	D_3	D_4	D_1	D_2	D_3	D_4
CW_1	32	32	32	32	32	32	32	32	32	32	32	32	32	32	32	32
CW_2		131	140	134		125	132	126		120	125	118		115	119	111
ete-delay(ms)	10.3	11.1	11.1	10.5	10.5	11.4	11.6	10.8	10.7	11.7	12.2	11.1	10.9	12.2	13.5	11.6

Table 7. $CW_i(= CW_{min})$ of the priority class of i and end-to-end (ete) delay vs. arrival rate

4.5 Extension to the case of coexisting of uplink and downlink streams

Next, we discuss whether two packet management schemes achieve location-independent delay of packets regardless of sources' locations in the case that both uplink and downlink streams coexist. First, let us consider the same linear WMN with downlink stream only. Let us consider zone D_i. Keep in mind that relay node R_i has i uplink buffers for uplink streams where the $k-$th uplink buffer in the relay node R_i stores packets originated from all end nodes in zone D_k. As shown in Fig.8, for downlink streams, the relay node R_{i+1} needs

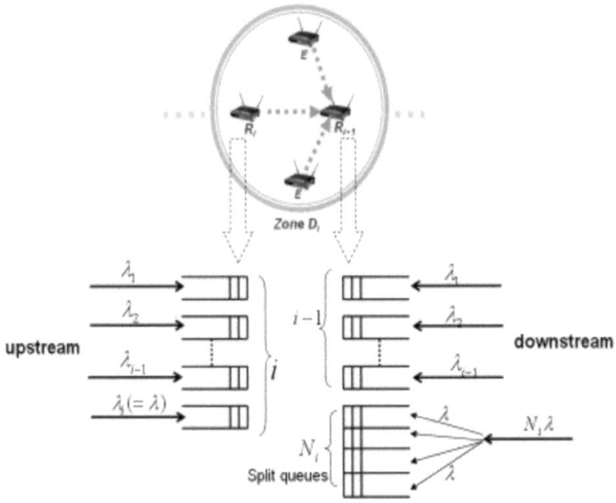

Fig. 8. Application of our proposed scheme to the case of coexisting up/ down streams

	$\lambda = 10/sec$			$\lambda = 11/sec$			$\lambda = 12/sec$		
	D_1	D_2	D_3	D_1	D_2	D_3	D_1	D_2	D_3
CW_1	32	32	32	32	32	32	32	32	32
CW_2		69	76		66	73		64	71
CW_3			192			180			171
ete-delay(ms)	$W_{end}^{(1)}$	$W_{end}^{(2)}$	$W_{end}^{(3)}$	$W_{end}^{(1)}$	$W_{end}^{(2)}$	$W_{end}^{(3)}$	$W_{end}^{(1)}$	$W_{end}^{(2)}$	$W_{end}^{(3)}$
simulation	8.3	7.9	7.4	8.8	8.7	8.7	9.3	9.6	9.3

	$\lambda = 13/sec$			$\lambda = 14/sec$			$\lambda = 15/sec$		
	D_1	D_2	D_3	D_1	D_2	D_3	D_1	D_2	D_3
CW_1	32	32	32	32	32	32	32	32	32
CW_2		61	67		59	65		57	63
CW_3			159			151			143
ete-delay(ms)	$W_{end}^{(1)}$	$W_{end}^{(2)}$	$W_{end}^{(3)}$	$W_{end}^{(1)}$	$W_{end}^{(2)}$	$W_{end}^{(3)}$	$W_{end}^{(1)}$	$W_{end}^{(2)}$	$W_{end}^{(3)}$
simulation	10.0	11.0	11.0	11.5	12.0	12.3	15.0	16.5	17.2

Table 8. $CW_k(= CW_{min})$ of the priority queue of class-k and end-to-end (ete) delay vs. arrival rate for differentiated CW policy

i downlink buffers where the $k-$th downlink buffer in the relay node R_{i+1} stores packets destined for the end nodes in zone D_k from gateway, $1 \leq k \leq i-1$, and the $i-$th downlink buffer stores packets destined for all the end nodes in zone D_i. Again, as depicted in Fig.8, we split packets in the $i-$th downlink buffer in relay node R_{i+1} into N_i queues as many as the number of end nodes in zone D_i equally. With these packet management for downlink stream, we have exactly symmetric structure between uplink and downlink schemes. We assume that the arrival rate of upstream packets originated from all the end node in a zone D_i is equal to that of downstream packets destined for end nodes in zone D_i. We assume that the priority queue of class$-k$ in relay node R_{i+1} uses the same CW_{min} as the priority queue of class-k in relay node R_i and also each split queue uses the same CW_{min} as an end node. With

	$\lambda = 10/\text{sec}$			$\lambda = 11/\text{sec}$			$\lambda = 12/\text{sec}$		
	D_1	D_2	D_3	D_1	D_2	D_3	D_1	D_2	D_3
CW_1	32	32	32	32	32	32	32	32	32
CW_2		144	195		137	182		130	170
ete-delay(ms)	$W_{end}^{(1)}$	$W_{end}^{(2)}$	$W_{end}^{(3)}$	$W_{end}^{(1)}$	$W_{end}^{(2)}$	$W_{end}^{(3)}$	$W_{end}^{(1)}$	$W_{end}^{(2)}$	$W_{end}^{(3)}$
simulation	8.3	7.8	7.8	8.8	8.6	9.0	9.0	9.1	9.7
	$\lambda = 13/\text{sec}$			$\lambda = 14/\text{sec}$			$\lambda = 15/\text{sec}$		
	D_1	D_2	D_3	D_1	D_2	D_3	D_1	D_2	D_3
CW_1	32	32	32	32	32	32	32	32	32
CW_2		124	160		118	151		113	142
ete-delay(ms)	$W_{end}^{(1)}$	$W_{end}^{(2)}$	$W_{end}^{(3)}$	$W_{end}^{(1)}$	$W_{end}^{(2)}$	$W_{end}^{(3)}$	$W_{end}^{(1)}$	$W_{end}^{(2)}$	$W_{end}^{(3)}$
simulation	9.8	10.1	10.8	9.8	11.0	10.5	14.8	15.4	16.6

Table 9. $CW_i(=CW_{min})$ of relay node and end node, respectively, and end-to-end (ete) delay vs. arrival rate for strict priority policy

these assumptions, in each zone, packet delay of priority queue of class-k for uplink stream is exactly same as that of the corresponding priority queue of class-k for downlink stream. Thus, in the case that only downlink stream exists, our proposed schemes can achieve the location-independent end-to-end delay of packets regardless of destination's location just as the case that only uplink stream exists.

In the case that uplink and downlink streams coexist, there are $N_i + i - 1$ contending entities for uplink transmission in zone D_i. Also, for downlink, there are $N_i + i - 1$ contending entities for downlink transmission in zone D_i. Thus, there are $2 \cdot (N_i + i - 1)$ contending entities in zone D_i for the case of differentiated CW policy. (For the case of strict priority policy, note that there are $2 \cdot (N_i + 1)$ contending entities.) In the case that uplink and downlink streams coexist, by symmetric structure between uplink and downlink schemes, we see that the end-to-end delay of upstream packet originated from an end node in each zone is exactly same as that of downstream packet destined for the corresponding end node in each zone.

In order to examine whether our proposed schemes provide location-independent end-to-end delay even in the case that both uplink and downlink streams coexist, we perform simulations using Matlab software. The parameters for simulations are set to the same as the case of uplink only: The number of hops is set to 3. The number of end nodes in each zone is set to 5. The downstream packet arrival process destined for each end node follows Poisson process with rate $\lambda(/\text{sec})$, which is the same as uplink packet arrival process of each end node. The priority queue of class-k for downlink at the relay node R_{i+1} uses the same CW_{min} of the corresponding priority queue of class-k for uplink, which are given by Table 2. As we see in Table 8 and 9, for a given packet arrival rate λ, our two proposed schemes ensure end-to-end delays of packets to be almost equal in the case that uplink and downlink streams coexist.

5. Acknowledgement

This research was partially supported by the MKE (Ministry of Knowledge Economy) under the ITRC (Information Technology Research Center) support program supervised by the NIPA (National IT Industry Promotion Agency) (NIPA-2010-C1090-1011-0007), Korea and in part by KT.

6. References

[1] Mesh Networking Forum, Building the business case for implementation of wireless mesh networks, Mesh Networking Forum 2004, San Francisco, CA, October 2004.

[2] P. Gupta, P. R. Kumar, "The capacity of wireless networks," *IEEE Transactions on Information Theory*, Vol. 46(2), 2000, pp. 388-404.

[3] J. Li, C. Blake, D.S.J. De Couto, H.I.Lee and R. Morris, "Capacity of ad hoc wireless networks", Proc. of ACM Mobicom 2001, pp.61-69, July 2001.

[4] Ian F. Akyildiz and Xudong Wang, " A Survey on Wireless Mesh Networks," IEEE Radio Communications, Sep. 2005

[5] K. Jain, J. Padhye, V. Padmanabhan and L. Qiu, "Impact on interference on multihop wireless network performance," Proceedings of *ACM MobiCom*, 2003, pp. 66-80.

[6] A. Raniwala, T. Chiueh, "Architecture and algorithms for an IEEE 802.11-based multi-channel wireless mesh network," Proceedings of *IEEE INFOCOM*, 2005, pp. 2223-234.

[7] A. Raniwala, K. Gopalan, T. Chiueh, "Centralized channel assignment and routing algorithms for multi-channel wireless mesh networks," *ACM Mobile Computing and Communications Review* (MC2R), Vol. 8(2), 2004, pp. 50-65.

[8] *IEEE Standard for Information technology. Telecommunications and information exchange between systems. Local and metropolitan area networks. Speci□c requirements Part 11: Wireless LAN Medium Access Control (MAC) and Physical Layer (PHY) speci□cations* Amendment 8: Medium Access Control (MAC) Quality of Service Enhancements (2005).

[9] *Draft Standard for Information Technology - Telecommunications and Information Exchange Between Systems - LAN/MAN Speci□c Requirements - Part 11: Wireless Medium Access Control (MAC) and physical layer (PHY) speci□cations: Amendment: ESS Mesh Networking*, IEEE Unapproved draft P802.11s/D1.02, Mar. 2007.

[10] V. Gambiroza, B. Sadeghi, and E. W. Knightly, "End-to-End Performance and Fairness in Multihop Wireless Backhaul Networks," *ACM MOBICOM*, Sep. 2004.

[11] Daji Qiao, and Kang G. Shin, "Achieving efficient channel utitlization and weighted fairness for data communications in IEEE 802.11 WLAN under the DCF." In Proceedings of Tenth IEEE International Workshop on Quality of Service, 2002. IWQOS 2006, 2002.

[12] J. Jun and M.L. Sichitiu, "Fairness and QoS in Multihop Wireless Networks," *IEEE VTC*, Oct. 2003.

[13] I.F. Akyildiz, X. Wang and W. Wang, "Wireless Mesh Networks: a survey", *Comput. Networks* 47 (4) (2005) 445-487

[14] R. Bruno, M. Conti, and E. Gregory, "Mesh networks:Commodity multihop ad hoc networks," *IEEE Communication Magazine*, pp.123-131, March 2005.

[15] Ashish Raniwala, and Tzi-cker Chieh, "Architecture and algorithms for an IEEE 802.11-based multi-channel wireless mesh network," In Proceedings of *IEEE INFOCOM 2005*. 24th Annual Joint Conference of the IEEE Computer and Communications, vol.3, pages 2223-2234, March 2005.

[16] L. Yang, "Issues for mesh media access coordination component in 11s," *IEEE 802.11-04/0968R13*, January 2005.

[17] J. Jun and M. L. Sichitiu, ąřThe nominal capacity of wireless mesh networks,ąś IEEE Wireless Communications, pp. 8.14, October 2003.

[18] S. Ganguly, V. Navda, K. Kim, A. Kashyap, D. Niculescu, R. Izmailov, S. Hong, and S. R. Das. Performance Optimizations for Deploying VoIP Services in Mesh Networks. IEEE Journal on Selected Areas in Communications, 24(11), November 2006.

[19] Jane-Hwa Huang, Li-Chun Wang, and Chung-Ju Chang, "Coverage Enhancement for a Multi-channel Ring-based Wireless Mesh Network with Guaranteed Throughput and Delay," *IEEE ICC 2006 proceedings*, vol.9, pp. 3903-3910, June 2006 .

[20] J. Lee and I. Yeom, "Achieving Throughput Fairness in Wireless Mesh Networks Based on IEEE 802.11," IEEE INFOCOM 2008.

[21] Ashish Raniwala, and Tzi-cker Chieh, "Architecture and algorithms for an IEEE 802.11-based multi-channel wireless mesh network", In Proceedings of *IEEE INFOCOM 2005*. 24th Annual Joint Conference of the IEEE Computer and Communications, vol.3, pp. 2223-2234, March 2005

[22] Y. H. Bae, K. J. Kim, M. N. Moon and B. D. Choi, "Analysis of IEEE 802.11 non-saturated DCF by matrix analytic method", Annals of Operations Research, Vol.162, Num.1, pp.3-18, Sept. 2008.

[23] Yun Han Bae, Kyung Jae Kim, Jin Soo Park and Bong Dae Choi,"Differentiated CW Policy and Strict Priority Policy for Location-Independent End-to-End Delay in Multi-hop Wireless Mesh Networks", IEICE TRANSACTIONS on Communications Vol.E93-B No.7 pp.1869-1880, July 2010.

[24] Tickoo, O. and Sikdar, B., "A queueing model for finite load IEEE 802.11 random access MAC," Communications, 2004 IEEE International Conference on, pp: 175- 179, Vol.1, June 2004.

[25] Hideaki Takagi, "Queueing analysis," North-holland, vol.1, 1993.

[26] Violeta Gambiroza, Bahareh Sadeghi, and Edward W. Knightly, ąřEnd-toend performance and fairness in multihop wireless backhaul networks.ąś In MobiCom ąŕ04: Proceedings of the 10th annual international conference on Mobile computing and networking, pages 287.301. ACM Press, 2004.

[27] Jangeun Jun, and Mihail L. Sichitiu, ąřFairness and QoS in multihop wireless network.ąś In Proceedings of Vehicular Technology Conference, 2003. VTC 2003-Fall. 2003 IEEE 58th., Volume 5, pages 2936.2940, October 2003.

[28] Bisnik, N. and Abouzeid, A., "Delay and Throughput in Random Access Wireless Mesh Networks," Communications, 2006. ICC '06. IEEE International Conference on, Vol. 1, pp. 403-408, June 2006.

[29] T. Liu and W. Liao, "Location-Dependent Throughput and Delay in Wireless Mesh Networks," Vehicular Technology, IEEE Transactions on, Volume PP, Issue 99, Page(s):1 - 1, 2007.

[30] C. Sarr and I. G. Lassous, "Estimating Average End-to-End Delays in IEEE 802.11 Multihop Wireless Networks," available on http://hal.inria.fr/inria-00166017/en/.

[31] G. Bianchi, "Performance analysis of the IEEE 802.11 distributed coordination function," *IEEE J. Select. Areas Commun.*, vol. 18, no. 3, pp. 535–547, Mar. 2000.

[32] N. S. Nandiraju et al., "A novel queue management mechanism for improving performance of multihop flows in IEEE 802.11s based mesh networks," In Proc. of IPCCC 2006, April 2006.

[33] B. Aoun, R. Boutada, G. Kenward, "Analysis of capacity improvements in multi-radio wireless mesh networks", In Proc. of VTC 2006-Spring.

[34] Duffy, K., Leith, D.J., Li, T. and Malone, D. "Improving Fairness in Multi-Hop Mesh Networks Using 802.11e", In Proc. of WiOpt, pp. 1-8, 2006.

[35] A. Hamed Mohsenian Rad and Vincent W. S. Wong, "Cross-Layer Fair Bandwidth Sharing for Multi-Channel Wireless Mesh Networks", IEEE Transactions on Wireless Communications, vol.7, no.9, pp.3436-3445, Sep. 2008.
[36] A. K. Das, H. M. K. Alazemi, R. Vijayakumar, and S. Roy, "Optimization models for fixed channel assignment in wireless mesh networks with multiple radios," in Proc. IEEE SECON, Santa Clara, CA, Sept. 2005.
[37] Y. Y. Chen, S. C. Liu, and C. Chen, "Channel assignment and routing for multi-channel wireless mesh networks using simulated annealing," in Proc. IEEE Globecom, San Francisco, CA, Nov. 2006.

4

Self-adaptive Multi-channel MAC for Wireless Mesh Networks

Zheng-Ping Li, Li Ma, Yong-Mei Zhang, Wen-Le Bai and Ming Huang
North China University of Technology
China

1. Introduction

In order to enhance the transmission rate, multiple channels and multiple transceivers are employed in wireless mesh networks (WMNs). However, the bandwidth utilization rate is still low, and it is hard to design efficiency MAC. There are mainly three reasons. The first reason is that there are different kinds of nodes in WMN: some nodes are single transceiver and some nodes are multiple transceivers or multiple radios (Ian & Wang, 2005). The MAC needs to be suitable for single transceiver nodes and multiple transceiver nodes at the same time. The second reason is that the MAC not only needs to control multiple nodes but also multiple transceiver or multiple radios to access multiple channels. How to coordinate all the nodes and the transceivers, radios of each node to access the channels and enhance the bandwidth utilization rate is a multi-parameter optimization problem. The third reason is that the traffic load on each link is varying. The channel allocation scheme need be adaptive to the load of links.

Common control channel (CCC) based multi-channel MAC is a representative proposal (Benveniste & Tao, 2006) for WMN. All the MAC control signals are exchanged on a common control channel, and the data are sent on data channel. This MAC scheme is very flexible to combine with existing channel allocation schemes. However, the handshaking is made on the control channel, which can't avoid the interference of the non CCC based MAC on data channel. The second problem is that the switching time on the data channel is longer than the transmission time of sending a data packet, which will reduce the efficiency of CCC. Moreover, when there is only one radio, CCC will have the hidden terminal problem (N. Choi, et al. 2003).

Based on CCC, a self-adaptive multi-channel MAC is proposed in this chapter. To keep the flexibility of CCC, common control channel still remains in our scheme. To reduce the channel switching delay and avoid interference from non-CCC based MAC, spreading code based channel division scheme is employed on the data channel. Our scheme can inherit the merits of CCC and remove its faults. Moreover, based on the common control channel framework, we proposed a self-adaptive channel allocation scheme which can adjust the medium access process according to the number of idle channels, and the load of the links on a node to maximize the bandwidth utilization rate of the WMN system.

The rest of this chapter is organized as follows: section 2 makes a survey of the existing multi-channel medium access schemes, section 3 proposes our scheme and makes theory analysis and simulation, and conclusion is drawn in the last section.

2. Related work

To enhance the transmission rate, multi-channels are introduced into WMN. To present our scheme, we need to introduce the related multi-channel MAC first.

2.1 Random medium access schemes

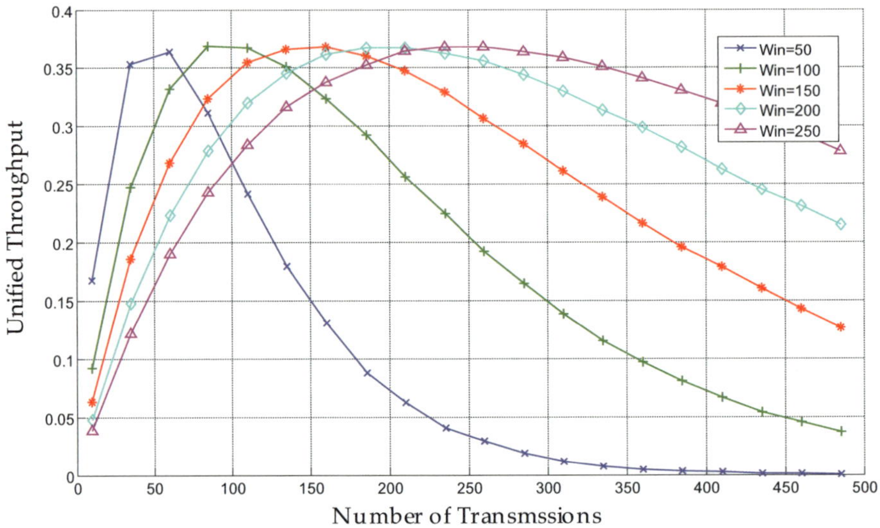

Fig. 1. The throughput vs. the number of transmissions of ALOHA based MAC

MAC controlling signals need be exchanged among the nodes within the communication range. Since no fixed channel is allocated to each node, the MAC controlling signals are sent following random medium access scheme which are mainly Carrier Sensing Multiple Access/Collision Avoidance (CSMA/CA) scheme. CSMA/CA is gotten from ALOHA based scheme. Fig.1 shows that the throughput of ALHOA varies with the number of transmissions. The system throughput can be maximized if the number of transmissions is properly set, which is the original idea of CSMA/CA. CSMA/CA can maximize the system throughput by controlling the number if transmissions in the contention window. CSMA/CA employs carrier sensing, and random back-off schemes to access the channels. There are many kinds of CSMA/CA schemes, and the one used in WLAN is the most common one. The CSMA/CA adopted by IEEE 802.11 works as follows. Before accessing the channel, a node needs to generate a random time which is uniformly distributed within contention window and sense the carrier to get the channel conditions: idle or busy. If the channel is idle, the node will do back-off with a back-off timer. If the channel is busy, the node will turn off the timer until the channel is idle again, and if the channel become idle, the node will turn on the timer after a distributed inter-frame space (DIFS) (IEEE 802.11-1999 (R2003), 2003). If the timer expires, the node can access the channel immediately. Fig.2 shows the CSMA/CA based medium access process of 4 nodes in IEEE 802.11. Node 2, 3, 4 are ready to send frame during node 1 sending frame. Then, node 2, 3, 4 generate back-off times respectively which are 5, 21, 12 seconds, and then begin deferring. After node 1 finish sending the frame, the three nodes start their back-off clocks after a short time DIFS and then begin back-off. This process is done in the contention window during which the three nodes compete for the channel. Since node 2's back-off time is the smallest one, this node's timer timeout first and sends its frame immediately. Node 3, 4's left back-off time are 16s and 7s respectively and this two nodes stop the back-off timers and then go to defer. After node 2 finish sending the frame, node 3, 4 follow the similar process like node 2.

2.2 Channel division scheme

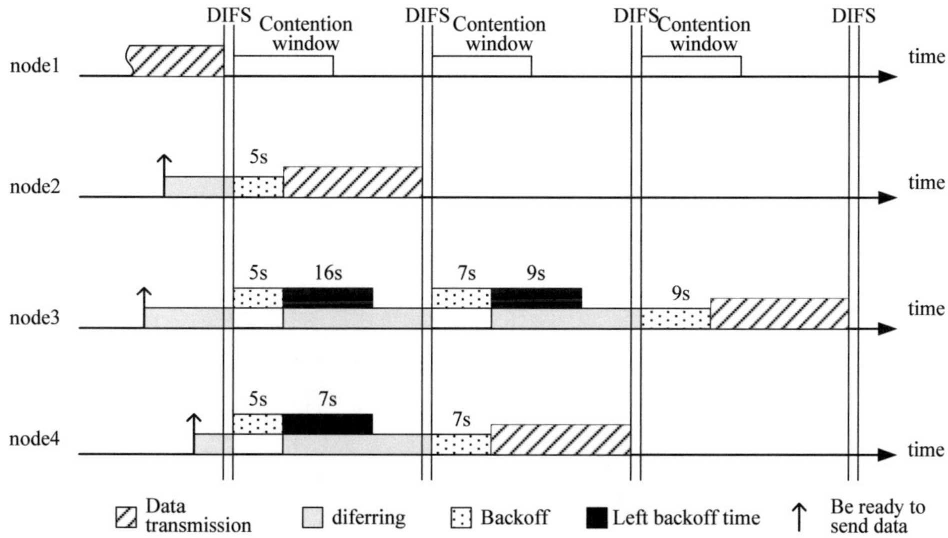

Fig. 2. The CSMA/CA based medium access process in IEEE 802.11

The wireless resource needs be divided into multiple channels, and the resource division scheme is very important. The wireless resource can be divided with frequency, time, space, and spreading code. These schemes have different features and can be used in corresponding conditions. Properly selecting the resource division schemes can enhance the system's performance, otherwise, the performance might be declined. Frequency based channel division scheme can effectively remove inter-channel interference through properly set the band of each channel, and the transmission on each channel can be done simultaneously. Once the band of each channel is set, the capacity of each channel will be constant, and can't be adjusted according to requirements. Time based channel division scheme divides a period of time into many time slots, and allocates each channel with several time slots. The capacity of each channel is based on the number of time slots, and can be arbitrarily changed through setting the number of time slots. To realize time based channel division scheme, synchronization among the nodes in a communication area must be needed, which could consume much wireless resource. Space based channel division scheme divides channels through setting the covering area of each antenna. By properly designing the antenna's covering area, frequency reuse rate can be increased, but space based channel division scheme can't realize full duplex transmission. Spreading code based channel division scheme divides channels through setting spreading code for each channel. The spreading codes have low cross correlation and high autocorrelation, which is employed to divide wireless resource. This scheme can arbitrarily adjust the transmission

rate according to the interference and data transmission requirements of services. Moreover, this scheme can weaken the interference from the same channel. However, this scheme is a interference constrained system, and too many channels can increase the interference and reduce the transmission rate.
The proposed scheme in this chapter employs frequency, and spreading code based channel division scheme which is shown in Fig. 3. To avoid interference from data channel and realize simultaneous transmission on these channels, control channel and data channel are divided with frequency. The data channel is divided into several sub-channels with spreading codes. Employing the spreading code channel division scheme can weaken the hidden terminal interference and make the transmission on sub-channel be adjustable according to the interference and the service requirements.

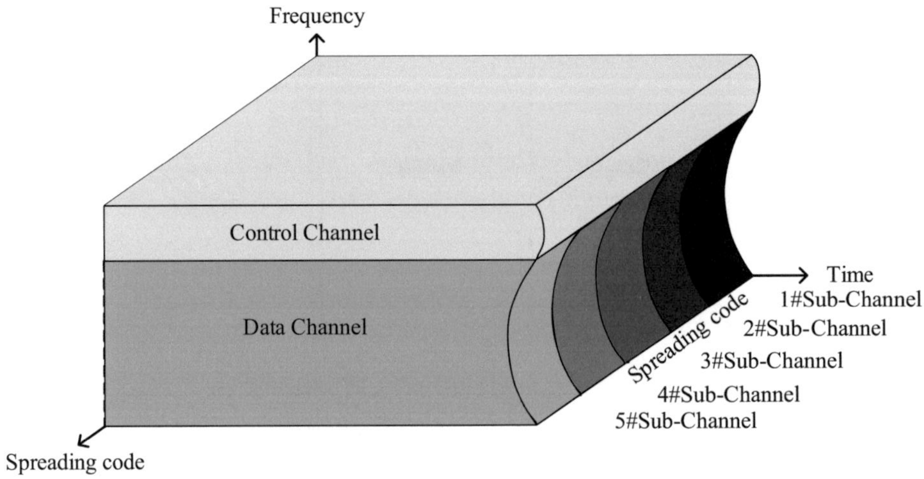

Fig. 3. Channel division scheme of the proposed scheme

2.3 Related medium access control schemes

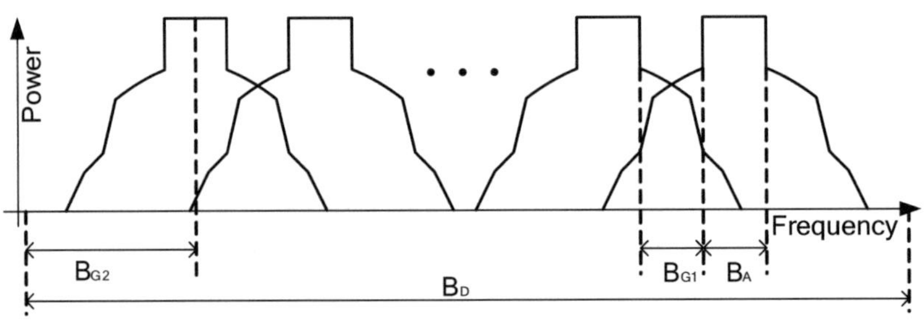

Fig. 4. Sub-band allocation of CCC's data channels

Since mesh has been employed in Wireless Personal Area Networks (WPAN), Wireless Local Area Networks (WLAN), Wireless Metropolitan Area Networks (WMAN), corresponding Medium Access Control (MAC) schemes have to be proposed to enhance the network performance. IEEE has set up IEEE 802.11s working group for the mesh networks in WLAN networks. The draft of IEEE 802.11s has been proposed in 2006 (802.11s Working Group, 2006), but there are still many issues demanding solutions (Wang & Lim, 2008).

During the drafting process, common control channel (CCC), a common control channel based MAC for multi-channel WMN, is a representative proposal (Benveniste and Tao, 2006). CCC divides the wireless resource with frequency which is shown in fig.4, and there are one control channel and several data channel. Control channel is used for the transmission of request-to-send (RTS) and clear-to-send (CTS) which are distributed coordinating signals. After the handshaking on control channel, the nodes can access the requested channel for data transmission. This scheme has two problems.

1. The first problem is that when all the data channels are occupied, the control channel will be idle, and this is a waste of wireless resource. Fig. 5 shows one control channel and two data channels of a CCC system, and node B accesses channel 1 and node C accesses channel 2 after handshaking on control channel with RTS and CTS respectively. When node B and node C occupy the two data channel, the control channel become idle until one of the data channel become idle again, and then node A can send handshaking signals on the control channel to make channel request. The idle time of the control channel is a waste of wireless resource.
2. The second problem is the hidden terminal interference. RTS and CTS handshaking process can constrain the hidden terminal interference, but the interference radius is larger than the communication radius, so the hidden terminal interference can't be entirely removed.

Moreover, CCC didn't proposed channel allocation scheme to enhance the bandwidth utilization rate.

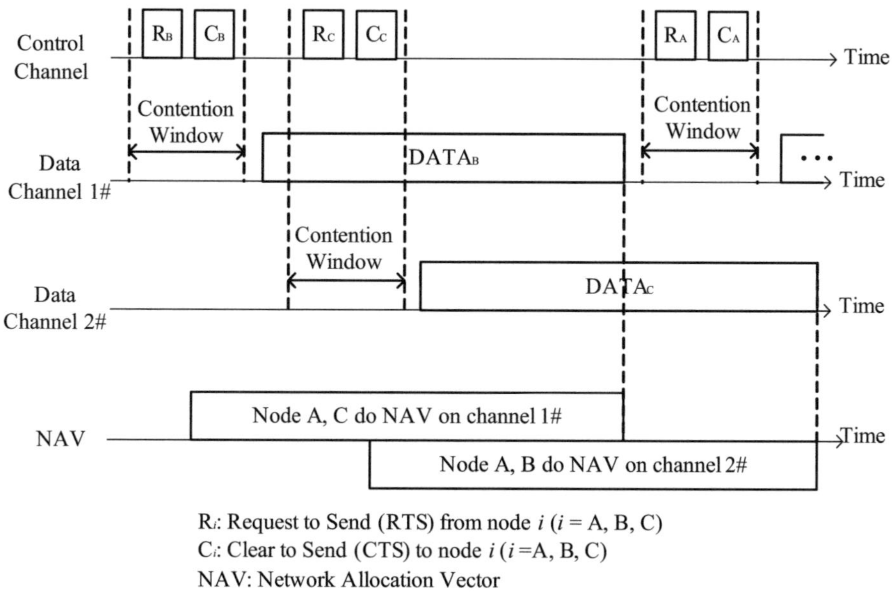

Fig. 5. DCF of CCC.

3. Self-adaptive multi-channel MAC for wireless mesh networks

This scheme employs the channel division scheme shown in fig. 3, does channel request with RTS-CTS handshaking scheme on the control channel, and sends data on the data sub-channels. The channel allocation process can adapt to the traffic load on each link to reduce congestions and maximize the bandwidth utilization rate.

3.1 Medium access control scheme

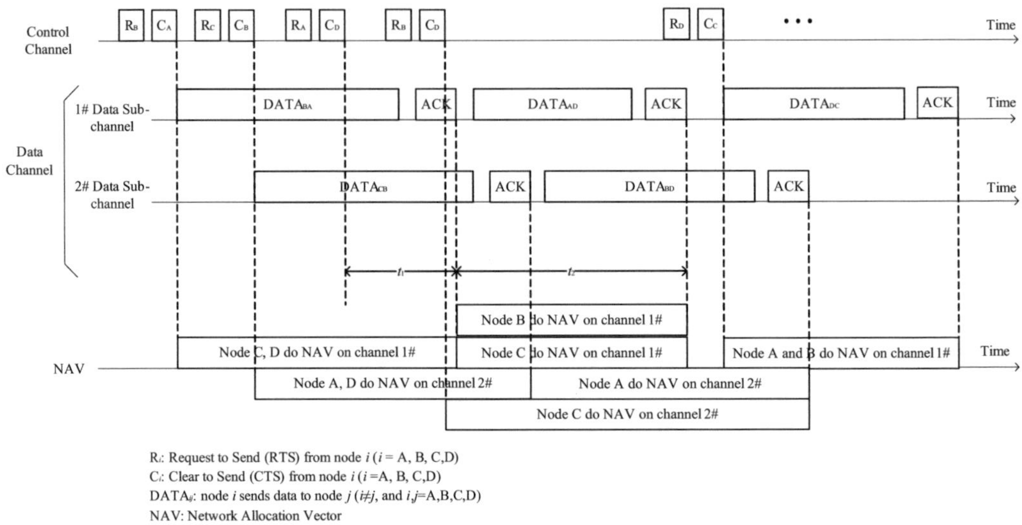

Fig. 6. DCF of the proposed scheme

The main goal of our scheme is to reduce the channel access delay and increase the system's throughput. Our scheme has a control channel and a data channel which are divided by frequency. The data channel is divided into data sub-channels with spreading codes. One spreading code corresponds to one data sub-channel. We use the sub-channel for short instead of data sub-channel in the following. RTS-CTS handshake is used on the control channel. The process of RTS-CTS handshake is designed as follows:

a. Node A sends RTS to node B for data transmission on a sub-channel. CSMA/CA is used as the medium access control scheme.
b. When node B receives RTS from node A, it sends CTS to node A after a short time interval, named short-inter-frame-space (SIFS).
c. If the sub-channel in Step 1 isn't occupied, node A will send data to node B immediately on the sub-channel. Otherwise, the data will be arranged for transmission on the sub-channel.

In our scheme, both RTS and CTS massages are of the same form which is composed of source ID, destination ID, spreading code ID, and duration. Source ID is the ID of the node which sends the message; destination ID is the ID of the node which receives the message; and spreading code ID is the ID of the selected spreading code for transmission. Duration is the length of time will be spent for the data transmission and is estimated according to the amount of data, the length of the spreading code, and the data rate of the system. In the data channel, data packet and ACK are transmitted.

Data channel reservation is proposed in our scheme. In IEEE 802.11 MAC, DCF was proposed for single channel system. Data, ACK, RTS and CTS are sent on the same channel. Data is transmitted immediately after a successful RTS-CTS handshake. In our scheme, RTS and CTS are transmitted in control channel; data and ACK are transmitted in data channel.

Data shall not be transmitted immediately after a successful RTS-CTS handshake as in Step 3. When there is no idle sub-channel, RTS-CTS handshake can still be done on control channel. After a successful RTS-CTS handshake, data is arranged for transmission on a sub-channel. When the time for transmission comes, the data will be transmitted immediately. In this process, data channel reservation is needed for the arrangement. It is realized with virtual carrier sensing, which was proposed in IEEE 802.11 MAC, and is realized with networks allocation vector (NAV). In our scheme, NAV records the start time and the duration of an arranged traffic on the sub-channel based on duration information in RTS/CTS message. When a node receives RTS/CTS with destination ID which is different from its own ID, it need conduct virtual carrier sensing. The virtual carrier sensing acts differently in the following three conditions:

a. A node receives RTS/CTS when it is idle. The start time of NAV is the current time. Duration of NAV is that in RTS/CTS. The node does NAV immediately.
b. A node receives RTS/CTS when it is doing NAV and the remaining duration is tn. The node needs to substitute the tn with the duration in RTS/CTS. The start time of NAV is the current time.
c. A node receives RTS/CTS when it is sending data, and the remaining transmission time is t_d. The start time of NAV is the sum of current time and t_d. Let t be the duration in TRS/CTS. The duration of NAV is t minus t_d. When this node finishes the transmission, it starts doing NAV.

For example, there are four nodes A, B, C and D within the radio coverage of each other, and there are two sub-channels and a control channel. The communication process is shown in fig. 6. At first, node B sends data to node A on sub-channel #1 (sub-channel #1 is simply named #c1 below). After a while, node C sends data to node B on sub-channel #2 (sub-channel #2 is simply named #c2 below). Since the two channels are all idle, these two transmissions start immediately after the CTS on #c1 and #c2. Nodes C and D do NAV on #c1, and nodes A and D do NAV on #c2. After a while, node A has data to send to D when the two sub-channels are all busy. Node A finds that the transmission on #c1 will finish soon, so node A makes a reservation for #c1. Node A sends RTS to node D. In RTS, the spreading code ID is that of #c1, and the duration D_{AD1} is t_1+t_2. t_1 is the remaining occupying time on #c1, and t_2 is the estimated data transmission length of time from nodes A to D. Then, nodes B and C should start doing NAV. Node C is in the state of NAV now. The remaining NAV time of node C on #c1 is t_1. Then, this node extends the NAV time by $t_2= D_{AD1}-t_1$. Node B is transmitting data to node A on #c1 now, and the remaining time of the transmission is t_1. In this case, node A arranges the NAV from the end of its transmission on #c1, and the NAV duration is $t_2= D_{AD1}-t_1$. After some time, node B has data to be sent to node D. At this time, the two data channels are all busy. Fig. 6 shows that #c1 has been reserved, and #c2 will be the first one to finish transmission. Therefore, node B makes reservation on #c2. Some time later, node D needs to send data to node C when #c1 is idle. In this case, node D needn't do channel reservation and requests for the idle channel as usual.

To enhance the throughput of the system, traffic flow adaptive channel allocation scheme is proposed. This scheme adaptively allocates the channels to the node according to the node's load level. The node with heavy load accesses the channel with high priority, and the node with light load accesses the channel with low priority. This scheme can help the node with heavy load occupy more channel than the node with light load. In this way, the node with heavy load can borrows idle channels from the node with low load, and the system's channel utilization rate and throughput is enhanced. The traffic load can be estimated through the self-similarity of the traffic (Crovella & Bestavros, 1997) (Leland, et. al., 1994). Then, each node's busy level is defined as follows:

Let $\psi(i)$ be the expected load of node i, and let h be the number of data channels. Suppose the capacity of each data channel is constant and it is denoted with C. Then, the capacity within node i's coverage is:

$$\Phi(i) = C^*h \tag{1}$$

Let $\rho(i)$ be the ratio of ratio of the espected load and the channel capacity within node i's coverage.

$$\rho(i) = \psi(i)/\Phi(i) \tag{2}$$

$\rho(i)$ is employed to denote the busyness degree of node i. Based on $\rho(i)$, the busyness level can be gotten with the following formula :

$$B(i) = \begin{cases} 1 & \rho(i) > (1-a) \\ k+1 & (1-a*k) \geq \rho(i) > (1-a-a*k) \\ h & \cdots\cdots \quad (1+a-a*h) \geq \rho(i) \end{cases} \tag{3}$$

In (3), k=1,2,…,(h-2), and a is the stem-length of business level and a*h<1. $B(i)$=1 is the highest busyness level and $B(i)$=h is the lowest busyness level. In this algorithm, all the data channels are being numbered first, and channels are allocated according to the numbered channel and busyness level of each node. The channel allocation process is as follows:

a. let Θ(k, m)(k=1,2,…, h; m=1,2,…, h) be the traffic load of the node with busyness level m on channel k. Firstly, node i needs to decide whether channel k satisfy following condition which is named condition 1:

$$\psi(i) < C - \sum_{m=1}^{h} \Theta(k,m) \tag{4}$$

(4) means that node i's expected load is smaller than channel k's available bandwidth. If only one channel satisfy condition 1, this channel will be selected by node i. If there is more than one channel satisfy condition 1, minimum interference hybrid channel allocation algorithm (Jeng & Jan 2006) shall be employed to select a channel from them.

b. If no channel satisfies condition 1, node i will search for channels satisfy condition 2:

$$\psi(i) < \left[C - \sum_{m=1}^{h} \Theta(k,m) \right] + \left[\sum_{m>B(i)} \Theta(k,m) \right] \tag{5}$$

In the right side of (5), the former part is the available bandwidth on channel k, and the latter part is the bandwidth occupied by the node with busyness level lower than $B(i)$. If there is one channel satisfies condition 2, this channel will be selected. If there are more than one channel satisfy condition 2, minimum interference hybrid channel allocation algorithm shall be employed to select a channel from them.

c. each node periodically estimate their traffic load, and search the channel with the step a and step b.

3.2 Performance analysis models

A. Throughput with hidden terminals

a. *Throuput of our scheme without hidden termianls*

Let N be the number of channels. Each channel i, $1 \le i \le N$, is assigned an n bit pseudo-random noise (PN) sequence. From (Hui, 1984), we can get the sum capacity of the CDMA channels in binary input Gaussian condition is:

$$C_{b/c} = \frac{N}{n}\left(\log_2 2\pi e - \int_{-\infty}^{\infty} P(y)\log_2 P(y)dy\right) \tag{6}$$

in which

$$P(y) = \left(P_1(y) + P_{-1}(y)\right)/2$$

and

$$P_m(y) = \exp\left[-(y - m\sqrt{n/N})^2 / 2\right] / \sqrt{2\pi}$$

This capacity is denoted with bits/chip. Let r_d be the chip rate of the data channel. Then we can get the capacity C_d denoted with bit/s.

$$C_d = C_{b/s} \bullet r_d$$

$$= \frac{Nr_d}{n}\left(\log_2 2\pi e - \int_{-\infty}^{\infty} P(y)\log_2 P(y)dy\right) \tag{7}$$

Let B be the bandwidth of the WMN. In our scheme, B is composed of two parts: control channel and data channel. The bandwidth of the control channel is B_C. Then the bandwidth of the data channel is $B_D = B - B_C$. According to Shannon formula, the capacity of WMN is:

$$C = B\log_2\left(1 + \frac{E}{n_0 B}\right) \tag{8}$$

in which E is the signal power. Let E(P) be the average length of the data packet. Since packet rate equals bit rate dividing by average packet length, the capacity of WMN denoted with packet rate is:

$$P_{WMN} = \frac{bits \cdot per \cdot \mathrm{second}}{\mathrm{packet} \cdot \mathrm{length}} = \frac{C}{\mathrm{E}(P)} \tag{9}$$

According to Shannon formula, the capacity of data channel r_{DHMA} of our scheme is:

$$r_{DHMA} = B_D \log_2\left(1 + \frac{E}{n_0 B_D}\right) \tag{10}$$

in which E, n_0 and B_D have been defined in the upper parts.
Replacing r_d in (7) with r_{DHMA} in (10), we can get the sum capacity of the sub-channels:

$$C_{DHMA} = \frac{Nr_{DHMA}}{n}\left(\log_2 2\pi e - \int_{-\infty}^{\infty} P(y)\log_2 P(y)dy\right) \tag{11}$$

in which N is the number of sub-channels and n is the length of the spreading code. Sum capacity of the sub-channels is the capacity of the data channel with CDMA access scheme, and the sum capacity of the sub-channels is named achievable data rate of the data channel. The achievable packet rate of the data channel is the ratio of the achievable data rate of the data channel C_{DHMA} to the average data packet length E[P]. Then the achievable packet rate of the data channel is:

$$P_{DHMA} = \frac{C_{DHMA}}{\mathrm{E}(P)} \tag{12}$$

According to Shannon formula, the capacity of control channel r_{CC} in our scheme is:

$$r_{CC} = B_C \log_2\left(1 + \frac{E}{n_0 B_C}\right) \tag{13}$$

in which E, n_0 and B_C have been defined in the upper parts.
Let E($RTSCTS$) be the average cycle of a success RTS-CTS two-way handshake. From (Liu, 2004), we can get E($RTSCTS$) as follows:

$$\mathrm{E}(RTSCTS) = T_{RTS} + SIFS + \delta + T_{CTS} + DIFS + \delta \tag{14}$$

in which T_{RTS} and T_{CTS} are the transmission time of RTS and CTS, and they are equal to $T_{RTS}=RTS/r_{CC}$, $T_{RTS}=CTS/r_{CC}$. RTS and CTS are the frame length of the RTS and CTS. SIFS is short inter-frame space, and DIFS is DCF inter-frame space. δ is propagation delay.
Then, the capacity of the control channel denoted with packet rate is the inverse of average cycle of a success RTS-CTS handshake, that is:

$$P_{CC}^{DHMA} = \frac{1}{\mathrm{E}(RTSCTS)} \tag{15}$$

From (Kleinrock & Tobagi, 1975), we can get the throughput of CSMA/CA on control channel under the offered load G:

$$S_{DHMA} = \frac{Ge^{-aG}}{G(1\text{-}2a)+Ge^{-aG}} \tag{16}$$

in which a is normalized propagation delay of the radio.
Since throughput can be denoted with the ratio of achievable packet rate to channel capacity (Liu, 2004), achievable packet rate can be denoted with the product of throughput and channel capacity. From (15) and (16), we can get the achievable packet rate of control channel under the offered load G:

$$R_{CC}^{DHMA} = P_{CC}^{DHMA} S_{DHMA} \tag{17}$$

Let L_{DHMA} be the achievable packet rate on data channel of DHMA under offered load G. In our system, every transmission of a data packet on data channel needs a success of RTS-CTS handshake on control channel. Therefore, L_{DHMA} is equal to the achievable packet rate on control channel when the achievable packet rate on control channel is smaller than the achievable packet rate on data channel. When the achievable packet rate on control channel is higher than the achievable packet rate on the data channels, the data channels are full loaded, and L_{DHMA} is equal to the achievable packet rate on data channel. Then, L_{DHMA} is as follows:

$$L_{DHMA} = \begin{cases} R_{CC}^{DHMA} & R_{CC}^{DHMA} \leq P_{DHMA} \\ P_{DHMA} & R_{CC}^{DHMA} > P_{DHMA} \end{cases} \tag{18}$$

The throughput of our scheme under offered load G is the ratio of L_{DHMA} to the capacity of the WMN P_{WMN}. From (9) and (18), we can get the throughput of our scheme under the offered load G:

$$T_{DHMA} = \frac{L_{DHMA}}{P_{WMN}} \tag{19}$$

b. *Throughput of CCC without hidden terminals*

CCC is a multi-channel MAC proposed for IEEE 802.11 mesh networks. In order to make a comparison of CCC and our scheme, we analyze the throughput of CCC. In CCC, there is control channel and data channels. The bandwidths of them are denoted with B_C and B_D. The data channels are divided by frequency. The bandwidth allocation of the data channels is shown in Fig. 5. In this figure, B_A is the allocated bandwidth for a data channel, B_{G1} is the guard band between data channels, and B_{G2} is the guard band at the edge of data channels. Supposing there are N data channels, we can get B_A by the following equation:

$$B_A = \frac{1}{N}\left[B_D - 2B_{G2} - (N-1)B_{G1}\right] \tag{20}$$

From Shannon formula, we can get the capacity on each data channel:

$$r_A = B_A \log_2\left(1 + \frac{E}{n_0 B_A}\right) \tag{21}$$

in which E and n_0 have been defined in the upper parts.

The bandwidth of the control channels in CCC and DHMA are equal. From (13), we can get the capacity on control channel in CCC. The access process of CCC in Fig.6 shows that the contention window is between two data packets. Then, the average cycle of a success RTS-CTS handshake $E'(RTSCTS)$ is the sum of the average cycle of a success RTS-CTS handshake in DHMA and the average data transmission interval. Let $E(P)$ be the average data packet length, and N be the number of data channels. Then the average data transmission interval is $E(P)/r_A/N$. From (15), we can get $E'(RTSCTS)$ as follows:

$$E'(RTSCTS) = E(RTSCTS) + E(P)/r_A/N \tag{22}$$

The capacity of control channel denoted with packet rate is the reverse of average cycle of a success RTS-CTS handshake $E'(RTSCTS)$:

$$P_{CC}^{CCC} = \frac{1}{E'(RTSCTS)} \tag{23}$$

Suppose the offered load of the system is G. Since contention window is embedded between two data packets in CCC, contention window can not be arranged at any place on the control channel. Therefore, the offered load G is converged in short intervals on the control channel. Then, the load in the contention window is:

$$G' = (E'(RTSCTS)^{*}G)/E(RTSCTS) \tag{24}$$

From (Kleinrock & Tobagi, 1975), we can get the throughput of CSMA/CA of control channel in CCC under the offered load G:

$$S_{CCC} = \frac{G' e^{-aG'}}{G'(1\text{-}2a) + e^{-aG'}} \tag{25}$$

Similar to (17), we can get the achievable packet rate on control channel in CCC from (23) and (25):

$$R_{CC}^{CCC} = P_{CC}^{CCC} S_{CCC} \tag{26}$$

Let L_{CCC} be the achievable packet rate on data channel of CCC under offered load G. Because each success of RTS-CTS handshake is followed by a data packet on the data channels immediately, L_{CCC} is equal to the achievable packet rate on control channel:

$$L_{CCC} = R_{CC}^{CCC} \tag{27}$$

Similar to (19), we can get the throughput of CCC under the offered load G:

$$T_{CCC} = \frac{L_{CCC}}{P_{WMN}} \tag{28}$$

c. *Thoughput of the two system with hidden terminals*

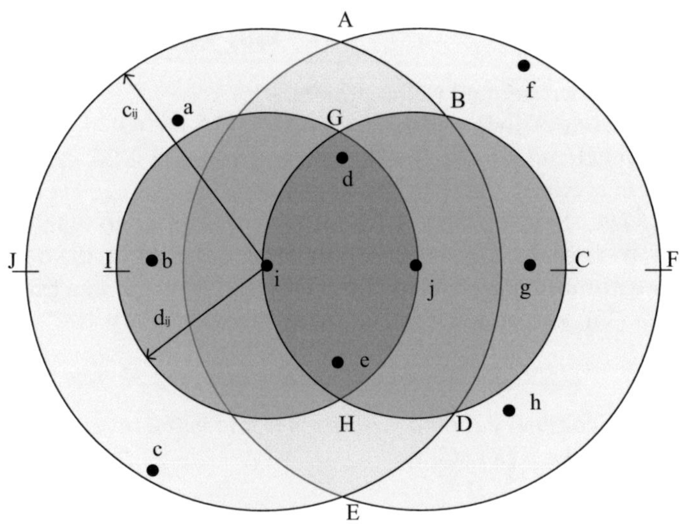

Fig. 7. The coverage areas and capture areas of node i and j.

When there are hidden terminals in the system as shown in Fig. 7, we estimated the throughput of our system and CCC.

In Fig. 7, capture area of node i is circle AJE, and is denoted with I_i. c_{ij} is the capture distance of node i, and is the radius of the capture areas of node i in Fig. 7. c_{ij} can be denoted with $c_{ij}=ad_{ij}$. a is the capture factor and $1\leq a$. d_{ij} is the distance from node i to node j. Coverage area of node i is circle GIH which is maximum communication area of node i, and is denoted with C_i. d_{ij} is the radius of node i's coverage area.

Let H_i be the aggregate of node i's hidden terminals. Let C(i,j) be the aggregate of the communication pair (m,n), and $m \in C_i \cup C_j - H_i$, $n \in C_i \cup C_j - H_j$. Suppose the data transmitted from node i to node j follow Poisson distribution. Then, the length of the data's inter-arrival time follows exponential distribution, and the average of the inter-arrival time is denoted with G(i,j). Let O_i be the aggregate of hidden terminals during the transmission from node i to node j with RTS-CTS handshake scheme. O_i is as follows:

$$O_i = I_i \cup I_j - I_i \cup C_j \tag{29}$$

In Fig. 7, the coverage of O_i equals to the area ABCDEF. Let β be the propagation delay from node i to node j, and T_{RTS} be the transmission time of RTS.

From (Liu, 2004), we can get the success probability of a transmission from node i to node j in hidden terminal condition as follows:

$$P_S(i,j) = \exp\left\{-\beta \sum_{(m,n)\in C(i,j)} G(m,n) - T_{RTS} \sum_{m\in O_i, n\in C(m,n)} G(m,n)\right\} \tag{30}$$

In our scheme, suppose the spreading code is m-sequence. Then, we can get the autocorrelation $\rho(\tau)$ of the spreading code from (Goldsmith, 2005).

$$\rho(\tau) = \begin{cases} 1 - \dfrac{|\tau|(1+1/N)}{T_C} & |\tau| \leq T_C \\ 1/N & |\tau| > T_C \end{cases} \tag{31}$$

in which τ is the delay offset of spreading code, T_C is the chip duration of the spreading code, and n is the length of the spreading code. Since the medium accesses of two nodes are independent, the delay offset τ is uniformly distributed on $[0,nT_C]$. The expectation of $\rho(\tau)$ is $E(\rho(\tau))=3(n-1)/(2n^2)$. Suppose the radio signal is transmitted in free space. From (Goldsmith, 2005), we can get the path loss:

$$P_L = \frac{G_t G_r \lambda^2}{(4\pi)^2 d^2} \tag{32}$$

in which G_t and G_r are the antenna gain of transmitter and receiver respectively. λ is the wave length of the radio, and d is the distance from transmitter to receiver. In CSMA/CA, there is a power sensing threshold of the capture area. Nodes with sensed power lower than the power sensing threshold take the channel as busy; otherwise take the channel as idle. From (32), we can denote the power sensing threshold P_0^{CCC} of CCC as follows:

$$P_0^{CCC} = \frac{G_t G_r \lambda^2}{(4\pi)^2 \left(c_{ij}^{CCC}\right)^2} P \tag{33}$$

in which c_{ij}^{CCC} is the capture distance of CCC, and P is the transmission power. Since spreading code is used in our scheme, the power sensing threshold P_0^{DHMA} of DHMA is the product of sensed power and the average autocorrelation of the spreading code. So, P_0^{DHMA} is as follows:

$$\begin{aligned} P_0^{DHMA} &= \left(\frac{G_t G_r \lambda^2}{(4\pi)^2 \left(c_{ij}^{DHMA}\right)^2} P \right) E(\rho(\tau)) \\ &= \frac{G_t G_r \lambda^2}{(4\pi)^2 \left(c_{ij}^{DHMA}\right)^2} \frac{3(n-1)}{2n^2} P \end{aligned} \tag{34}$$

in which c_{ij}^{DHMA} is the capture distance of DHMA, and P is the transmission power. Supposing the power sensing threshold of CCC and DHMA is equal, from (33) and (34), we can get the ratio η_C of capture distances of CCC and our scheme.

$$\eta_C = \frac{c_{ij}^{CCC}}{c_{ij}^{DHMA}} = \frac{n}{\sqrt{3(n-1)/2}} \tag{35}$$

From (35), it can be seen that $\eta_C>1$, which indicates that the capture distance of CCC is longer than that of our scheme. Fig.7 shows that the longer of capture distance, the heavier of hidden terminal interference. Moreover, η_C is in direct ratio with n. When the length of spreading code is increased, η_C is increased. Supposing the capture distance of CCC is constant, we can see that the capture distance of our scheme is decreased when the length of spreading code is increased. So, the hidden terminal interference in CCC is heavier than that in our scheme, and the hidden terminal interference in our scheme can be removed when the spreading code is long enough.

Suppose node *i*'s coverage areas in our scheme and CCC are equal, and they are denoted with C_i. In Fig.7, C(*i,j*) is the aggregate of the communication pair (*m,n*) and $m \in C_i \cup C_j - H_i$, $n \in C_i \cup C_j - H_j$. Therefore, C(*i,j*) in our scheme and CCC are same. Suppose the transmission power of every node in our scheme and CCC is equal. From (35), it can be seen that node *i*'s capture areas in our scheme and CCC are different, and they are denoted with I_i^{DHMA} and I_i^{CCC} respectively. From (29), we can get O_i of our scheme and CCC:

$$O_i^{DHMA} = I_i^{DHMA} \cup I_j^{DHMA} - I_i^{DHMA} \cup C_j \tag{36}$$

$$O_i^{CCC} = I_i^{CCC} \cup I_j^{CCC} - I_i^{CCC} \cup C_j \tag{37}$$

O_i^{DHMA} and O_i^{CCC} are corresponding to the area ABCDEF in Fig. 7. Suppose the nodes are uniformly distributed in the WMN with density ζ. From Fig. 7, we can get O_i^{DHMA} and O_i^{CCC} as follows:

$$O_i^{DHMA} = \zeta\left(\pi - 2\arccos\frac{d_{ij}}{2c_{ij}^{DHMA}}\right)\left(c_{ij}^{DHMA}\right)^2 + \zeta d_{ij}c_{ij}^{DHMA}$$

$$-\zeta\left(4d_{ij} - 2\left(c_{ij}^{DHMA}\right)^2\right)\arccos\frac{c_{ij}^{DHMA}}{2d_{ij}} - \zeta c_{ij}^{DHMA}\sqrt{d_{ij}^2 - \left(c_{ij}^{DHMA}/2\right)^2} \quad (38)$$

$$O_i^{CCC} = \zeta\left(\pi - 2\arccos\frac{d_{ij}}{2c_{ij}^{CCC}}\right)\left(c_{ij}^{CCC}\right)^2 + \zeta d_{ij}c_{ij}^{CCC}$$

$$-\zeta\left(4d_{ij} - 2\left(c_{ij}^{CCC}\right)^2\right)\arccos\frac{c_{ij}^{CCC}}{2d_{ij}} - \zeta c_{ij}^{CCC}\sqrt{d_{ij}^2 - \left(c_{ij}^{CCC}/2\right)^2} \quad (39)$$

Let $P_S^{DHMA}(i,j)$ and $P_S^{CCC}(i,j)$ be the success transmission probability of our scheme and CCC respectively. From (29), we can get $P_S^{DHMA}(i,j)$ and $P_S^{CCC}(i,j)$:

$$P_S^{DHMA}(i,j) = \exp\left\{-\beta \sum_{(m,n)\in C(i,j)} G(m,n) - T_{RTS} \sum_{m\in O_i^{DHMA}, n\in C(m,n)} G(m,n)\right\} \quad (40)$$

$$P_S^{CCC}(i,j) = \exp\left\{-\beta \sum_{(m,n)\in C(i,j)} G(m,n) - T_{RTS} \sum_{m\in O_i^{CCC}, n\in C(m.n)} G(m,n)\right\} \quad (41)$$

in which T_{RTS} and T_{CTS} are the same with that in (14).

The throughputs of our scheme and CCC with hidden terminals are the product of throughput without hidden terminals and the transmission success probability with hidden terminals, and they are as follows:

$$T_H^{DHMA} = T_{DHMA}P_S^{DHMA}(i,j) \quad (42)$$

$$T_H^{CCC} = T_{CCC}P_S^{CCC}(i,j) \quad (43)$$

n	N	*e*	B	B_D
100	10	2.718	100MHz	99MHz
B_C	B_{G1}	B_{G2}	β	n_0
1MHz	2MHz	4.9MHz	0.005	70dBm
RTS	*CTS*	*E(P)*	*a*	*SIFS*
20Byte	14Byte	3000Byte	0.1	0.03
DIFS	E	G_t	G_r	
0.14	90dBm	1	1	

Table 1. The evaluation parameter

B. Access delay

Access delay is the length of time from a node sending RTS to this node starting to send data. Access delay D_A is the sum of contention access delay D_{CA} and data channel waiting delay D_W, that is,

$$D_A = D_{CA} + D_W \tag{44}$$

Contention access delay is the length of time from a node sending RTS to this node receiving CTS. Data channel waiting delay is the length of time from a node receiving the CTS to this node starting to send data.

a. *Access delay of our scheme*

In our scheme, RTS-CTS handshake is used on the control channel, so the collision happens only at RTS period. Since CSMA/CA is used during RTS, according to (Kleinrock & Tobagi, 1975), we can get the average contention access delay on control channel:

$$D_{CA}^{DHMA} = (G / S_{DHMA} - 1)(1 + 2a + \alpha + \delta) + 1 + a \tag{45}$$

in which G is the offered load for the WMN, S_{DHMA} is the throughput of the control channel defined in (16), a is the propagation delay normalized by the transmission time of RTS T_{RTS}, α is the ratio of frame length of CTS to that of RTS, and δ is ratio of the average length of backoff time E[B] to the transmission time of RTS. From (Cali, et. al., 1998), we can get the average length of backoff time E[B]=(E[CW]-a* T_{RTS})/2. Here, a* T_{RTS} is the propagation delay. Then, δ= E[B]/ T_{RTS}.

When the RTS-CTS handshake is succeeded, transmission of a data packet will be arranged on the data channels. This process can be modeled as an M/M/1 queuing process. The average packet arriving rate of the queue is R_{CC}^{DHMA} defined in (17), and the average packet serving rate is the product of the throughput of DHMA and the capacity of WMN system, that is, $P_H^{DHMA} = T_H^{DHMA} P_{WMN}$. Then, the average waiting delay D_W is the average waiting delay in the queue. According to Little formula, we can get the average waiting time:

$$D_W^{DHMA} = \frac{R_{CC}^{DHMA}}{P_H^{DHMA}(P_H^{DHMA} - R_{CC}^{DHMA})} \tag{46}$$

Replacing D_{CA} and D_W in (44) with D_{CA}^{DHMA} and D_W^{DHMA}, we can get the access delay of our scheme:

$$D_A^{DHMA} = D_{CA}^{DHMA} + D_W^{DHMA} \tag{47}$$

b. *Access delay of CCC*

Suppose G is the offered load of the WMN. From (24), we can get the offered load G' in the contention window of CCC. According to (Kleinrock & Tobagi, 1975), the contention access delay is:

$$D_{CA}^{CCC} = (G' / S_{CCC} - 1)(1 + 2a + \alpha + \delta) + 1 + a \tag{48}$$

G' and S_{CCC} are defined in (24) and (25). All the other parameters are the same with that of (45). In Fig.6, data packet is transmitted on data channel immediately after the success of RTS-CTS handshake on control channel. Then, the data channel waiting delay is equal to zero in CCC. From (44), we can get that the access delay in CCC is equal to the contention access delay on control channel, that is:

$$D_A^{CCC} = D_{CA}^{CCC} \tag{49}$$

3.3 System evaluation and comparisons

Table 1 offers the required parameters of our system and CCC during the evaluation. Suppose the nodes are uniformly distributed as shown in Fig. 7. C(*i,j*) is {(b,i), (d,i), (d,j), (e,i), (e,j), (j,g)}. From (35), it can be seen that the capture distance of our scheme is smaller than that of CCC. When n=100, the ratio of the two capture distances η_C is 8.2. Therefore, c_{ij}^{CCC} is much larger than c_{ij}^{DHMA}. Since c_{ij}^{CCC} and c_{ij}^{DHMA} are the radiuses of I_i^{CCC} and I_i^{DHMA}, I_i^{CCC} is larger than I_i^{DHMA}. From (36) and (37), we can see that O_i^{CCC} is larger than O_i^{DHMA}. In Fig. 7, suppose O_i^{CCC} ={h, f}, and O_i^{DHMA} ={h}.

Firstly, the throughput of DHMA is estimated. From (19), we can get the throughput of DHMA without hidden terminals. The estimation results are shown in Fig.8. When the offered load is increased, the throughput of DHMA is increased at first and decreased when the control channel is overloaded. The estimation curve for CCC without hidden terminals is similar to that of our scheme. However, because the contentions on control channel of CCC are more serious than that of ours, the throughput of CCC is lower when the offered load is over 23. When there are hidden terminals, the throughputs of our scheme and CCC are decreased. Because of the hidden terminal immune property which is analyzed in (35), our scheme has smaller throughput reduction than CCC does.

Then, the access delay of DHMA is estimated. The access delay of our scheme can be estimated with (47). Fig.9 shows the estimation results which imply: when the offered load is bellow 20, the access delay increases slowly; when the offered load is over 60, the access delay increase quickly. Comparing with CCC, the access delay of our scheme is shorter. When the offered load is increased, the access delay of our scheme increases slower than

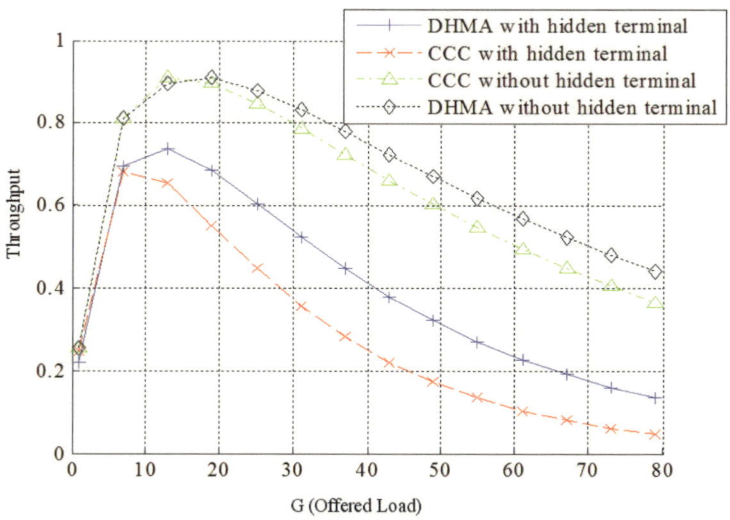

Fig. 8. Throughput vs. offered load of DHMA and CCC.

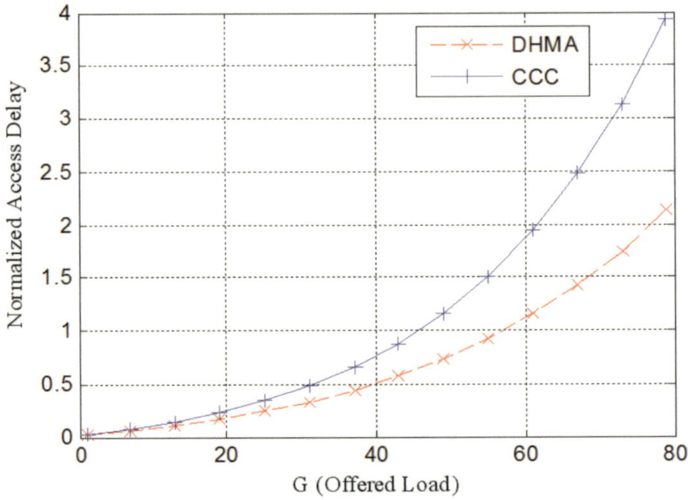

Fig. 9. Throughput vs. offered load of DHMA and CCC.

that of CCC. The reason is that the contention window of our scheme can be arranged at any place on the control channel while that of CCC must be arranged between the two data transmissions. So, when the offered loads in DHMA and CCC are equal, there are fewer RTSs within each contention window in our scheme than those in CCC.

4. Conclusions and future works

This chapter proposes a medium access scheme for multi-channel WMN. Our scheme is designed to reduce the channel access delay and increase the system's throughput. Since CDMA is used to divide sub-channels on data channel, data-rate on each sub-channel can be adjusted by changing the length of spreading code. Because the transmissions between different communication pairs aren't synchronized, hidden terminal interference can be reduced. From the theory analyses and performance evaluations, we can see that our scheme has high throughput and short access delay and outperforms CCC.

However, there are still problems needed to be solved. This scheme can be employed for the nodes with multiple transceivers. For the system with single transceiver, the node needs to switch transceiver from control channel and data channel frequently. When the node switches to the data channel, the handshaking information on the control channel might be missed, and cause system malfunction. Therefore, transceiver adaptive MAC need be further studied. Cognitive radio has been introduced into WMNs (Chen et al., 2008), which can relax the conflict between wireless resource supply and demand. But cognitive radio makes the medium access control more difficult. In cognitive radio based WMNs, wireless resources are unlicensed and licensed users' actions will cause mesh node's frequent channel switching, which makes the MAC of WMNs more complicated. QoS guarantee of MAC in WMN need be studied. Channel allocation and channel switching processes might cause delay for transmission which will reduce the QoS of real-time services, especially the real-time multimedia services. Because of frequent channel switching, QoS guarantee in cognitive radio based WMN is much more difficult. If the QoS problem isn't solved, WMN won't be widely accepted by the consumer.

5. Acknowledgment

This work was supported by 2009 Scientific Research Fund of NCUT and 2008 Scientific Research Platform and Team Construction Fund of NCUT.

6. References

Ian, F. & Wang, X.(2005). A Survey on Wireless Mesh Networks, *IEEE Communications Magazine*, Vol. 43, No. 9, (Sept. 2005), s23-s30, ISSN 0163-6804.

Benveniste, M. & Tao, Z. (2006). Performance Evaluation of a Medium Access Control Protocol for IEEE 802.11s Mesh Networks, *Proceedings of IEEE Sarnoff Symposium* 2006, pp. 1-5, Princeton, NJ USA, ISBN 978-1-4244-0002-7, Mar. 2006.

Wang, X. & Lim, Azman O. (2008). IEEE 802.11s Wireless Mesh Networks: Framework and Challenges, *Ad Hoc Networks Journal (Elsevier)*, 2008, 6(6): 970-984.

Jeng, A. A.-K. & Jan, R.-H. (2006). Optimization on Hybrid Channel Assignment for Multi-channel Multi-radio Wireless Mesh Networks. *Proceedings of IEEE GLOBECOM* 2006, pp. 1-5, San Francisco, California, USA, Nov.-Dec. 2006.

IEEE 802.11-1999 (R2003), Information Technology-Telecommunications and Information Exchange between Systems-Local and Metropolitan Area Networks-Specific Reqirements-Part11: Wireless LAN Medium Access Control (MAC) and Physical Layer (PHY) Specitications, Jun. 2003

IEEE 802.11 Working Goup, Draft Standard for Inormation Technology-Telecommunications and Information Exchange Between Systems-LAN/MAN Specitific Requirements-Part 11: Wireless Medium Access Control (MAC) and physical layer (PHY) specifications: Amendment: ESS Mesh Networking, IEEE P802.11s/D1.0, Nov. 2006.

Crovella, M. E. & Bestavros, A. (1997). Self-Similarity of World Wide Web Traffic: Evidence and Possible Causes, *IEEE/ACM Transactions on Networking*, 1997, 5(6): 835-846, ISSN 1063-6692.

Leland, W. E.; Taqqu, M. S.; Willinger, W.; Wilson, D. V. (1994). On the Self-Similar Nature of Ethernet Traffic (Extended Version), *IEEE/ACM Transactions on Networking*, 1994, 2(1): 1-15, ISSN 1063-6692.

Tzamaloukas, A. & Garcia-Luna-Aceves, J. J. (2001). A Receiver-Initiated Collision-Avoidance Protocol for Multi-Channel Networks, *Proceedings of IEEE INFOCOM* 2001, vol. 1, pp. 189-198, Anchorage, AK USA, ISBN: 0-7803-7016-3, Apr. 2001.

Choi, N.; Seok, Y. & Choi, Y. (2003). Multi-Channel MAC Protocol for Mobile Ad Hoc Networks, *Proceedings of IEEE VTC* 2003-*Fall*, vol. 2, pp. 1379-1382, Orlando, Florida USA, ISBN: 0-7803-7954-3, Oct. 2003.

Hui, J. Y. N. (1984). Throughput Analysis for Code Division Multiple Accessing of the Spread Spectrum Channel, *IEEE Journal on Selected Areas in Communications*, 1984, SAC-2(4): 482-486, ISSN 0733-8716.

Liu, N. (2004). *Wireless Local Area Networks* (*WLAN*): *Principle, Technique and Application*, Press of Xidian University, ISBN 7560613624, Xi'an China.

Kleinrock, L. & Tobagi, F. A. (1975). Packet Switching in Radio Channels: Part I—Carrier Sense Multiple-Access Modes and Their Throughput-Delay Characteristics, *IEEE Transactions on Communications*, 1975, 23(12): 1400-1416, ISSN 0090-6778.

Goldsmith, A. (2005). *Wireless Communications*, Press of Cambridge University, ISBN 978-0-521-83716-3, New York, NY, USA.

Cali, F.; Conti, M. & Gregori, E. (1998). IEEE 802.11 Wireless LAN: Capacity Analysis and Protocol Enhancement, *Proc. of IEEE INFOCOM* 1998, vol. 1, pp. 142-149, San Francisco, CA, USA, ISBN 0-7803-4383-2, Mar.-Apr. 1998.

Chen, T.; Zhang, H.; Matinmikko, M.; Katz, M. D. (2008). CogMesh: Cognitive Wireless Mesh, *Proceedings of GLOBECOM* 2008, pp. 1-6, ISBN: 978-1-4244-3061-1, New Orleans, LO, USA, Nov.-Dec. 2008.

5

A Layered Routing Architecture for Infrastructure Wireless Mesh Networks

Glêdson Elias, Daniel Charles Ferreira Porto and Gustavo Cavalcanti
Federal University of Paraíba
Brazil

1. Introduction

Wireless Mesh Networks (WMN) is a new technology that promises improved performance, flexibility and reliability over conventional wireless networks. WMNs are easy to deploy and have self-configurable and self-healing capabilities. In essence, a WMN is a dynamic, multi-hop wireless network in which the nodes automatically establish and maintain connectivity among them. Thus, routing protocols have a fundamental role by providing paths to allow communication between non-neighbor nodes and so keep up best routes. One of the most important goals for routing protocols developed for WMNs is to reduce the routing overhead and improve network scalability.

The WMN's architecture defines two types of nodes: mesh client (MC) and mesh router (MR). They can play different roles in the network, forwarding packets in behalf of other ones or just using the network resources. Depending on such roles, three types of WMNs can exist: client, infrastructure and hybrid (Akyldiz et al., 2005).

A client WMN is just an ad hoc network built only by MCs. The infrastructure WMN (IWMN) is the most common type, being formed by a fixed, dedicated group of MRs, which builds a wireless backbone, providing a coverage area for keeping connected mobile MCs, even when they are moving (Fig. 1).

In IWMNs, MCs cannot forward packets and besides cannot communicate directly with each other. Finally, in a hybrid WMN, the backbone is built by mobile and fixed devices. Hence, both MCs and MRs can forward packets, although only MRs can connect the backbone to other networks.

The routing facilities required by WMNs are already present in protocols developed for ad hoc networks. So, ad hoc routing protocols like DSR (Johnson et al., 2004), AODV (Perkins, C. et al., 2003) and OLSR (Clausen, T. & Jacquet, P., 2003) have been applied in several WMN projects (Chen, J. et al., 2006) (Bicket, J. et al., 2005) (Tsarmpopoulos, N. et al., 2005). However, such protocols do not perform very well in WMN and the throughput drops as the number of nodes increases (Akyldiz, I. F. et al., 2005). One of the major problems of such routing protocols is that they do not use properly the infrastructure provided by WMNs. Therefore, taking into account WMN features, research efforts have been focused on enhanced them or designing new protocols such as RA-OLSR (Bahr, M., 2006), HWMP (Bahr, M., 2006) and AODV-ST (Ramachandran, K. et al., 2005).

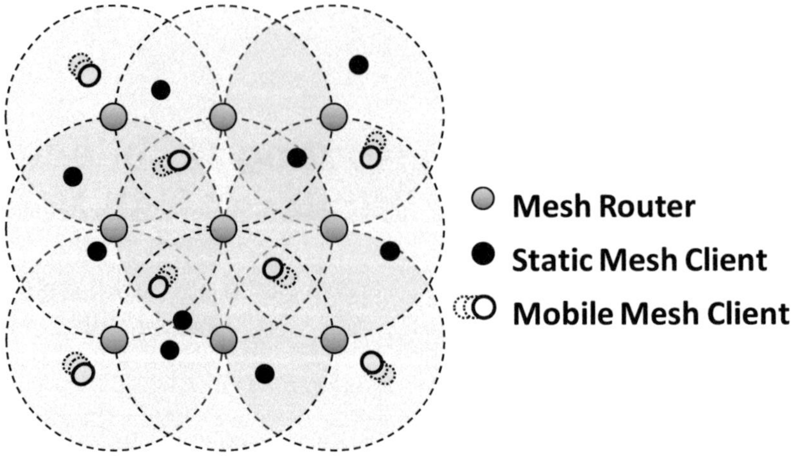

Fig. 1. Infrastructure wireless mesh network

As evinced in (Chen, J. et al., 2006) and (Hossain, E. & Leung, K., 2008), improved scalability in terms of the number of nodes in the network may be achieved reducing routing overhead. Hence, a scalable routing protocol can be applied to small as larger number of nodes without exhaust network resources with excessive sending of control messages.

An interesting approach to address routing problems is to split routing capabilities into a layered routing architecture. So a specialized strategy can be applied to address problems for each layer to improve routing protocol´s scalability.

In such a context, this chapter presents the efforts of network research group at Federal University of Paraíba in Brazil on specifying a scalable, layered routing architecture, called Infrastructure Wireless Mesh Routing Architecture (IWMRA) (Porto, D.C.F. et al., 2009), which is specifically designed considering IWMN's features. The proposed architecture allows separating routing concerns into a three-layered architecture and designing of a specialized protocol for each layer. The main strengths and innovations of the proposed architecture are the separation of routing concerns in three independent layers and the differentiation of routing strategies for MR and MCs to reduce signaling overhead, adopting proactive and reactive strategies for static and mobile nodes, respectively.

The remainder of this chapter is organized as follows. The related works and context are presented in Section 2. Then, the proposed three-layered routing architecture is presented in Section 3. Afterward, the main features of protocols applied in each layer are a briefly described in Sections 4, 5 and 6. The initial results of performance evaluations are described in Section 7. Finally, the Section 8 presents the concluding remarks and future work.

2. Related work and context

As WMNs are essentially a dynamic multihop wireless network, the topology can change very fast. Thus, the routing protocols play an important role providing needed paths to allow communication among the nodes. The wireless routing protocols have to be aware to topological changes caused, for instance, by node movement. These topological changes may happen in the neighborhood of the nodes or in the links of path between them. Then, the routing protocol has to restore or compute a new path for keeping the communication.

Among of a variety of routing protocols applied to WMNs, the OLSR´s first version (here, simply indicated as OLSR) is an example of modular core architecture with well defined neighborhood discovery and topology dissemination processes. Nevertheless, these processes are integrated in OLSR´s specification but not as independent protocols. However, for the OLSR´s second version (OLSRv2) (Clausen T. et al., 2010), the neighborhood discovery process was separated from its specification as an independent protocol called Neighborhood Discovery Protocol (NHDP)(Clausen T., C. Dearlove & J. Dean, 2010). The NHDP is intended to be used for routing protocols to provide continued tracking of neighborhood changes and allows routing protocols to access neighborhood information.
The OLSRv2 specification retains the same basic mechanisms and algorithms of OLSR (topology dissemination and routing calculation process), while using a more flexible signaling framework that refers NHDP as responsible for manage neighborhood information.
It must be emphasized that OLSR´s neighborhood process is basically identical to NHDP, except that NHDP uses a new packet structure and address compression technique defined by the packetbb (Clausen, T., et al., 2009) specification.
Due the clear separation of OLSR´s processes, it is not too difficult to make a performance evaluation between OLSR and IWMRA´s protocols. Taking into account that NHDP and OLSRv2 are not available for the adopted simulator yet, the presented performance evaluation has just compared the protocols of IWMRA and the processes of OSLR.
In order to make possible to understand the reasoning presented in the performance evaluation, a brief description of the OLSR processes (neighborhood discovery and topology dissemination) is presented at this point.
In OLSR, in all nodes, the neighborhood discovery process periodically sends HELLO messages in broadcast at a regular time interval (2 seconds, by default). Note that MRs and MCs periodically send HELLOs but they do not forward them. A given node X declares other node Y as neighbor whenever X receives a HELLO from Y. In complement, a given node X declares the neighborhood with other node Y as lost when X does not hear three HELLOs from Y (6 seconds by default).
To disseminate the neighborhood data through the network OLSR uses an optimized link state algorithm. Each node in the network employs an algorithm to select a set of neighboring nodes to retransmit its Topology Control (TC) messages. This set of nodes is called the multipoint relays (MPR) of that node. Any node which is not in the set can read and process each TC but do not retransmit. Note that, MRs and MCs can be selected as MPR of a node, according to MPR´s selection algorithm. Thus the OLSR reduces the number of rebroadcasting nodes over conventional flooding. The node sends its TC messages in broadcast at a regular time interval, 5 seconds by default, but the MPRs have to rebroadcast it in up to 0.5 seconds.

3. Infrastructure Wireless Mesh Routing Architecture

The Infrastructure Wireless Mesh Routing Architecture (IWMRA) splits routing concerns into a layered routing architecture specifically designed taking IWMNs features.
An application scenario, already depicted in Fig. 1, includes a set of fixed MRs, planned to provide a continuous coverage area, and also a set of fixed or mobile MCs. In this initial version of the architecture, all nodes have just one wireless interface and links are bidirectional.

As already mentioned, in IWMN's architecture, the MRs and MCs play different roles where only MRs are responsible to build a wireless backbone and forward network traffic, while MCs just uses network resources. Since the MRs are fixed devices, they can be connected directly to power source, unlike the MCs which are mobile devices and have constrained power supply provided by batteries (Akyldiz et al., 2005)(Zhang, Y. et al., 2006). These IWMN's features are explored by IWMRA to reduce control message overhead and increase network scalability.

To achieve its goals, the IWMRA splits routing functionality into three independent layers: neighborhood, topology and routing (Fig. 2). In each layer, an independent protocol has been designed to handle specific features of IWMNs. Each protocol provides to the upper layer a couple of well defined services. By separating the functionality in layers, the architecture enables further adaptations.

The neighborhood layer is defined by SNDP protocol (Elias, G. et al., 2009). Briefly, the neighborhood layer is responsible to detect the presence and status of directly reachable neighbors, keep track of neighborhood changes and alert the topology layer whenever a change is detected. The neighborhood layer may also detect the metric of the link, which is used by upper layers to calculate the overall path cost and select the best routes.

The topology layer is defined by MLSD protocol (Porto, D. C. F., 2010). Based on a flooding approach, the topology layer efficiently disseminates neighborhood information to all MRs over the network, allowing the MRs to build a topological map of the network. The topology layer is responsible to keep accurate topological information and synchronize databases among the MRs. It also alerts the routing layer whenever a topological change is detected for the routes to be updated.

Finally adopting a proactive and a reactive approaches, the routing layer compute and configure the best routes for all nodes.

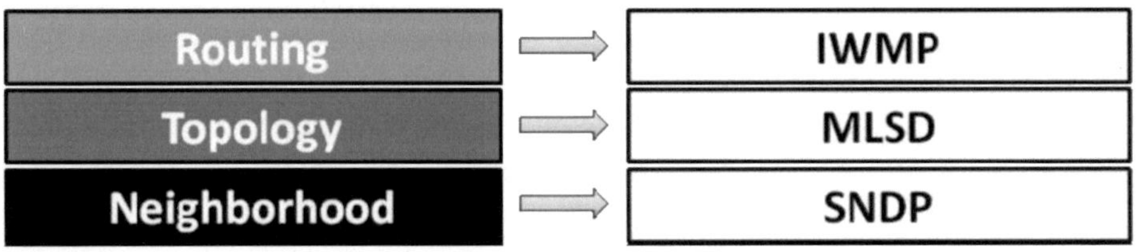

Fig. 2. IWMRA layers and its respective protocols

SNDP adopts a hybrid, collaborative signaling strategy, in which MRs employ a proactive, timer-based signaling approach, whereas MCs make use of a reactive, event-based signaling approach.

MLSD is a low-overhead link-state dissemination protocol. Unlike current proactive routing protocols applied in WMNs, such as OLSR, MLSD employs an event-based approach with a reliable message delivery strategy and a flooding control in order to reduce the message overhead.

In the routing layer, IWMP is a multiple routing, hybrid protocol, which is under refinement. IWMP makes use of information provided by topology layer to build a graph and to calculate the best paths using the SPF algorithm (Dijkstra, E.W., 1959).

It is important to emphasize that topological information is only stored and handled by MRs. Thus, MCs have to request routes to neighbor MRs, which can promptly answer to such requests.

As a proof of concept, the following sections introduce the main insights and concepts of protocols that compose the IWMRA. Then, simulation results of the neighborhood and topology layers are presented and compared to similar functionalities provided by the OLSR protocol.

4. Neighborhood layer - SNDP (Scalable Neighborhood Discovery Protocol)

SNDP is a scalable neighborhood discovery protocol, which has been specifically designed taking into account architectural features of IWMNs. Based on such architectural features and in order to reduce control message overhead, SNDP adopts a hybrid, collaborative signaling strategy. On the one hand, the proposed signaling strategy is said to be hybrid because MRs and MCs adopt distinct signaling approaches. On the other hand, the proposed signaling strategy is said to be collaborative because MRs and MCs work together to detect the presence and absence of nodes.

Considering that MRs have unlimited power supply, they employ a proactive, timer-based signaling approach, which uninterruptedly and periodically sends messages even when there does not exist any node in their transmission ranges. In contrast, as MCs have limited power supply, they adopt a reactive, event-based signaling approach, which sends messages as a consequence of receiving other ones from MRs in their transmission ranges.

The next sections briefly describes the signaling approaches adopted by MRs and MCs, and also how they work together to manage neighborhood among nodes. Note that, the SNDP can only operate on IWMNs that adopt bidirectional links among all nodes and provide continuous connectivity within the coverage area of the wireless backbone.

4.1 Neighborhood discovery

SNDP is employed to detect the presence and status of neighbor nodes in IWMNs. As previously mentioned, in IWMNs, MCs do not communicate directly with each other. In such scenario, the communications among MCs are mediated by MRs. Thus, MCs do not need to detect other ones as neighbors. Therefore, MCs have to detect MRs as neighbors, while MRs ought to detect MRs and MCs. Due to such distinct neighborhood discovery requirements, SNDP adopts a hybrid, collaborative signaling strategy.

The MRs adopts a proactive, timer-based approach, where periodically they send HELLO messages in broadcast even when do not exist nodes in their transmission ranges. Such an approach allows an MR to be promptly detected as neighbor by any other MR or MC that comes into its transmission range. The MR signaling rate is regulated by a protocol parameter, which by default is 2 seconds.

Notwithstanding, the MCs adopts a reactive, event-based approach, where they send HELLO messages in broadcast as responses to other ones, previously received from MRs in their transmission ranges. Such an approach allows an MC to be detected as neighbor by any MR in its transmission range. Note that a given MC only generates a HELLO immediately after detecting a given MR as neighbor. Thus, although MRs send periodic HELLOs, MCs only react to the first HELLO detected from neighbor MRs.

Considering the proactive, timer-based approach, two MRs require the exchange of a pair of HELLOs in order to recognize their neighborhood in both directions. Hence, as illustrated in Fig. 3a, each one declares the other one as neighbor after receiving the first periodic HELLO message from the other one.

In a similar way, the neighborhood between MRs and MCs are established exchanging HELLOs. Although, the MCs only reacts to first HELLO sent by the MR. Usually, as also illustrated in Fig. 3b, the MR proactively sends a HELLO (arrow 1), and, in turn, the MC reactively sends a HELLO as response (arrow 2).

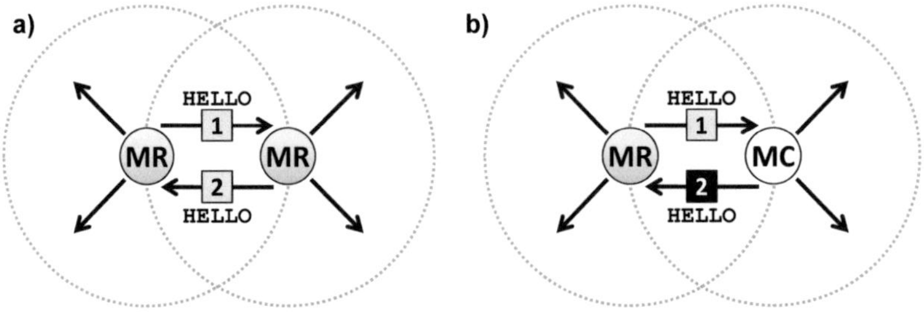

Fig. 3. MR-MR and MR-MC discovery process

Beside of that, the signaling approaches adopted by MRs and MCs have to integrate mechanisms to handle transmission problems that causes message loss. In MRs, the proactive, timer-based signaling approach just handles transmission errors by simply resending the HELLO message in the next time interval.

In MCs, the reactive, event-based signaling approach deals with transmission errors by adopting a confirmed service, in which MRs must acknowledge in their succeeding HELLO the reception of HELLOs sent by MC, as illustrated in Fig. 4.

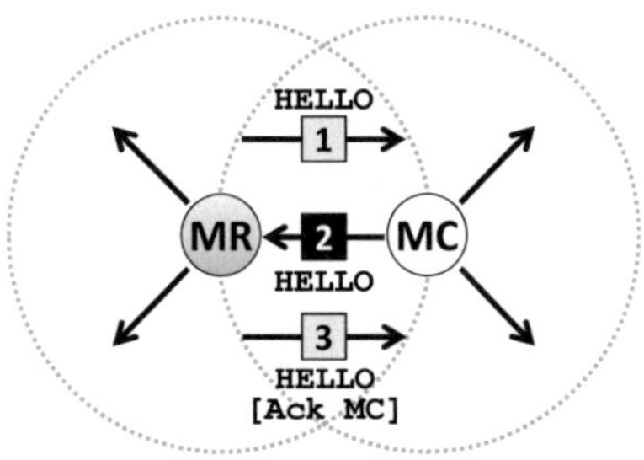

Fig. 4. MR-MC discovery process with acknowledgement

Thus, in case of message loss, the robustness of the process relies on immediately after detecting a neighbor MR, an MC must reply with a HELLO message for each one received from that MR, until it receives an acknowledgment sent by the MR. Note that, the acknowledgement is indicated by just including the MC's address in the MR's HELLO message, which contains the list of MCs from which the MR has received HELLOs during its last signaling interval (around 2 seconds).

4.2 Neighborhood loss

When a node is declared as neighbor, SNDP needs to monitor the neighbor node in order to detect the instant in which the neighborhood is lost. Once more, SNDP adopts a hybrid strategy for detecting and managing neighborhood loss. On the one hand, as MRs periodically sends HELLOs, MCs adopt a timer-based approach. On the other hand, as MCs reactively send HELLOs, MRs adopts a notification-based approach.

As MCs adopts the timer-based approach, when an MC declares an MR as neighbor, it also configures an expiration time, which by default is 2 seconds. Then, whenever an MC receives a HELLO from the neighbor MR, it just updates the expiration time. If an MC goes out of the transmission range of a neighbor MR, it will not receive HELLOs from that MR, and so, the neighborhood entry associated with that MR expires. At this moment, the MC declares the neighborhood as lost.

In the notification-based approach, as shown in Fig. 5, MCs ought to notify MRs about the neighborhood that has been lost. To do that, immediately after detecting the neighborhood loss, the MC broadcasts a HELLO, including the notification that the neighborhood with the MR has been lost. However, since the connectivity between the source MC and the target lost MR is no longer available, the notification-based approach requires collaboration among intermediary MRs, which forwards the notification to the lost MR.

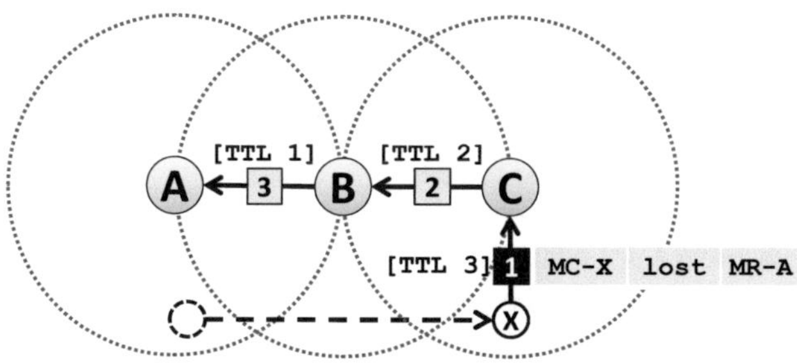

Fig. 5. Notification process

SNDP employs a bounded flooding technique to limit the notification area. Note that notifications do not generate additional signaling messages because it piggybacks on periodic HELLOs of intermediary MRs.

To limit the notification area, each notification has a TTL (Time to Live) field, which indicates the number of hops that the notification can reach. By default, the TTL field is 3 hops. When an intermediary MR receives a notification, it must decrement the TTL before broadcasting the notification in its next HELLO. If the TTL reaches zero, the intermediary MR does not forward the notification. When the notification reaches the target lost MR, it just declares the source MC as lost.

As a result of rebroadcasting notifications, the bounded flooding can make intermediary MRs and the target lost MR to receive replicated notifications. Hence, each notification generated by a given source MC to a lost MR has a sequence number field that enables the MRs detect and discard replicated ones.

Due transmission errors, a given MC may not receive a HELLO broadcasted by its neighbor MR. In such a case, as a mean to avoid erroneously declaring the neighborhood as lost, the MC and MR have to cooperate, as depicted in Fig. 6.

On the MC's side, after expiring the neighborhood entry associated with its neighbor MR due to error transmission (arrow 1), the MC broadcasts a HELLO with the notification (arrow 2), but internally it does not declare the neighborhood as lost. Instead of that, the MC just waits for a hold time interval (default 0.5 seconds). The MC can only declare the neighborhood as lost if it does not receive a HELLO from the MR during the hold time interval.

On the MR's side, after receiving the notification directly from the MC (arrow 2), it immediately broadcasts in advance its HELLO (arrow 3), making possible to the MC to keep its neighborhood with the MR. Hence, when HELLOs sent by MRs are subjected to transmission errors, the notification process avoids MCs to erroneously declare the neighborhood with MRs as lost.

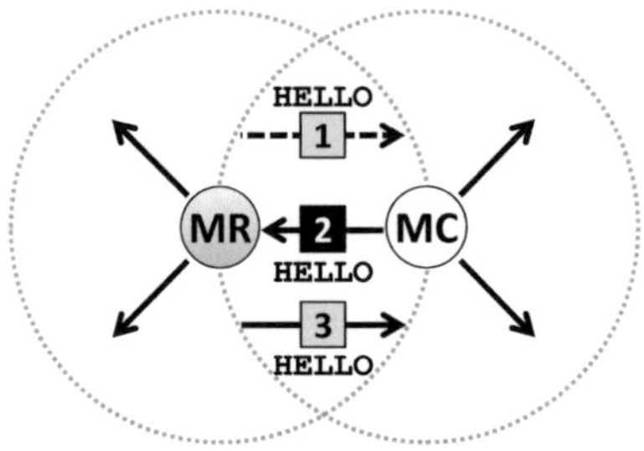

Fig. 6. Avoiding the neighborhood loss

4.3 Additional improvements for SNDP

When the density of MCs is relatively low throughout the wireless backbone, it is common some MRs do not have MCs as neighbors. In such cases, MRs broadcast HELLOs only with the purpose of keeping the neighborhood with other MRs.

In order to reduce signaling load, SNDP specifies a low signaling rate, which by default has a time interval of 32 seconds. The low rate is only adopted by a given MR when it and its neighbor MRs do not have neighbor MCs. To do that, a flag in MRs' HELLO informs when a given MR has MC neighbors. As a consequence, each MR can adopt two signaling rates. The low signaling rate (each 32 seconds) when the own MR and its neighbor MRs do not have neighbor MCs, and otherwise, the high signaling rate (each 2 seconds).

5. Topology layer – MLSD (Mesh Network Link State Dissemination Protocol)

MLSD is a low-overhead link state dissemination protocol, which has also been designed taking into account architectural features of IWMNs. It defines how to spread and maintain consistent and updated information about network topology, allowing the MRs to build a topological map of the network making possible to routing layer build best routes.

Considering that in IWMNs the backbone is built only by MRs, MLSD defines that the topological information is only managed by them. As a consequence, only MRs can send and process link state update messages (LSU). Despite of the MCs do not process or send LSUs, they also store topological information. However, the topological information maintained by MCs is just the links with its neighbor MRs, which are informed by neighborhood layer. As already mentioned, when an MC needs to communicate to other nodes, it must use the services provided by routing layer to request and configure a route.

In order to reduce the message overhead caused by link state messages, MLSD employs an event-based approach with a reliable message delivery strategy and a flooding control.

By adopting an event-based approach, the MRs sends small incremental update messages only when topology changes. Therefore, to ensure the consistency of topological information in all MRs the MLSD also adopts reliable flooding, which uses a positive implicit acknowledgment with retransmission strategy to deploy the update throughout the backbone and a synchronization process to fully update new MRs. Beside of that, MLSD controls flooding by adopting time-slots which are automatically configured among neighbors MRs to help avoiding exhaust network resources due excessive events or retransmissions.

The next sections presents a concise description of dissemination process with positive implicit acknowledgment, synchronization and time-slot configuration approaches adopted by MLSD to manage topology among the nodes.

Likewise SNDP, MLSD can only work on IWMNs that adopt bidirectional links among all nodes and provide continuous connectivity within the coverage area of the backbone.

5.1 Topology dissemination – positive implicit acknowledgment

Once the neighborhood layer makes available neighborhood information, based on a flooding approach, the topology layer is responsible for disseminating such information (called link state advertisement - LSA) to all MRs over the network. Each MR broadcasts in their LSUs one or more LSAs (by default, up to 128). The LSAs flooded by all MRs are employed to derive the network topological database, which is identical for all MRs. Each LSA may define one of two operation types, link discovery (ADD) or link loss (REM) and it is assigned with unique sequence number to allow identifying if it is duplicated or outdated. In the event-based approach, the MRs broadcasts each LSA in the LSU only once. Due transmission errors, a given MR may not receive a LSU broadcasted by its neighbor. In such a case, as a mean to avoid inconsistencies in topological database, the MRs adopts a flooding with positive implicit acknowledgment with retransmission as illustrated in Fig. 7.

A given MR-A that broadcasts a LSA in its LSU (Fig. 7a) assures that it has been effectively delivered to a neighbor MR-B, which is indicated as forwarder for such LSA, when the MR-B rebroadcasts the same LSA, in its own LSU, to another neighbor MR-C (Fig. 7b). Since the LSU broadcasted by MR-B is received by all neighbors, it can also work as an acknowledgment to MR-A. Therefore, MR-C must also rebroadcast the LSA, at least once, in order to acknowledge the MR-B, even though it has no other neighbors MRs (Fig. 7c).

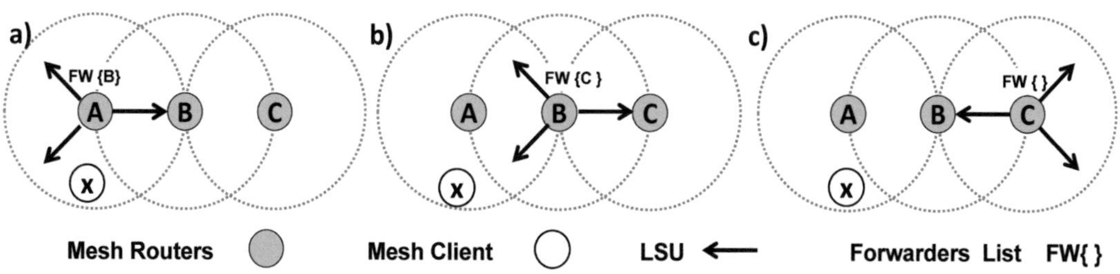

Fig. 7. Flooding with positive implicit acknowledgment

Each LSA have a list of forwarders which are address of the neighbors MRs that must rebroadcast it. As also illustrated in Fig. 7, when a LSA is generated in response to neighborhood layer update of a given MR-A, it defines all its neighbors MRs as forwarder to such LSA. When MR-B receives a LSA from neighbor MR-A, it defines all its MR neighbors as forwarders to such LSA, except the one from which the LSA was received (MR-A).

It is important to emphasize that when all LSA are successfully delivered and acknowledged, all MR broadcasts its LSU only once. Thus, no additional message is needed

and the total LSUs employed is the same that a conventional flooding. However, the positive implicit acknowledge avoid need flooding LSA throughout the backbone again when transmission problems causes message loss.
As depicted in Fig. 8, when the MR-B broadcasts a new LSA in its LSU, for instance adding a new link with a given MC-Y, it indicates all its neighbors MRs as forwarder to such LSA. Nevertheless, transmission problems may cause message loss to a given neighbor MR-C (Fig. 8a). After broadcasting the LSU, MR-B internally configures an expiration time to retransmit the LSA which is sufficient to all its neighbors of MR-B also rebroadcast it. The section 5.3 describes how retransmission time is calculated.
During the time waited for retransmit the LSA, the MR-B receives the acknowledgment by MR-A, however, as MR-C lost the LSU sent from MR-B, it will not rebroadcast the LSA (Fig. 8b). When the retransmission time expires the MR-B rebroadcasts the LSA, although only the MR-C is indicated as forwarder to LSA (Fig. 8c). As a consequence, MR-C must rebroadcast the LSA. However, despite of the MR-A also receives the LSU, it is not identified as forwarder to such LSA and do not sends the message again. Consequently, only the MR-C rebroadcasts the LSA and acknowledges the MR-B (Fig. 8d).

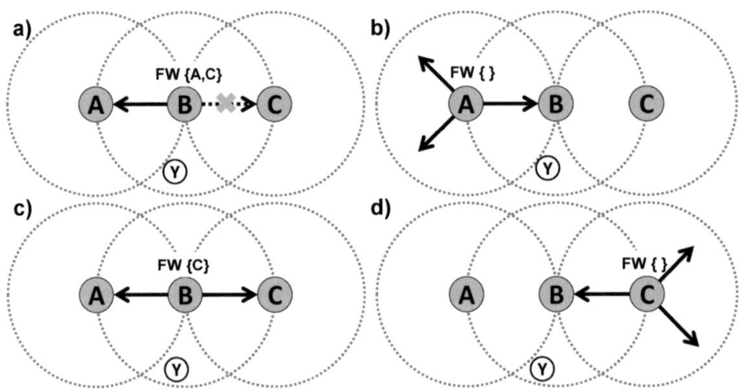

Fig. 8. Message loss causes retransmission.

As already mentioned, each LSU may carry up to 128 LSAs. Since each update has a list of forwarders that has to acknowledge the MR source, the LSA must adopt a compressed packet format to avoid LSU get too large due repetition of MRs' address list.
In the LSU, instead of a list of forwarders for each update there is only one list, which may includes the address of all neighbors MRs indicated as forwarder (usually up to 4) for at least one update carried in the packet. Besides, each LSA can carry more than one update, which are set with unique sequence number generated by its MR source to make possible detect and discard outdated ones. The updates with identical MR source and identical operation code (ADD/REM) are grouped per LSA. Therefore, each LSU actually can carry up to 128 updates, regardless if all of them belong to only one LSA, or if there are 128 LSAs with one update. Thereafter, a bitmap is built to match each update to the forwarders list, enabling the receiving neighbors MRs to derive if they are forwarder for each update.
A concise view of most important fields in LSU is presented in Fig. 9. When a given MR-B has to broadcast a LSU, it builds a forwarder list based on updates to send. Then, it also includes compacted LSAs with all updates from the same MR and same operation (ADD).

Finally a bitmap matches each update in LSA with the forwarders list. As the forwarders list has two elements, each update must be represented for two bits in bitmap. Hence, for the first update (B ADD X #3), the bitmap defines that it must be forwarded by MR-A (first element in forwarders list) but not by MR-C (second element in forwarders list). The second update (B ADD Y #4) must be forwarded only by MR-C and, the third (B ADD Z #5), must be forwarded by both MR-A and MR-C. The bitmap is built in a group of octets and the remaining bits must be filled with 0's.

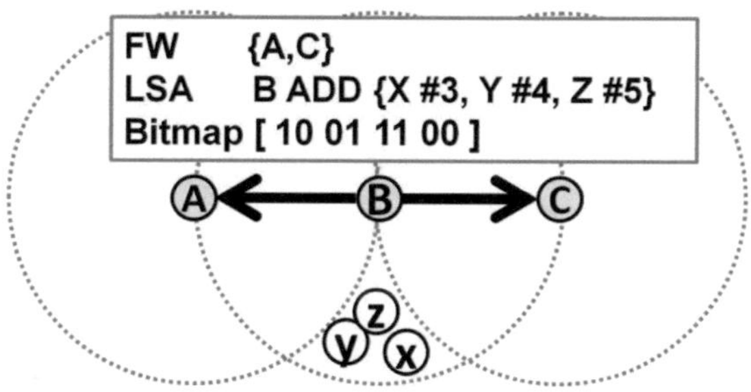

Fig. 9. Concise view of compacted fields in MR-B's LSU.

5.2 Synchronization process

The incremental updates are mostly applied to links between MRs and MCs. Notwithstanding, whenever links among neighbors MRs are removed or discovered it also triggers a database cleaning or database synchronization respectively, to ensure that all MRs' database reflect the current topology state.

On the one side, a database cleaning is trigged whenever a link between two MRs is lost. When a given MR crashes, its neighbors MRs have to broadcast an update removing the link lost. However, such link loss may also split the backbone into distinct sets of nodes. Thereafter, each neighbor MR of the crashed node disseminates an update across all reachable MRs.

Whenever an update removing a link between two MRs is processed, internally, each MR performs a connectivity test building a connected set, enabling the MR detect and clean all links of unreachable MRs.

On the other side, a synchronization of topological database is trigged whenever a link between two neighbors MRs is discovered. When a given MR-B adds a new link with a given MR-A, beside the discovered neighbor, several others nodes may become reachable. Therefore, the neighbors MRs must exchange their topological databases in order to let know possible new links reachable through each other.

To exchange the databases, each MR retrieves all stored updates and set the discovered neighbor MR as forwarder for all of them in the next LSU. Hence, at the moment of a given MR-B discovers a neighbor MR-A, it retrieves all links stored in its database, including the one just discovered, and sets MR-A to forward them in the next LSU to send (Fig. 10a).

In turn, MR-A may receives the LSU sent by MR-B regardless the neighborhood layer has been detected the neighbor MR-B yet. As MLSD assumes that all links are bidirectional, the MR-A adds the link with MR-B in advance, retrieves all links stored in its topological database and sets MR-B to forward them in the next LSU to send (Fig. 10b). Note that, the updates retrieved from MR-A's database and the forwarding updates that acknowledges MR-B goes together in the same LSU broadcasted by MR-A. Finally, the MR-B broadcasts

another LSU with the new updates received but without forwarder set for them, as a mean to acknowledge MR-A (Fig. 10c).
It's important to note that due the dissemination process forwards new links across all reachable nodes, they also will be added to all other MRs, synchronizing all databases.

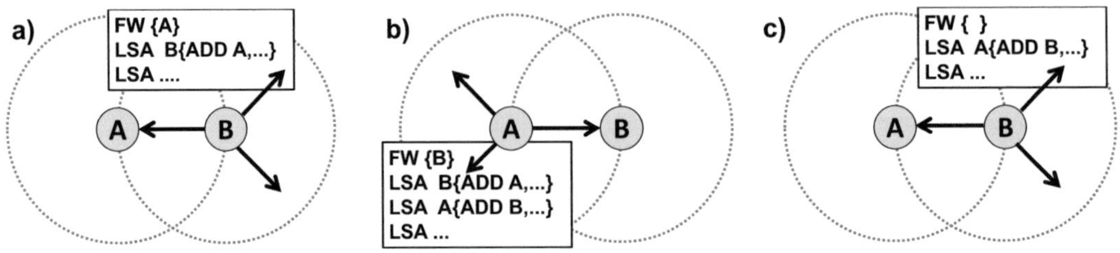

Fig. 10. Topological database exchanging.

5.3 Time-slot based flooding control

The IWMNs also supports mobile MCs. Consequently mobile MCs often cause changes of the MRs' neighborhood. Moreover, transmission problems like collisions may lead MRs to retransmit messages. For instance, the hidden terminal problem rises when a given MR broadcast a LSU with updates and all of its neighbors MRs has to forward them. In such a case, all neighbors MRs will receive the LSU close to same moment and may broadcast them very close or at the same time causing collisions in MR source, and therefore, retransmissions which can drastically increase the overhead generated by link state updates. To handle excessive events or retransmissions the MLSD employs time-slots automatically configured among neighbors MRs. In such an approach, a LSU sent by a given MR configures its neighbors MRs to broadcast their LSUs in distinct moments (Fig. 11a.).

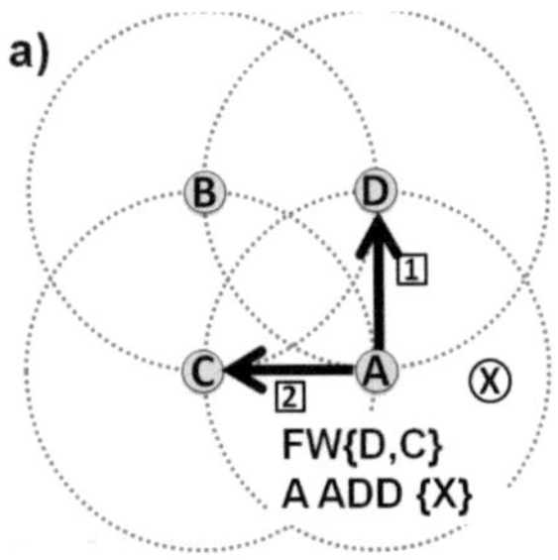

b)

MR	1	2	3	4	5
A	TRANSMISSION				RETRANSMISSION
D		TRANSMISSION			
C			TRANSMISSION		
B					

TRANSMISSION → FORWARDERPOSITION*TIME_SLOT_INTERVAL
RETRANSMISSION→(#FORWARDERS+2)*TIME_SLOT_INTERVAL*#LSU

Fig. 11. Time-slot configuration.

As also illustrated in Fig. 11, when a given MR-A detects a neighborhood change, it schedule the update to broadcast within a LSU in the first slot (Fig. 11b). During this time, all events detected will be scheduled to be sent in the same LSU.
When the LSU is broadcasted, the forwarder list is built based on neighbors MRs and the updates to send. Then, both neighbors MR-D and MR-C receive the LSU and evaluate the order they appear in forwarder list to schedule the slot they must use to transmit the update. Besides, after broadcasts a LSU an MR-A schedules a retransmission time for the updates sent, considering the time expected to all neighbors rebroadcast them and a multiplier consisting of number of LSUs need to disseminate all updates (Fig. 11b), which is usually one. To avoid the retransmission time get too long the multiplier can only grows up to 5.

Note that, when an update is acknowledged, its retransmission is unscheduled. Hence, if all updates were successfully delivered and acknowledged no additional message is sent.
It is important to cite that the time slot value is a configurable parameter in MLSD and the value adopted is calculated taking into account the features of wireless technology. For instance, in 802.11 we suggest 0.03125s which is obtained considering bit rate, inter-frame times and maximum message size. In addition, by defining different moments for neighbors MRs broadcast the LSU, MLSD reduces the chance of collision in the source MR and so retransmissions, leveraging the chance of the message being successfully delivered, while handle successive events in a short time interval grouping them in the same message.

6. Routing layer – IWMP (Infrastructure Wireless Mesh Protocol)

In the routing layer, a multiple routing, hybrid protocol called IWMP is under refinement. To discover routes, the topology and routing layers have to cooperate.
On the MR's side, IWMP makes use of proactively information provided by the topology layer to build routes. On the MC's side, the MCs do not have full topology information, and therefore, they have to reactively request routes to neighbor MRs, which can promptly answer to such requests.
Since MRs are the only allowed to forward packets in IWMN, each one gets a graph constituted only by MR nodes from topology layer and then uses the SPF algorithm (Dijkstra, E.W., 1959) to build best routes to all other MRs, afterwards the links with neighbors MC are added. Hence, when topology layer notify the routing layer of a new update related to a link with an MC, it can be processed without recalculate all routes. However, whenever MRs' graph is updated, then all the best routes have to be recalculated.
On the MC's side, in IWMP, all MCs adopt a reactive approach. Thus, when an MC needs a route to other nodes it broadcasts a route request for its neighbor MRs (Fig 12a). Hence, all neighbors MRs responds immediately, in unicast, because they already have the route proactively configured (Fig. 12b). Thereafter, an MC can choose the best route (Fig. 12c).

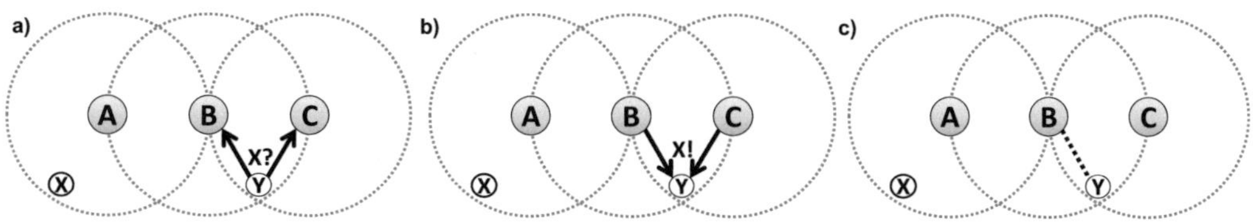

Fig. 12. MC's reactive route setup process.

As already mentioned, the IWMP is an ongoing work, and therefore, its facilities still under refinement. For instance, how to connect the IWMN to Internet through multiple gateways and load balance scheme to adopt are topics under development.

7. Performance evaluation

The performance evaluation of SNDP and MLSD was contrasted to OLSR's processes by adopting a simulation-based performance evaluation. All simulation scenarios consider a full mesh topology defined by a grid of 10×10 stationary MRs, which defines a rectangular coverage area of 1.04 Km × 1.04 Km, where stationary or mobile MCs move around

adopting the random waypoint model without thinking time and all nodes have an 802.11b wireless interface configured to 100m of range.
Besides, the simulation scenarios have varied the number and the speed of MCs. In all protocols their speed ranges from 0 to 20 m/s. For SNDP, the number of MCs varies from 0 to 500 nodes totalizing 340 scenarios while MLSD has initial results from 0 to 100 MCs in sum of 48 scenarios. For each simulation scenario, average values of the evaluated performance metrics were calculated based on several simulation experiments, considering a relative estimation error of 5% and a confidence interval of 95%. Together, all simulation scenarios required around 4800 simulation experiments, which were conducted using NS-2 (Fall, K. et al. 2008) together with the UM-OLSR implementation (Ros, J. F., 2008).
The performance metric evaluated was the message overhead, which considers the total messages sent by all nodes during the evaluation time of simulation (2840s). The smaller the message overhead, the more scalable is the corresponding protocol because the transmission channel will not be saturated with control messages. As a way to show the general SNDP and MLSD behaviors, this chapter only presents the performance gains in scenarios with four speed configurations including an average of 10 m/s, varying uniformly between 0 and 20 m/s.

7.1 Performance evaluation of neighborhood layer

In OLSR's neighborhood discovery process all nodes (MR/MC) periodically send HELLOs in a constant rate. Hence, the neighborhood message overhead raises as the number of nodes increases, because more nodes will send HELLOs. However, OLSR's message overhead is independent of the speed of the nodes and, in the Fig. 13, all its curves are overlapped.

a)
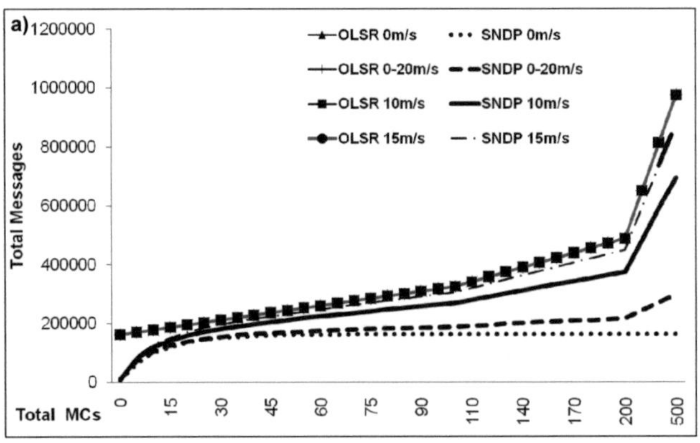

b)

Relative gains %		Number of MCs			
		0	100	200	500
Speed (m/s)	0	94	50	67	83
	10	94	18	23	29
	15	94	6	8	10
	0-20	94	42	56	70

Fig. 13. Message overhead of SNDP and OLSR's neighborhood process.

Nevertheless, the performance of SNDP evinced by the curves in Fig. 13a, reveals interesting outcomes of the hybrid strategy adopted. As MRs periodically send HELLOs in a high or low rate, depending on the density of MCs in the network, in SNDP, the MRs message overhead is strongly reduced in scenarios with a small number of MCs (less than 30). In such scenarios, several MRs adopt a low signaling rate due to the absence of neighbor MCs.
Hence, as the number of MCs increases, the probability of MRs adopting a high signaling rate increases. Once the SNDP high signaling rate and the OLSR signaling rate are equal, the MRs message overhead of both tend to be similar as the number of MCs increases up to 30 MCs.

Analyzing the message overhead generated by stationary MCs (0m/s), in OLSR, like MRs, all MCs also periodically send HELLOs. In contrast, in SNDP, MCs do not periodically send HELLOs. In SNDP, message overhead is basically constant and independent of the number of MCs. When contrasted with OLSR, as can be seen in Fig. 13b, SNDP reduces in almost 83% the message overhead in a presence of 500 MCs.

Conversely, as evinced by level of the curves in 10m/s and 15m/s, the MCs speed has an important impact on the SNDP message overhead. The higher is the speed, more neighborhood detection and loss events are generated, and so, the higher is the SNDP message overhead. Such a behavior is a consequence of the reactive signaling approach adopted by MCs. Simulations evince that SNDP reaches the OLSR message overhead when MCs speed exceeds 20 m/s.

An important result was revealed when the speeds varies from 0 to 20m/s. As also showed in Fig. 13b, the curve has values close to best performance of protocol, and for 500 MCs the gain compared to OLSR reaches 70%.

7.2 Performance evaluation of topology layer

In OLSR's topology management process, the MPRs selection algorithm reduces the number of rebroadcasting nodes. However, the total of MPR nodes are not essentially minimal to coverage all backbone (Clausen, T. & Jacquet, P., 2003). Hence, MPRs selection algorithm may chooses more rebroadcasting nodes than the minimum to provide the coverage area.

Besides, each TC can carry up to 64 links per MR but, by default, only 4 TCs can be sent per packet, totalizing up to 256 updates per OLSR packet. Hence, when a MPR has to forward TCs from more than 4 nodes, more packets are immediately sent, in order to deliver all TCs. Therefore, as evinced in Fig. 14a, the OLSR's message overhead at 0m/s rises as the total of nodes increases, because more MPR can be selected, then will periodically send and forward more OLSR packets with TCs.

It is important to evince that, to conduct a fair comparison, only OLSR packets with at least one TC was considered in performance evaluation. Even when the packet carries 4 TCs it was counted as only one message.

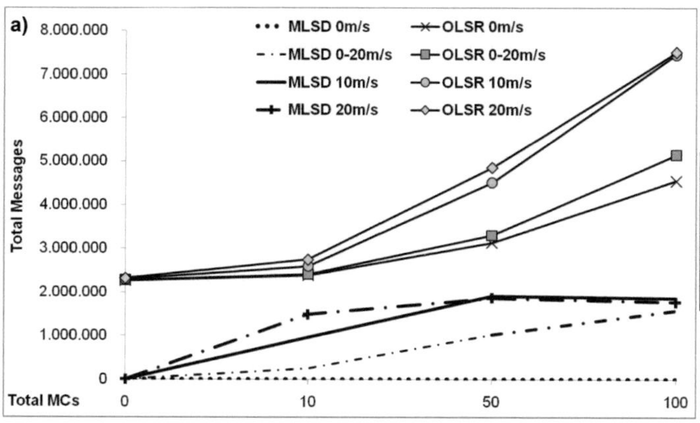

b)

Relative gains %		Number of MCs			
		0	10	50	100
Speed (m/s)	0	100	100	100	99
	10	100	63	58	75
	20	100	46	62	76
	0-20	100	90	69	69

Fig. 14. Message overhead of MLSD and OLSR's topology dissemination process.

The speed of MCs has also a large impact on OLSR's topology message load, because as the node moves around even more MPR are selected, then more nodes sends and forward packets.

Analyzing the curve of MLSD when the MCs are stationary (0m/s), it shows that the number of stationary MCs has an insignificant impact in message overhead. The reason is the event-based approach adopted by MLSD. When events are not detected no message are sent, resulting in gains of 100% compared to OLSR, as illustrated in the Fig. 14b.

Nevertheless, when mobile MCs are present, the faster the MCs moves, the more events are detected increasing the message overhead, as shown by the level of the curves in 10m/s and 20m/s, and therefore, the number of mobile MCs and their speed impacts in message overhead.

However, unlike OLSR, the message overhead grows slowly in higher speeds and many MCs, presenting gains up to 76% even with 100 mobile MCs moving at 20m/s, with similar overhead at 10m/s. Such a behavior is an effect of the time-slot approach adopted. When new updates are sent, the MRs will wait for its neighbors MRs rebroadcast them in their LSU before retransmit another message, as explained. However, the retransmission time calculated considering the number of LSUs to send. Hence, when the number of events grows the MR may use more then one LSU disseminate all of them. Consequently the MR will also wait more time before retransmit them again, accumulating new updates per packet and avoiding disseminate many messages with few updates in a short time.

When MCs moves with speeds varying from 0 to 20m/s, the curve still reveals a good performance of MLSD with gain of at least 69% compared to OLSR, as showed in Fig 14b.

8. Concluding remarks and future work

The simulation results considering the message overhead evince that both protocols SNDP and MLSD have excellent performance when contrasted with OLSR, especially considering static scenarios, unveiling gains of 94% and 100% for SNDP and MLSD respectively.

On the neighborhood layer, considering mobile scenarios, the hybrid collaborative approach of SNDP shows a good performance in average mobility, when the speeds varies from 0 to 20m/s with at least 42% of gain, and a comparable performance when MCs adopt speeds superior to 15 m/s. However, it is also important to note that nowadays it is uncommon to find real scenarios with a large number of highly mobile MCs. Therefore, considering the evaluated metrics, SNDP has an excellent performance in typical IWMNs scenarios.

On the topology layer, the performance evaluation turns out expressive gains of MLSD's event-based approach, in all evaluated scenarios. Indeed, even the worst case of MLSD (20m/s) still has better outcomes, in terms of message overhead, than OLSR's best case (0m/s).

The results evince the effectiveness of the strategies adopted by both protocols SNDP and MLSD. Such results show a well-tuned, layered routing architecture has the potential to drastically reduce message overhead, and so, improve scalability of IWMNs.

As future work, in neighborhood layer, new simulations considering new metrics are still need. For instance, to evaluate the load in terms of bytes. Moreover, in topology layer, although rigorous experiments have been realized to validate the convergence and consistence of databases, a formal proof still important and must be conducted to further studies. Additionally, a detailed study about convergence time and also new simulations considering scenarios with many MCs are needed. At last, as already mentioned, in routing layer additional features are still under investigation. When the protocol stack is fully implemented, a performance evaluation contrasting other protocols and considering a new set of metrics like aggregate throughput and routing overhead will be conducted.

9. References

Akyildiz, I. F.; Xudong, W. & Wang, W. (2005). Wireless mesh networks: a survey, *Computer Networks,* Vol. 47, No. 4, March 2005, pp. 445-487, ISSN 1389-1286

Johnson, D.B., Maltz, D.A. & Hu, Y.C. (2003). The Dynamic Source Routing Protocol for Mobile Ad Hoc Networks (DSR), April 2003, pp. 1-111. IETF draft

Perkins, C., Belding-Royer, E. Das, S. (2003). Ad-Hoc On-Demand Distance Vector (AODV) Routing, July 2003, pp. 2-37. IETF RFC 3561.

Clausen, T. and Jacquet, P. (2003). Optimized Link State Routing Protocol (OLSR), October 2003, pp 1-75. IETF RFC 3626

Chen, J. and Lee, Y.Z. Maniezzo, D. & Gerla, M. (2006). Performance Comparison of AODV and OFLSR in Wireless Mesh Networks, *MedHocNet'06,* pp 271-278, Lipari, Italy, June 2006, IEEE

Bicket, J. et al., (2005). Architecture and evaluation of an unplanned 802.11b mesh network. *Proceedings of 11th annual international conference on Mobile computing and networking,* pp 31-42, ISBN 1-59593-020-5, Cologne Germany, September 2005, ACM, New York

Tsarmpopoulos, N., Kalavros, I. & Lalis, S. (2005). A low-cost and simple-to-deploy peer-to-peer wireless network based on open source Linux routers, *Tridentcom 2005,* pp. 92-97, ISBN 0-7695-2219-X, Trento Italy, February 2005, IEEE

Bahr, M., (2006). Proposed Routing for IEEE 802.11s WLAN Mesh Networks. *2nd Annual International Workshop on Wireless Internet,* pp. 5, ISBN 1-59593-510-X, Boston, USA, August 2006, ACM, New York, NY, USA

Ramachandran, K. et al., (2005). On the design and implementation of infrastructure mesh networks", *IEEE Workshop on Wireless Mesh Networks (Wimesh),* pp. 12, Santa Clara, CA, September 2005, IEEE

Hossain, E. & Leung, K., (2008). *Wireless Mesh Networks: Architectures and Protocols,* Springer Science, ISBN 978-0-387-68838-1, New York, USA

Porto, D.C.F. et al., (2009). A Layered Routing Architecture for Infrastructure Wireless Mesh Networks. *Proceedings of the 2009 Fifth International Conference on Networking and Services,* pp 366-369, ISBN 978-1-4244-3688-0, Valencia Spain, April 2009, IEEE

Clausen T., C. Dearlove & P. Jacquet (2010). The Optimized Link State Routing Protocol version 2, April 2010, pp 1-82, IETF draft

Clausen T., C. Dearlove & J. Dean (2010). Mobile Ad Hoc Network (MANET) Neighborhood Discovery Protocol (NHDP), July 2010, pp 1-84, IETF draft

Clausen T., C. Dearlove, J. Dean & C. Adjih (2009). Generalized Mobile Ad Hoc Network (MANET) Packet/Message Format, February 2009, pp 1-57, IETF draft

Zhang, Y. and Luo, J. & Hu, H. (2006). *Wireless mesh networking: architectures, protocols and standards,* Auerbach Pub, ISBN 0-8493-7399-9, Boca Raton FL, USA

Elias, G., Novaes, M. Cavalcanti, G. & Porto, D. (2009). A Scalable Neighborhood Discovery Protocol for Infrastructure Wireless Mesh Networks, *Advances in Mesh Networks 2009,* pp 132 - 137, ISBN 978-0-7695-3667-5, Athens Greece, August 2009, IEEE

Porto, D. C. F., (2010). A Link State Dissemination Protocol for Infrastructure Wireless Mesh Networks, *Master Thesis*, March 2010, Federal University of Paraíba (in Portuguese)
Dijkstra, E.W., (1959). A Note on Two Problems in Connexion with Graphs, *Numerische Mathematik*, Vol. 1, pp. 269-271, 1959, Springer
Fall, K. et al., (2008). The NS manual, available in http://www.isi.edu/nsnam/ns/doc
Ros, J. F., (2008). Masimum UM-OLSR, http://masimum.dif.um.es/?Software:UM-OLSR

6

Trends and Challenges for Quality of Service and Quality of Experience for Wireless Mesh Networks

Elisangela S. Aguiar[1], Billy A. Pinheiro[1],
João Fabrício S. Figueirêdo[1], Eduardo Cerqueira[1],
Antônio Jorge. G. Abelém[1] and Rafael Lopes Gomes[2]
[1]*Federal University of Pará*
[2]*State University of Campinas*
Brazil

1. Introduction

Wireless communications have received a lot of attention from both industry and academic groups. Wireless access allows independency between the user's position and the physical bearer used to access services from the network, as well as, it supports the delivery of multimedia content ubiquitously (Chiti et al., 2008) (Akyildiz & Wang, 2009). Nowadays, the Wireless Mesh Network (WMN) model is one of the most relevant approaches to provide last-mile access in emerging communication systems (Held, 2005) (Zhang et al., 2006) (Hossain & Leung, 2009), such as The Institute of Electrical and Electronic Engineers (IEEE) 802.11s (802.11s, 2010). Another standard that enables mesh mode is the IEEE 802.16 (802.16, 2010), which is used in Worldwide Interoperability for Microwave Access (WiMAX) systems.

WMNs are a special case of ad hoc networks, which allow multiple hops, increase the coverage area, and have low implementation cost and support ubiquitous features for Internet access. A WMN consists of clients (Mesh Clients (MC)), routers (Mesh Routers (MR)) and gateways, where routers provide connectivity to a set of fixed and/or mobile users and gateways assure Internet connectivity as presented in Figure 1.

Multimedia applications, such as video streaming, Voice over IP (VoIP), and Internet Protocol Television (IPTV), will be abundant in future wireless mesh systems and, consequently, the end-to-end quality level support for these services is a major requirement for a near future. In this context, new Quality of Service (QoS) (Bok-Nyong Park et al. 2006) and Quality of Experience (QoE) (Jain, 2004) approaches are needed to optimize the usage of (scarce wireless) network resources and increase the user´s satisfaction.

QoS-based schemes define a set of network level (and packet level) measurement and control operations to guarantee the distribution of multimedia content, in wired and wireless networks, with an acceptable quality level. Traditional QoS metrics, such as packet loss rate, packet delay rate and throughput, are typically used to indicate the impact on the multimedia quality level from the network's point of view, but do not reflect the user's experience. The receiving (by user devices), presenting (by displaying units) and perceiving (by end-users) of the applications are not considered.

In order overcome the limitations of current QoS schemes, QoE approaches have been introduced. Trends in QoE-based solutions are creating a new paradigm regarding human-based quality level support in WMNs. QoE measurement operations can be used as an indicator of how a WMN environment meets the end-user needs. Researches on assessment schemes, control mechanisms and wireless resource management approaches based on QoE are being developed and will used as an extension to current QoS solutions (Yamada et al., 2007) (Monteiro & Nunes, 2007) (De Vleeschauwer et al., 2008).

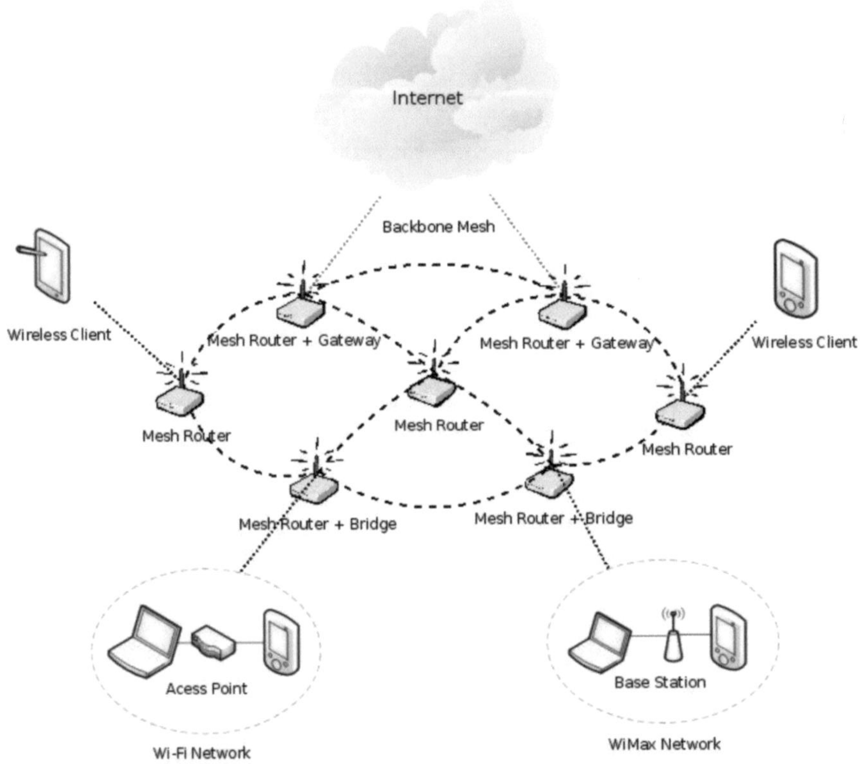

Fig. 1. A Generic Wireless Mesh Network

In this context, the development of an efficient and wise QoS/QoE-based routing scheme is one important challenge for the success of emerging multimedia-aware WMNs. Additionally, novel routing solutions must support a cross-layer approach to improve the overall system performance. Other issues that aim to provide quality level support in WMNs include new admission control, load-balance, resource reservation, over-provisioning, and cognitive radios mechanisms.

In this chapter, an overview of the most relevant challenges and trends in WMNs (focused on IEEE 802.11s) in terms of routing, cross-layer, QoS, and QoE support will be addressed. Due to the importance of routing schemes on optimization operations, a particular attention will be given in this area. To assist on explaining of such challenges, different approaches on QoE and QoS issues will be discussed. In order to show the benefits of cross-layer routing solutions on WMNs, simulation experiments were carried out to present the impact of a new routing scheme on the network´s and user´s point of view.

The remainder of this chapter is organized as follows. Section 2 discusses quality of level issues in WMNs. Cross-layer schemes are described in Section 3. The implementation and validation of a WMNs routing solutions are presented in Section 4. Finally, Section 5 presents the final considerations.

2. Quality of level support in wireless mesh networks

This section discusses the main approaches to assure quality level support for multimedia applications in wireless mesh networks with focus on QoS and QoE issues.

2.1 Quality of service issues

End-to-end quality of service control for fixed and mobile users is a core requirement for the success of emerging wireless systems. This control aims to increase the user satisfaction, while enlarging the revenue to network operators. With this goal in mind, the Internet has been a heavily researched topic in QoS networking for more than one decade. Several QoS models have been proposed with the goal of enriching the Internet with QoS guarantees that the current best effort model cannot support. Each approach defines its own mechanisms and parameters for traffic control and resource management, although usually at different granularities. It is common for a QoS model to be based on the notion of a class as supported by well-know QoS models, such as Differentiated Service (DiffServ), IEEE 802.11e, IEEE 802.16d and Universal Mobile Telecommunication System (UMTS).

Trends in last-mile Internet access require new QoS control mechanisms for IEEE 802.11s (mesh networks). However, the end-to-end QoS support in such scenarios is not trivial and is a research challenge. First of all, they must assure the high capacity needs of the access nodes that have to forward the accumulated traffic of their underling users. Moreover, WMNs have to cope with multiple strict QoS requirements of a large number of multimedia applications, including packet delay, throughput, and packet-error-rate. Finally, they must provide a large enough effective communication range to ensure that no Access Point (AP)s (or groups of APs) are isolated from the Internet gateways. In order to satisfy the above requirements, a set of novel QoS techniques needs to be exploited, such technology enablers include but not limited to multi-hopping, various multiple antennas techniques, novel Medium Access Control (MAC), resource reservation, over provisioning, admission control schemes and routing, where the last one will be explored in this chapter.

Another issue to be investigated in WMNs is that most of current works on QoS-aware protocols for WMNs are mainly based on a layered approach. This layered model led to the robust scalable protocols in the Internet and it has become the de facto architecture for wireless systems. However, the spatial reuse of the spectral frequency, the broadcast, unstable and error prone nature of the channel and different operational time scales for protocol layers, make the layered approach sub-optimum for the overall system performance of WMNs.

For instance, bad resource scheduling in MAC layer can lead to interference that affects the PHY layer performance due to reduced signal-to-interference-plus-noise-ratio (SINR) and ultimately deteriorates the overall network performance. Local capacity optimization with opportunistic scheduling techniques that exploit the multi-user diversity may increase the overall outgoing transceiver's throughput but they can also generate new bottlenecks in several routes in the network. Moreover, imprecise impact estimation of newly admitted applications on existing ones running in the network may jeopardize all ongoing QoS-aware services.

As described above, limitations of layered architectures are stimulating the development of a new WMN cross-layer design. In a cross-layer paradigm, the joint optimization of control over two or more layers can yield significantly improved performance. In general, QoS implementations for WMNs can be classified based on network layered schemes (Gavrilovska & Atanasovski, 2005).
Each layer has a set of mechanisms to provide quality level support for applications as following:

- MAC/LL Layer: Extensions of MAC mechanisms aim to provide QoS assurance in WMNs, such as IEEE 802.11e;
- Network Layer: Extensions of routing protocols and resource reservation schemes aim to provide QoS assurance in WMNs, such as The Optimized Link State Routing Protocol (OLSR) (Clausen & Jacquet, 2006) routing protocol with QoE support as will be evaluated in Section 4 and presented in next section;
- Application Layer: Application layer QoS schemes aim to improve the distribution of multimedia content, by adapting sessions to the current network conditions;
- Cross-layer: Improves the overall system performance, by optimizing wireless resources and services based on information about more than one layer.

Recent advances in WMNs have introduced routing schemes as attractive solutions to provide end-to-end quality level control in such multi-hop scenarios. The number of hops was the first criteria adopted by traditional routing protocols. However, it is clear that these approaches are not suitable for multimedia applications, such as real-time video streaming, which require strict QoS guarantees.
Routing protocols need to be aware of the overhead caused by information exchange. This process generates traffic to gather routing information and therefore, it consumes bandwidth. In addition, mainly in the WMNs, the interference caused by the data and control frames transmitted also consumes bandwidth. Therefore, WMNs routing protocols need to minimize the amount of state information exchanged and also maximize the network throughput by using an appropriate selection process of the best path.
The best path can be defined by using a set of QoS and QoE parameters. There are three strategies to gather such information in WMNs:

- Proactive: each node maintains updated information about network topology in routing tables through the constant exchange of routing information. This information is transmitted by flooding on the network. When a source node needs to establish a route to the destination node, the route is selected by an appropriated algorithm based on the exchanged information;
- Reactive: protocols that belong to this category do not exchange routing information periodically, but gather routing information on-demand when it is required and therefore, a process of route discovery is started among involved nodes;
- Hybrid: this approach has advantages from both proactive and reactive protocols. Therefore, it reaches a good balance between proactive and reactive protocols. In addition, a hybrid protocol can be adaptive to a wide range of network characteristics (e.g., mobility and traffic patterns) and to optimize routing layer parameters for the different applications.

The implementation of routing schemes in WMNs is a hard task, but several solutions have been proposed. Among them, the OLSR protocol (Clausen & Jacquet, 2006) is an adaptation of the traditional link-state algorithm for ad hoc networks. OLSR is a proactive protocol which uses a routing table obtained through the exchange of messages, between nodes, about the network conditions. A benefit of the OLSR protocol from the QoS perspective is its proactive nature that allows routes to be available before the source need to start a packet flow control to a destination. Another advantage of the OLSR protocol is that route computation is performed by using the knowledge about the entire network.

However, the hop count metric used by OLSR is unable to support QoS, because paths are selected based only on the number of hops (no well-know QoS metrics are used) along the session path. In this context, some extensions were developed for OLSR protocol, which are based on other link quality metrics. Among them, the OLSR Expected Transmission Count (ETX) and Minimum Delay (MD) are well-know metrics and will be presented below.

The OLSR extension based on ETX metric proposed in (De Couto et al., 2003) aims to find routes with the lowest expected number of transmissions that are necessary to ensure that a package can be delivered and has its arrival confirmed by the final destination. Other approach is the OLSR-MD (Cordeiro et al., 2007) that measures the link delay, calculating it through the Ad hoc Probe technique. Therefore, the calculation of the routing table can be based on the delay calculated to each neighboring node. Hence, in the OLSR-MD protocol the route selection between the current node and any other node in the network will have as criteria the lowest sum of the different transmission delays of all links along the path.

The OLSR-Dynamic Choice (OLSR-DC) extension (Gomes et al., 2008) aims to provide QoS support, giving different treatment to traffic from applications that use TCP and UDP, using the ETX metric for routing TCP packets and MD metric for routing UDP of packets. The protocol can also decouple the routing of TCP and UDP packets, this is achieved due to each packet be routed according to the metrics that best reflect their needs. This protocol was used as basis for the OLSR-FLC (Fuzzy Link Cost), since the proposed FLC is based on metrics that express the characteristics relevant to multimedia traffic. We can also configure FLC to route only UDP packets usually used for multimedia applications.

Comparing the previous solutions, OLSR-FLC seems to be the most suitable approach to guarantee the quality level support for multimedia applications in WMNs and will be explored and evaluated in this chapter. This novel cross-layer version uses a fuzzy logic to build a fuzzy system that aims to solve the problem of using multiple metrics for routing. The proposed fuzzy system has as base the values of the ETX and MD metrics collected from the network to define the FLC, which are used to route packets. TCP packets are still routed based on the ETX metric, as occurs with the OLSR-DC protocol. A detailed description of the OLSR-FLC can be found in (Gomes et al., 2009).

A major challenge regarding QoS-aware systems, including WMN routing schemes, is the lack of solutions to assure quality level control for applications according to the user´s perception. Traditional measurement schemes on the network can be used to estimate the impact of the quality of a media, such as video, but do not represent the entire set of metrics that will enable the management end-to-end quality-focused user's experience. Network statistics alone do not represent the perception of the user (Siller & Woods, 2003).

Current techniques that aim to maximize the quality level of multimedia services on a network are centralized in the aspects of QoS-based schemes that define a set of control operations and measurement, at the network level and packages to ensure the distribution of multimedia content in wired and wireless, with an acceptable level of quality (Zapater & Bressan, 2007). However, existing QoS metrics such as package loss rate, delay and throughput, are typically used to indicate the impact of the quality of a video (or any media) from the viewpoint of the network, but not reflects the situation experienced by the user. Consequently, these QoS parameters fail to capture the subjective aspects associated with human perception.

In order to overcome the current limitations of networks in their schemes of QoS for multimedia applications, considering the aspects of human perception and subjectivity related to the approach of QoE has been introduced (Takahashi et al., 2008), as characteristics of feelings, perceptions, views of users and how they interact with their environments and can be enjoyable and fun or annoying and frustrating (Patrick et al., 2004).

2.2 Quality of experience Issues

QoE issues have been creating a new assessment and management paradigm in multimedia systems and gaining a special attention in WMNs. QoE metrics have considered important metrics to measure the quality level of multimedia content based on the user´s perspective (Rowe & Jain, 2005) (De Vleeschauwer et al., 2008). QoE approaches aim to overcome the limitations of current QoS-aware schemes regarding human perception and subjective-related aspects (Jain, 2004) (Klein, 2007).

The emerging of QoE issues required the inclusion of a new user-level (abstraction) layer on Open Systems Interconnection (OSI) and Internet architectures (Siller & Woods, 2003). This layer can be seen as an extension of the application layer with user´s perception (Bauer & Patrick, 2004). Therefore, the results of QoE procedures can be used as an extension to the traditional QoS in the sense that QoE provides information regarding the delivered multimedia service from the user's point of view. Hence, QoE procedures can be explored to improve the accuracy of QoS control plane operations and to ensure smooth transmission of audio and video over WMNs.

It is important to highlight that QoE results are widely dependent on subjective aspects related with human perception, as well as, user´s location, screen size, hardware and applications (Valerdi et al., 2009) (Bhatti et al., 2000). For instance, video sequences with different complexities, motions and frame rates will produce different QoE results (Greengrass et al., 2009).

QoE measurement operations can be used as an indicator of how a networking environment meets the end-user needs. The QoE applicability scenarios, requirements, evaluations and assessment methodologies in multimedia systems have been investigated by several researchers and working groups, such as International Telecommunication Union - Telecommunication Standardization Sector (ITU-T) (ITU-T, 2010), Video Quality Experts Group (VQEG) (VQEG, 2010) and European Technical Committee for Speech, Transmission, Planning, and Quality of Service (ETSI STQ) (ETSI, 2010).

Advances in QoE-aware systems will allow the deployment of new QoS/QoE-sensitive services as well as provide new paradigms for the creation of new protocols, assessment solutions, objective and subjective metrics, routing approaches and overlay networks, such as the deployment of QoE routing schemes and user-aware packet controllers. Nowadays, QoE operations are not fully implemented in end-to-end networking systems due to the high CPU and memory consumption required by current QoE schemes, as well as to the lack of accuracy of in-service quality assessment methods. Usually, only QoE out-service measurement procedures are accomplished to evaluate the quality level of multimedia services WMNs and other systems.

Regarding QoE assessment issues, matching the multimedia quality level by computerized measurement is a research challenge and needs to take as input many factors related with the user´s perception. Multimedia quality evaluation approaches are classified into two main orthogonal criterions as described in the remainder subsections and presented in Figure 2: (i) the amount of the reference information required to assess the quality and (ii) the measured features based on objectivity/subjectivity (i.e. the way the quality is expressed).

2.2.1 Classification based on objectivity and subjectivity

In general, there are main methods to assess the quality level of multimedia contents, namely objectivity, subjectivity and hybrid. The output of these schemes is useful for QoE-aware billing/accounting procedures, assessment solutions and management issues.
Subjective metrics assess how audio and/or video streams are perceived by users (Kishigami, 2007), i.e., what is their opinion on the quality of particular audio/video sequences, as described in ITU-T recommendation BT 500 (ITU-R, 1995). The most popular subjective metric is called Mean Option Score (MOS). The quality level of a video (or audio) sequence based on MOS model is rated on a scale of 1 to 5, where 5 is the best possible score as presented in Table 1.
The MOS values are achieved based on subjective tests and methodologies performed with a set of viewers. For instance, the Single Stimulus Continuous Quality Evaluation (SSCQE) test allows viewers to dynamically rate the quality of an arbitrarily long video sequence using a slider mechanism with an associated quality scale. The drawback of subjective metrics is the fact that they are neither practical nor scalable for real-time multimedia environments. Other approaches are Double Stimulus Impairment Scale (DSIS), Double Stimulus Continuous Quality Scale (DSCQS), Single Stimulus Continuous Quality Evaluation (SSCQE), Simultaneous Double Stimulus for Continuous Evaluation (SDSCE) and Stimulus Comparison Adjectival Categorical Judgment (SCACJ) (Bocca-Rodríguez et al., 2007).

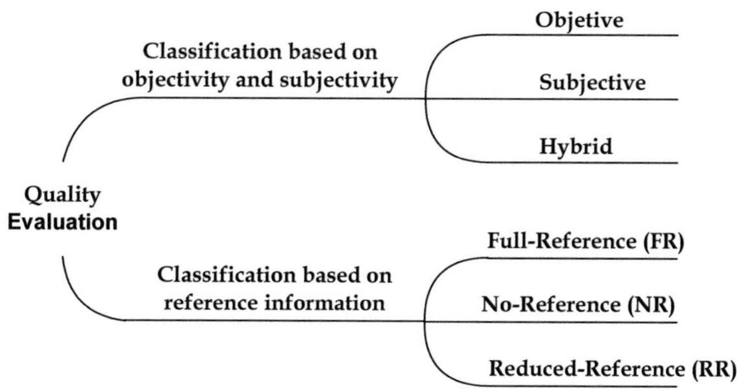

Fig. 2. Quality evaluation (Romaniak et al. 2008)

Several objective QoE metrics have been developed to estimate/predict (based on mathematical models) the quality level of multimedia services according to the user's perception. Among them, the Peak Signal to Noise Ratio (PSNR) is a traditional objective metric used to measure, in decibels, the video quality level based on original and processed

MOS	Quality	Impairment
5	Excellent	Imperceptible
4	Good	Perceptible but not annoying
3	Fair	Slightly annoying
2	Poor	Annoying
1	Bad	Very annoying

Table 1. Mean Option Score

video sequences. Typical values for the PSNR in lossy videos are between 30 dB and 50 dB, where higher is better. The PSNR of a video is defined through the Mean Square Error (MSE) metric. Considering the luminance (Y) of the processed and original frames and assuming frames with MxN pixels, the MSE is obtained using the Equation 1.

$$MSE = \frac{1}{MxN} \sum_{i=0}^{M-1} \sum_{j=0}^{N-1} \| Ys(i,j) - Yd(i,j) \|^2 \quad (1)$$

In Equation 1, while Ys(i,j) designates the pixel in the position (i, j) of the original frame, the Yd(i,j) represents the pixel located in the position (i, j) of the processed frame. Based on the MSE definition and on 8bits/sample, the PSNR, in a logarithmic scale, is achieved using the Equation 2.

$$PSNR = 20\log_{10}\left(\frac{255}{\sqrt{\frac{1}{MxN} \sum_{i=o}^{M-1} \sum_{j=0}^{N-1} \| Ys(i,j) - Yd(i,j) \|^2}} \right) \quad (2)$$

The MSE and PSNR metrics only provide an indication of the difference between the received frame and a reference signal, and do not consider any other important aspects which can strongly influence the video quality level, such as Human Visual System (HVS) characteristics (a detailed analysis of HVS can be found in (Wang et al., 2004).

The Structural Similarity Index Metric (SSIM) improves the traditional PSNR and MSE, which are inconsistent with HVS characteristics, such as human eye perception (Wang et al., 2004). The SSIM metric is based on frame-to-frame measuring of three components (luminance similarity, contrast similarity and structural similarity) and combining them into a single value, called index. The SSIM index is a decimal value between 0 and 1, where 0 means no correlation with the original image, and 1 means the exact same image.

The Video Quality Metric (VQM) method defines a set of computational models that also have been shown to be superior to traditional PSNR and MSE metrics (Revés et al., 2006). The VQM method takes as input the original video and the processed video and verifies the multimedia quality level based on human eye perception and subjectivity aspects, including blurring, global noise, block distortion and color distortion. The VQM evaluation results vary from 0 to 5 values, where 0 is the best possible score.

The Moving Picture Quality Metric (MPQM) evaluates the video quality using HVS modeling characteristics (Lambrecht, & Verscheure, 1996). The input to the MPQM metric is an original video sequence and a distorted version of it. The distortion is first computed as the difference between the original and the distorted sequences. The original and the error sequences are then decomposed into perceptual channels segmented using uniform areas, textures and contours classification.

2.2.2 Reference-based classification

Three different approaches are used to classify video quality assessment methods, based on reference-related video procedures, namely Full Reference (FR), Reduced Reference (RR) and No Reference (NR) (Engelke & Zepernick, 2007) (Garcia et al., 2009).

The FR approach assumes unlimited access to the original multimedia sequence. This approach uses the video reference to predict the quality level (degradation) of the processed video, by comparing the difference of every pixel in each image of the distorted video with its corresponding pixel in the original video. As consequence, it provides, in general, superior quality assessment performance. The FR method is difficult to implement in real-time networking systems (QoE-aware equipment/monitoring agent) because it always requires the original sequence during the evaluation process (common for offline experiments). Examples of metrics based on an FR approach are PSNR, SSIM and MPQM.

For in-service video quality measurements, RR and NR approaches are generally more suitable. The RR approach differs from the FR approach only selected multimedia parameters (or characteristics) are required during quality evaluation process, such as motion information. The set of reference parameters can be transmitted piggy-backed with the multimedia flow or by using a secondary channel. The objective of RR is to be as accurate as the full reference model, although using less network and processing resources. An example of an RR scheme is Video Quality Model (VQM), developed by the National Telecommunications and Information Administrative (NTIA) and reported in (Pinson & Wolf , 2004).

The NR approach tries to assess the quality of a distorted multimedia service without any reference to the original content. This approach is usually used when the coding method is known. NR-based metrics can be used in in-service network monitoring/diagnostic operations, when the original multimedia sequence is not available. The drawbacks of NR metric are the following: (i) low correlation with MOS; (ii) high CPU and memory consumption; (iii) time limitation. An example of NR schemes is the V-Factor model (V-Factor 2010) that outputs MOS.

3. Cross-layer design

The methodology of layered protocol design has been applied for decades in different types of network, for instance, OSI and Internet architectures. In this model, protocols, services and applications are designed without being constrained by each other. Many advantages such as scalability of network size, portability of protocols in different layers, flexibility in protocol design, and so on can be easily obtained in layered architectures. However, advances in emerging networks and heterogeneous systems are changing the traditional layered model.

There are many reasons behind the improvement of the layered design as follows: (i) the requirement of service quality is ever-increasing; (ii) the network heterogeneity is much higher than years ago; (iii) the conventional layered architecture is effective for integrating them into the same network, but tile performance is not optimized; (iv) many networks today, especially wireless networks have no dedicated links between nodes. In a wireless network, transmission between two nodes also interferes with other nodes in the neighborhood. Thus, the meaning of "link" pertained to a conventional wired network does not exist anymore. The capacity of a link is variable and can be fully cross-related with other links. Such inter-dependence in fact breaks the transparency between different protocol layers, where a multi-hop network, such as IEEE 802.11s, is concerned, this problem becomes much more obvious.

QoS and QoE support for multimedia application is a good example to explain the need of a cross-layer design, where the end-to-end quality level support over emerging systems involves the cooperation of three layers, namely physical, MAC and network. Therefore, a cross-layer paradigm must be applied to allow a tight communication between layers and improve the system performance (Kozat et al., 2004).

Several networking proposals have created to explore the benefits of cross-layer architectures to increase the network performance (Kawadia & Kumar, 2005) (Bhatia & Kodialam, 2004) (Chiang, 2004) (Kozat et al., 2004). The design of cross-layer models can be done by using two main approaches as presented in Figure 3, namely loosely coupled cross-layer design and tightly coupled cross-layer design.

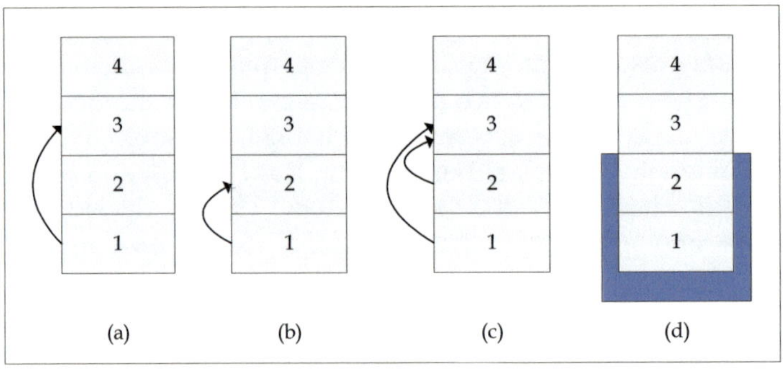

Fig. 3. Loosely coupled cross-layer design and rightly coupled cross-layer design

In the loosely coupled cross-layer design, the optimization is carried out without crossing layers, but focusing on one protocol layer. In order to improve tine performance of this protocol layer, parameters in other protocol layers are taken into account. Thus, information to one layer must be passed to another layer. Typically, parameters in the lower protocol layers are reported to higher layers. For example (Figure 3a), the packet loss rate in the MAC layer, or channel condition in the physical layer can be reported to the transport layer so that a TCP protocol is able to differentiate congestion from packet loss. As another example (Figure 3b), the physical layer can report the link quality to a routing protocol as an additional performance metric for the routing algorithms.

It should be noted that information from multiple layers can be used on another layer to perform cross-layer design (Figure 3c). There are two different ways of utilizing cross-layer information. The first one is the simplest case of cross-layer design, in which the information in other layers works just as one of the parameters needed by the algorithm in a protocol layer. The performance of this algorithm is improved because a better (more accurate or reliable) parameter is used, but the algorithm itself does not need a modification. For example, the physical layer can inform the TCP layer of the channel quality so that TCP can differentiate real congestion from channel quality degradation, and thus carry out congestion control more intelligently. In the second method, based on the information from other layers, the algorithms of a protocol have to be changed. For example, if a MAC protocol can provide a routine protocol information about its performance, the routing can perform multipath routing to utilize spatial diversity. However, the change from single-path routing to multipath routing needs a significant modification to the routing protocol rather than just parameter adaptation.

In the tightly coupled cross-layer design (Figure 3d), merely information sharing between layers is not enough. In this scheme, the algorithms in different layers are optimized together as one optimization problem. For example (Figure 3d), for MAC and routing protocols in a multichannel TDMA WMN, timeslots, channels, and routing path can be determined by one single algorithm. Using optimization across layers, it can be expected that much better performance improvement can be achieved by the tightly coupled cross-layer design than the loosely coupled scheme. However, the advantage of the loosely coupled design is that it does not totally abandon the transparency between protocol layers.

An extreme case of tightly coupled cross-layer design is to merge different protocol layers into one layer. According to the concept of "layering as optimization decomposition", this kind of design tries to improve network performance by re-layering the existing protocol stack. Merging multiple protocol layers into one layer keeps the advantage of tightly coupled cross-layer design. Furthermore, it can also eliminate the overhead in cross-layer information passing and is a trend in WMNs.

Interestingly, merging multiple protocol layers is not just a theoretical concept, but has been seriously considered in real practice. For example, in the IEEE 802.11s, the routing protocol is being developed as one of the critical modules in the MAC layer. Such a merging between routing and MAC layers provides great potential for carrying out optimization across MAC and routing, based on the same algorithm. Recent advances in wireless optimization are attracting researchers and industry to study cross-layer issues for future networks (Chen et al., 2007). Other cross-layer implementations supporting QoS in wireless system, as well as, load balance techniques can be found in (Pahalawatta et al., 2007), (Wu et al., 2007) and (Villalon et al., 2007).

It is clear that the cross-layer schemes will be predominant in WMNs, where new quality level mechanisms will be designed and implemented to increase the satisfaction of costumers and optimize the usage of network resources, such as routing protocols.

4. Performance evolution

As presented before, routing protocols based on QoE aims to optimize the usage of network resources, the system performance and the quality level of multimedia applications in WMNs. Novel QoE-aware cross-layer solutions will be essential for the success of next generation wireless system. In order to show the impact of this kind of solution on the user´s experience, this section presents the behavior of the OLSR-FLC protocol in WMNs, as well as to show the benefits, comparing it with the main well-known extensions of the OLSR protocol, OLSR-ETX, OLSR-MD, OLSR-DC.

This cross-layer OLSR extension is based on the dynamic choice of link quality metrics and in a FLC to decide on paths for multimedia packages. We analyzed the performance of the proposal through simulations on Network Simulator (NS-2) (Fall & Varadhan, 2010), using the scenario shown in Figure 4, which represents the WMN backbone partiality deployed in the Federal University of Para (UFPA) campus.

Table 2 describes the simulation parameters, which try to bring the simulation as close as possible to the considered real network scenario, representing the characteristics of the region and the used equipments. Path Loss Exponent and Shadowing Deviation parameters were used according to the measurements presented in (Moreira et al., 2008). The routers' carrier sense threshold and transmit power parameters were based on the IEEE 802.11 standard. The other values were used to represent the antennas and the routers used in the WMN at UFPA.

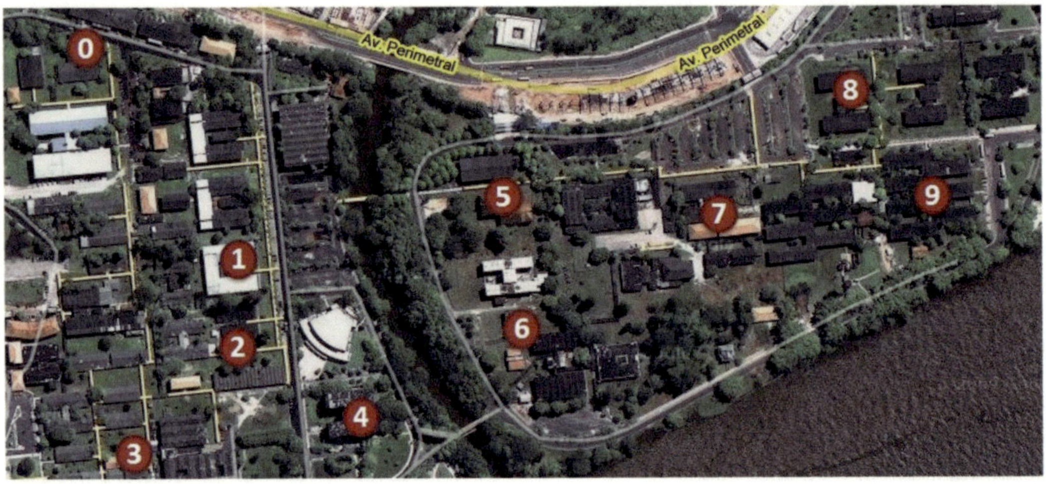

Fig. 4. Simulated scenario

Parameter	Value
Standard IEEE	802.11g
Propagation Model	Shadowing
Antenna	Omnidirectional 18dB
Router's Carrier Sense Threshold	-76dBm
Router's Transmit Power	-80dBm
Tansmission Power	17 dBm (WRT54G)
Frequency	2.422GHz (Channel 3)
Path Loss Exponent	1.59
Shadowing Deviation	5.4dB

Table 2. Simulation Parameters

Twenty simulations were performed using different seeds for each protocol: OLSR, OLSR-ETX, OLSR-MD, OLSR-DC, and OLSR-FLC. Table 3 shows the flow configuration used. All simulations were run for 50 seconds.

The configuration of flows aims to balance the flow over the MWN topology and to create a higher competition scenario, between data, audio, and video traffics. Hence, it brings the simulation to a common situation in WMNs, i.e., competition among all kind of flows where each flow has its own characteristics and requirements.

The simulation experiments comprised 3 VoIP (Voice over IP) calls, that on simulations are characterized by two flows, i.e. 6 UDP flows, 5 TCP-Reno flows and 3 video traffic. The video traffics were evaluated from the experience that the user obtained, through the QoE evaluation metric. The UDP flows have a bit rate of 8Kb/s and 40 bytes (RTP + UDP + Payload) of packet size, in order to represent the G.729 codec (Balam & Gibson, 2007). The TCP flows were characterized as FTP applications, following the Pareto model with a rate of 200k, 210 bytes of packet size and 500 ms burst duration.

Flow	Source	Destiny	Begin	End	Traffic
1	1	8	10	40	TCP – Reno
2	9	2	11	41	TCP Reno
3	7	4	12	42	TCP Reno
4	5	0	13	43	TCP Reno
5	6	4	14	44	TCP Reno
6	0	5	10	45	Video Paris
7	3	6	14	29	Video Foreman
8	3	6	30	45	Video News
9	2	9	6	46	UDP - CBR
10	9	2	6	46	UDP - CBR
11	1	8	7	47	UDP - CBR
12	8	1	7	47	UDP - CBR
13	4	7	8	48	UDP - CBR
14	7	4	8	48	UDP - CBR

Table 3. Flow Configuration

The video traffic was simulated through the Evalvid tool (Evalvid, 2010) that allows the control of the video quality in a simulation environment. Well-know real videos sequences were used, namely "Paris", "Foreman" and "News". These videos have frames in YUV format, which are compressed by MPEG-4 codec and sent at a rate of 30 frame/s. Each frame was fragmented into blocks of 1024 bytes where the packet has a size of 1052 bytes.

Well-know objective and subjective QoE metrics are used in the experiments, following the tests proposed by ITU-R (ITU-R, 1995). The subjective QoE metrics evaluate the quality of multimedia applications based on the receiver's opinion, where the MOS was used. The videos were analyzed using the MSU Video Quality Measurement Tool Software (MSU, 2010). The value of PSNR is expressed in dB (decibels). For a video to be considered with good quality it should have an average PSNR of at least 30dB.

The following tables and figures will demonstrate the simulation results of real video sequences collected from all protocols and based on QoS metrics. The tables show the average values, the highest value, the lowest value and standard deviation values for each protocol. Tables 4 and 5 show the values for the video "Foreman", where Table 4 presents the values of VQM and SSIM metrics and Table 5 the values of PSNR and MOS metrics. The same results are illustrated in Figure 5, 6 and 7 respectively.

Foreman			VQM				SSIM	
	Higher	Lower	Average	Standard Deviation	Higher	Lower	Average	Standard Deviation
OLSR	5	4,8	4,96	0,07	0,7	0,5	0,58	0,07
OLSR-MD	5	4	4,72	0,32	0,71	0,62	0,66	0,04
OLSR-ETX	5	4,5	4,86	0,23	0,77	0,5	0,61	0,08
OLSR-DC	5	2,3	4,5	0,83	0,87	0,61	0,67	0,08
OLSR-FLC	4,8	2,4	4,27	0,71	0,83	0,68	0,73	0,04

Table 4. VQM and SSIM Values of Video Foreman

Foreman		PSNR			MOS
	Higher	Lower	Average	Standard Deviation	
OLSR	18	14	15,8	1,62	Bad
OLSR-MD	23	16	19,1	2,26	Bad
OLSR-ETX	20	13	17,4	2,63	Bad
OLSR-DC	25	17	19,3	2,58	Bad
OLSR-FLC	25	22	22,9	0,88	Poor

Table 5. PSNR and MOS Values of Video Foreman

The transmission of the "Foreman" video, flow 7, begins after all the flows start their transmissions, and it starts in a moment of convergence of the protocols, which results in a high difficulty transmission with network congestion. These facts become clear from the data shown in the tables for the "Foreman" Video. Therefore, OLSR-FLC has the best values of QoE metrics, and is the only one which achieves the "Poor" quality while the other protocols obtain a quality considered "Bad", although the protocols OLSR-DC and OLSR-MD have values close to being considered as "Poor".

The Tables 6 and 7 show the values for the video "News", where Table 6 presents the values of VQM and SSIM values and Table 7 the values of PSNR and MOS metrics. The "News" video, flow 8, has the same destination and source as flow 7, "Foreman" video, however, it starts at a different time of the simulation. At this moment the protocols had already passed by the period of convergence, allowing a better choice of routes.

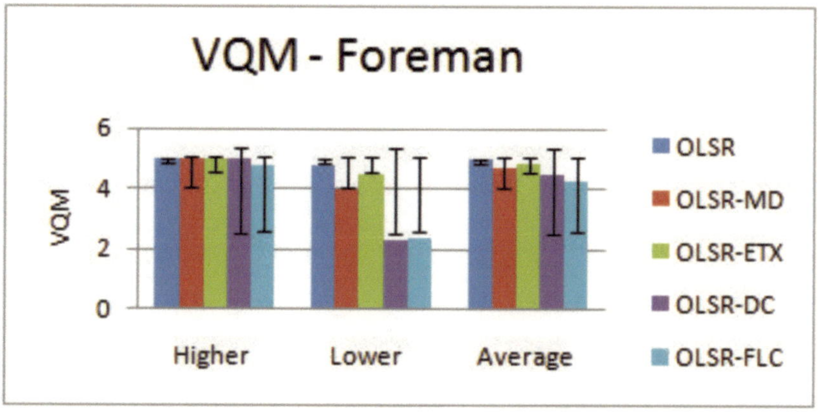

Fig. 5. VQM results of the Foreman video sequence for all protocols

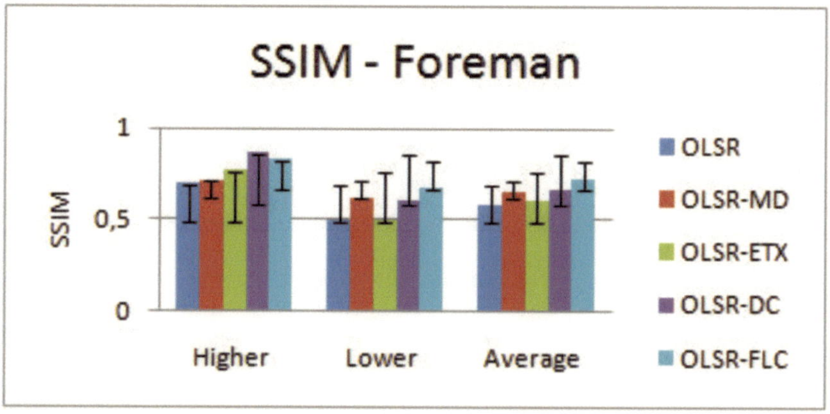

Fig. 6. SSIM results of the Foreman video sequence for all protocols

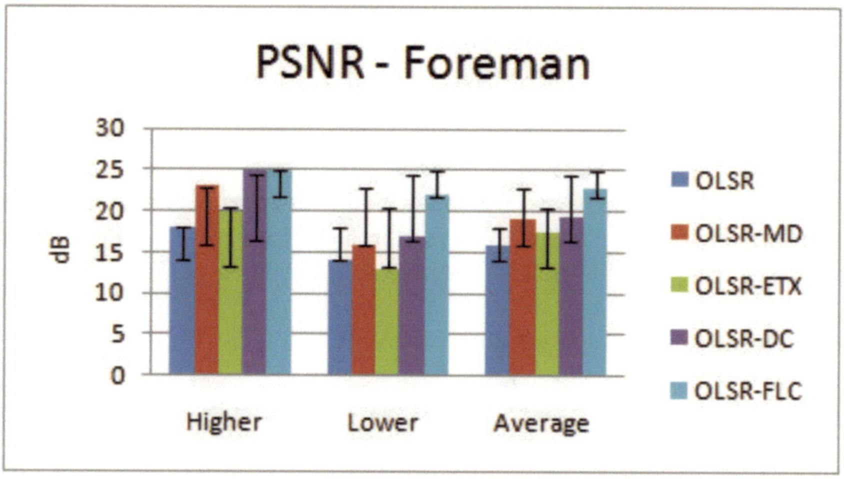

Fig. 7. PSNR results of the Foreman video sequence for all protocols

News			VQM				SSIM	
	Higher	Lower	Average	Standard Deviation	Higher	Lower	Average	Standard Deviation
OLSR	5	1,7	4,43	1,08	0,86	0,71	0,79	0,05
OLSR-MD	5	2,2	3,71	1,05	0,95	0,83	0,89	0,04
OLSR-ETX	5	2,2	4,09	1,01	0,89	0,78	0,85	0,04
OLSR-DC	4,9	1,4	3,44	1,17	0,97	0,84	0,89	0,05
OLSR-FLC	4	0,5	3,08	1,01	0,98	0,85	0,91	0,04

Table 6. VQM and SSIM Values of Video News

News		PSNR			MOS
	Higher	Lower	Average	Standard Deviation	
OLSR	25	17	19,7	2,91	Bad
OLSR-MD	27	19	22,5	3,21	Poor
OLSR-ETX	29	17	20,7	4,22	Poor
OLSR-DC	27	19	23,6	2,84	Poor
OLSR-FLC	44	20	25,7	7,01	Regular

Table 7. PSNR and MOS Values of Video News

We note this by comparing the data of "News" video with the data of "Foreman" video, where the "News" video has better values of QoE metrics. Again OLSR-FLC achieves the best video quality, having a quality considered "Regular", while the other protocols obtain qualities ranging from "Poor" to "Bad". Despite of having a better video quality, OLSR-FLC has a high standard deviation, showing a degree of instability in the quality of the transmitted videos, obtaining values better and of similar quality to the other protocols.

The Figure 8 and 9 show the values for the video "Paris", where Table 8 presents the values of VQM and SSIM values and Table 9 the values of PSNR and MOS metrics. Since the "Paris" video, flow 6, is longer than the other videos, it is transmitted during almost the entire simulation, this means that the transmission has a hard time during the convergence of the protocols at the beginning of its transmission, but most of the transmission occurs after the convergence period.

Unlike the other video transmissions, flows 7 and 8, the nodes involved in flow 6 have a clear line of sight, however, with a higher distance between the nodes. This makes that the use of a single hop increase the chance of packet loss, as well as, the use of multiples hops increase the end-to-end delay of the package. Within this reality, the usage of a single metric turns out to be insufficient to find the most appropriate route, because a good video transmission depends not only on small losses, but also on a small delay and jitter.

Paris			VQM				SSIM	
	Higher	Lower	Average	Standard Deviation	Higher	Lower	Average	Standard Deviation
OLSR	5	4,4	4,93	0,19	0,75	0,63	0,69	0,04
OLSR-MD	3,8	3	3,35	0,31	0,9	0,83	0,86	0,02
OLSR-ETX	4,9	3	4,09	0,59	0,83	0,73	0,79	0,04
OLSR-DC	3,5	2,3	2,94	0,41	0,93	0,87	0,88	0,02
OLSR-FLC	3,1	2,3	2,75	0,32	0,93	0,87	0,91	0,02

Table 8. VQM and SSIM Values of Video Paris

Paris		PSNR			MOS
	Higher	Lower	Average	Standard Deviation	
OLSR	17	14	15,2	1,23	Bad
OLSR-MD	27	23	24,8	1,14	Poor
OLSR-ETX	23	20	21,4	1,17	Poor
OLSR-DC	29	23	26,5	2,07	Regular
OLSR-FLC	31	25	29,2	2,15	Regular

Table 9. PSNR and MOS Values of Video Paris

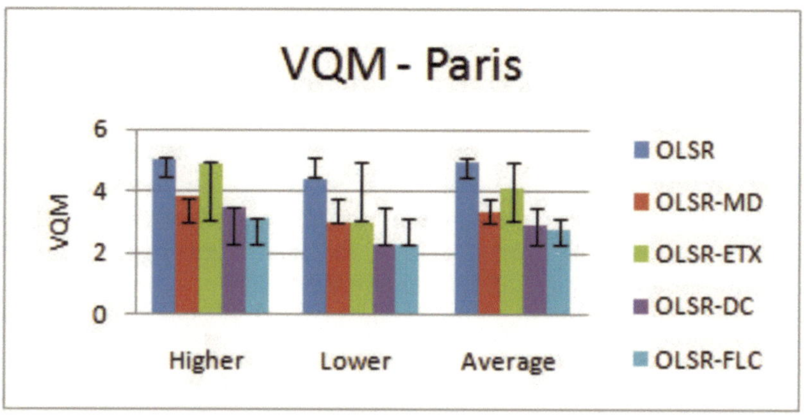

Fig. 8. VQM values for Paris video sequence

Therefore, we note that OLSR-FLC can adapt to this reality of multiple requirements, which is visible in the tables for the "Paris" video. The OLSR-FLC protocol, as well as, the OLSR-

DC protocol, have a video quality considered "regular", however OLSR-FLC reaches values close to "Good" quality level. In other words, because it is based on the OLSR-DC protocol, the OLSR-FLC protocol can better distribute the traffic, but it uses a fuzzy link cost, based on delay and quality of links. This enables the protocol to obtain a better video quality, against the protocols that use only one metric for routing.

In some of the existing works, as in (Moreira et al., 2008), (Gomes et al., 2008) and (Gomes et al., 2009a), the evaluation of the protocols occurred in scenarios of competition, however, the competition present in these works is small, when compared with the scenario used in this work, due to the quantity of flows used. This evaluation shows that only one metric for routing data and multimedia traffics may not be sufficient to reach acceptable QoS and QoE levels to answer the needs of the traffics, since each one has different requirements.

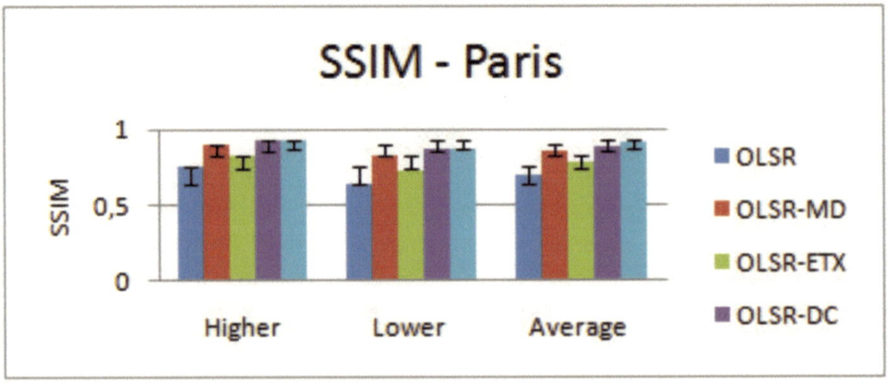

Fig. 9. SSIM values for Paris video sequence

5. Final considerations

Currently, wireless systems, multimedia distribution and quality level control continue to be strong research areas. Trends in these areas are expected to continue with various challenges emerging as a result of new services, protocols, wireless mesh networks, cross-layer schemes, emerging portable devices, changing user and terminal requirements. It is clear that recent advances on QoS/QoE have been allowed the creation of a new ubiquitous wireless multimedia approach in the Internet.

WMN routing solutions with QoE support will be essential for the success of multimedia communications, where users will be more satisfied with the received content and mobile operators will be able to increase their billing with the new attractions to clients and the operational costs reduction. However, one of the main challenges is to develop and implement new in-service routing QoE solutions in WMNs.

This chapter was intended to highlight important topics in WMNs, QoS, QoE, routing and cross-layer areas that need attention to address some of the most pressing challenges associated with them. We focus on four key areas, where the first one was on QoS, the second one on QoE, the third one on routing and the last on cross-layer issues. Furthermore, in order to demonstrate the behavior of an implementation of a QoE-aware cross-layer routing in WMNs, simulations were carried out. The results show the benefits of the proposed scheme on the user´s perspective, by using well-know QoE metrics.

We hope that this work will help improve our understanding of the issues and challenges that lie ahead in wireless mesh networks and QoS/QoE issues will serve as a catalyst for designers, engineers, and researchers to seek innovative solutions to address and solve those challenges.

6. Acknowledgements

Eduardo Cerqueira was supported by CNPq 476202/2009-4 & 557.128/2009-9, PROPESP UFPA and FAPESPA 5183.UNI319.4107.07072009 - 5467.UNI317.1279.31082009. We would like also thanks André Riker and Patricia Araujo de Oliveira for their value contributions for this chapter.

7. References

802.11s (2010). *IEEE Mesh Networking,* Task Group S. Status of Project IEEE 802.11s, accessed in March 2010, of http://www.ieee802.org/11/Reports/tgs_update.htm

802.16 (2010). *IEEE Standard.* Working Group on Broadband Wireless Access, accessed in March 2010, of http://grouper.ieee.org/groups/802/16/

Akyildiz, I. & Wang, X. (2009). *Wireless Mesh Networks (Advanced Texts in Communications and Networking),* Wiley

Balam, J. & Gibson, J (2007). Multiple descriptions and path diversity for voice communications over wireless mesh networks. *IEEE Transactions on Multimedia,* pages 1073–1088, August 2007.

Bauer, Ben; Patrick, Andrew S. (2004). *A Human Factors Extension to the Seven-Layer OSI Reference Model,* January 2004, accessed in April 2010, of http://www.andrewpatrick.ca/OSI/10layer.html

Bhatia, R. & Kodialam, M. (2004). On power efficient communication over multi-hop wireless networks: joint routing, scheduling and power control, *Twenty-third AnnualJoint Conference of the IEEE Computer and Communications Societies,* Vol.2, pp. 1457- 1466, March 2004, INFOCOM 2004

Bhatti, Nina; Bouch, Anna; Kuchinsky, Allan (2000). Integrating user-perceived quality into Web server design. *Proceedings of the 9th international World Wide Web conference on Computer networks: the international journal of computer and telecommunications networking.* Amsterdam, Vol. 33, No. 1-6, (June 2000) page numbers (1-16)

Bocca-Rodríguez, Pablo; Cancela, Héctor; Rubino, Gerardo (2007). Video Quality Assurance in Multi-Source Streaming Techniques, *Proceedings of the 4th international IFIP/ACM Latin American conference on Networking, pp. 83-93,* Applications, Technologies, Architectures, and Protocols for Computer Communication, 2007, San José

Bok-Nyong Park et al. (2006). QoS-driven wireless broadband home networking based on multihop wireless mesh networks. *Consumer Electronics,* IEEE Transactions on, Vol. 52, No. 4, (November 2006) page numbers (1220-1228)

Chen, B.; Lee, M.J. & Sun, Y. (2007). A Framework for Crosslayer Optimization from Physical Layer to Routing Layer on Wireless Ad Hoc Networks, *Global Telecommunications Conference,* pp. 3678-3683, November 2007, GLOBECOM '07

Chiang, M (2004). To layer or not to layer: balancing transport and physical layers in wireless multihop networks, *Twenty-third AnnualJoint Conference of the IEEE Computer and Communications Societies,* Vol.4, pp. 2525- 2536, March 2004, INFOCOM 2004

Chiti, F.; Fantacci, R.; Maccari, L.; Marabissi, D. & Tarchi, D. (2008). A broadband wireless communications system for emergency management, *Wireless Communications IEEE*, Vol.15, No.3, (June 2008) page numbers (8-14)

Clausen, T. & Jacquet, P. (2006). Optimized link state routing protocol (OLSR), RFC 3626, 2006

Cordeiro, W.; Aguiar, E.; Moreira, W.; Abelem, A. & Stanton, M. (2007). Providing quality of service for mesh networks using link delay measurements, *16th International Conference on Computer Communications and Networks*, pp. 991 -996, 2007

De Couto, D.; Aguayo, D.; Bicket, J. & Morris, R. (2003). A high-throughput path metric for multi-hop wireless routing, *9th Annual International Conference on Mobile Computing and Networking*, pp. 134 -146, 2003

De Vleeschauwer, B. et al. (2008). End-to-end QoE Optimization Through Overlay Network Deployment. *Information Networking, 2008. ICOIN 2008. International Conference on*, pp. 1-5, January 2008

Engelke, Ulrich; Zepernick, Hans-Jurgen (2007). Perceptual-based Quality Metrics for Image and Video Services: A Survey. *Next Generation Internet Networks, 3rd EuroNGI Conference on*, pp.190-197, May 2007.

ETSI (2010). *European Technical Committee for Speech, Transmission, Planning, and Quality of Service,* accessed in April 2010, of http://portal.etsi.org

Evalvid (2010). *Evalvid: A Video Quality Evaluation Tool-set*, accessed in March 2010, of http://www.tkn.tu-berlin.de/research/evalvid/

Fall, K. & Varadhan, K. (2010). *The network simulator – ns-2*, accessed in August 2010, of http://www.isi.edu/nsnam/ns/

Garcia, M. et al. (2009). A QoE Management System for Ubiquitous IPTV Devices. *Mobile Ubiquitous Computing, Systems, Services and Technologies, Third International Conference on*, pp.147-152, October 2009, UBICOMM '09

Gavrilovska, L.M.; Atanasovski, V.M. (2005). Ad hoc networking towards 4G: Challenges and QoS solutions. *International Conference on Telecommunications in Modern Satellite, Cable and Broadcasting Services*, September 2005

Gomes, R. L.; Ferreira Júnior, J.; Moreira Júnior, W. & Abelém, A. J. (2009). QoE and QoS in Wireless Mesh Networks. *IEEE Latin-American Conference on Communications,* 2009

Gomes, R.; Moreira Junior, W.; Cerqueira, E. & Abelem, A. (2009). Using a Fuzzy Link Cost and Dynamic Choice of Metrics to Achieve QoS and QoE in Wireless Mesh Networks. *Journal of Network and Computer Applications*, pp. 1084-8045, November 2009

Gomes, R.; Moreira, W.; Nascimento, V. & Abelem, A. (2008). Dynamic metric choice routing for mesh networks, *7th International Information and Telecommunication Technologies Symposium (I2TS)*, 2008

Greengrass, J.; Evans, J. & Begen, A.C. (2009). Not All Packets Are Equal, Part I: Streaming Video Coding and SLA Requirements. *IEEE Internet Computing,* Vol. 13 , No. 1, page numbers (70 – 75)

Held, G. (2005). *Wireless Mesh Networks,* Auerbach Publications

Hossain, E. & Leung, K. (2009). *Wireless Mesh Networks: Architectures and Protocols*, Springer US

ITU-R (1995). International Telecommunication Union - Radiocommunication Sector. Technical Report.Series BT: Broadcasting service (television). *Recommendation BT.500-7: Methodology for the Subjective Assessment of the Quality of Television Pictures,* October 1995

ITU-T (2010). International Telecommunication Union - Telecommunication Standardization Sector, accessed in April 2010, of http://www.itu.int/ITU-T/

Jain, R. (2004). Quality of experience. *IEEE Multimedia*, Vol. 11, No.1, 2004, page numbers (95- 96)

Kawadia, V. & Kumar, P.R. (2005). A cautionary perspective on cross-layer design. *Wireless Communications IEEE*, Vol.12, No.1, (February 2005) page numbers (3- 11)

Kishigami, Jay (2007). The Role of QoE on IPTV Services style, *Multimedia, ISM, Ninth IEEE International Symposium on*, pp. 11-13, Taichung

Klein, Anja (2007). Incorporating Quality Aspects in Sensor Data Streams, *Proceedings of the ACM first Ph.D, Conference on Information and Knowledge Management*, pp. 77-84, Lisboa, workshop in CIKM

Kozat, U.C.; Koutsopoulos, I. & Tassiulas, L. (2004). A framework for cross-layer design of energy-efficient communication with QoS provisioning in multi-hop wireless networks, *Twenty-third Annual Joint Conference of the IEEE Computer and Communications Societies*, Vol. 2, pp. 1446- 1456, March 2004, INFOCOM 2004.

Lambrecht, C. & Verscheure, O. (2006). Perceptual Quality Measure Using a Spatio-Temporal Model of the Human Visual System. *Proceedings of SPIE*, Vol. 2668, San Jose, USA.

Monteiro, Janio M.; Nunes, Mario S. (2007). A Subjective Quality Estimation Tool for the Evaluation of Video Communication Systems. *Computers and Communications, ISCC 2007, 12th IEEE Symposium on*, p. MW - 75 - MW – 80.

Moreira, W.; Aguiar, E.; Abelém, A. & Stanton, M. (2008). Using multiple metrics with the optimized link state routing protocol for wireless mesh networks. *26th Brazilian Symposium on Computer Networks and Distributed Systems*, May 2008

MSU MSU Video Quality Measurement Tool. accessed in August 2010, of http://compression.ru/video/quality measure/index en.html

Pahalawatta, P.; Berry, R.; Pappas, T. & Katsaggelos, A. (2007). Content-Aware Resource Allocation and Packet Scheduling for Video Transmission over Wireless Networks, *Selected Areas in Communications, IEEE Journal on*, Vol.25, No.4, (May 2007) page numbers (749-759)

Patrick, Andrew S. et al. (2004). A QoE sensitive architecture for advanced collaborative environments. *First International Conference on Quality of Service in Heterogeneous Wired/Wireless Networks*, pp. 319-322, October 2004, QSHINE 2004

Pinson, Margaret H.; Wolf, Stephen. (2004). *A new standardized method for objectively measuring video quality*. IEEE Transactions on Broadcasting, Vol. 50, No. 3, (September 2004) page numbers (312 - 322)

Revés, X.; Nafisi, N.; Ferrús, R. & Gelonch, A. (2006). User perceived Quality Evaluation in a B3G Network Testbed. *IST Mobile Summit*, Mykonos, Greece

Romaniak, P.; Mu, M.; Leszczuk, M. & Mauthe, A. (2008). *Framework for the Integrated Video Quality Assessment*. Blekinge, Institute of Technology, April 2008, Karlsrona, Sweden

Rowe, Lawrence A. & Jain, Ramesh (2005). ACM SIGMM Retreat Report on Future Directions in Multimedia Research. *ACM Transactions on Multimedia Computing*, Communications, and Applications (TOMCCAP), Vol. 1, No. 1, (February 2005) page numbers (3-13)

Siller, M. & Woods, J. (2003) Improving Quality of Experience for Multimedia Services by QoS arbitration on QoE Framework. *Proceedings of the 13th Packed Video Workshop 2003,* April 2003, Nantes, France

Takahashi, A.; Hands, D. & Barriac, V. (2008). Standardization activities in the ITU for a QoE assessment of IPTV. *IEEE Communication Magazine,* Vol. 46, No. 2, (February 2008)

Valerdi, J.; Gonzalez, A. & Garrido, F.J. (2009). Automatic Testing and Measurement of QoE in IPTV Using Image and Video Comparison. *Digital Telecommunications, ICDT '09, Fourth International Conference on,* pp.75-81, July 2009

V-Factor (2010). V-Factor Quality of Experience Platform, accessed in July 2010, of http://www.pevq.org/

Villalon, J.; Cuenca, P.; Orozco-Barbosa, L.; Seok, Yongho & Turletti, T. (2007). Cross-Layer Architecture for Adaptive Video Multicast Streaming Over Multirate Wireless LANs, *Selected Areas in Communications, IEEE Journal on,* Vol.25, No.4, pp.699-711, May 2007

VQEG (2010). *Video Quality Experts Group,* accessed in April 2010, of http://www.its.bldrdoc.gov/vqeg/

Wang, Zhou; Lu, Ligang; Bovik, Alan C. (2004). Video Quality Assessment based on Structural Distortion Measurement. *Signal Processing: Image Communication, Special Issue on Objective Video Quality Metrics,* Vol. 19, No. 2, pp. 121-132, February 2004

Wu, D.; Ci, S. & Wang, H. (2007). Cross-Layer Optimization for Video Summary Transmission over Wireless Networks, *Selected Areas in Communications, IEEE Journal on,* Vol.25, No.4, pp.841-850, May 2007

Yamada, H. et al. (2007). A QoE based service control scheme for RACF in IP-based FMC networks. *E-Commerce Technology and the 4th IEEE International Conference on Enterprise Computing, E-Commerce, and E-Services, CEC/EEE,* pp. 611-618

Zapater, Marcio Nieblas; Bressan, Graça (2007). A Proposed Approach for Quality of Experience Assurance of IPTV. *First International Conference on the Digital Society,* pp. 25-25, Guadeloupe, January 2007, ICDS '07

Zhang, Y.; Luo, J. & Hu, H. (2006). Wireless Mesh Networking: Architectures, Protocols and Standards, In: *Wireless Networks and Mobile Communications Series,* Auerbach Publications.

Part 2

Administrative Technical Issues in Wireless Mesh Networks

7

On the Capacity and Scalability of Wireless Mesh Networks

Yonghui Chen
Dept. of Electronics and Information Engineering of HUST
& Wuhan National Laboratory for Optoelectronic
Hubei University of Technology

1. Introduction

In practicable multi-user wireless networks, the communication should do among any nodes over the coverage. Since the nature of wireless channel is fading and share, the interferences and the collision becomes unable to avoid. It is difficult to balance reuse and interference while communications, location and mobility of each node are almost random. Considerate the cost, a practicable multi-user networking should have to be interference limited. Even though the Shannon capacity limitation for the single channel could be achieved by Turbo Coding(Berrou, Glavieux et al. 1993) or the MIMO (G.J.Foschini 1996) (E.Telatar 1999) technologies. In the other words, the capacity is always determined by the SIR or SINR. The flourishing cellular system and IEEE 802.11 networks are typical interference limited systems also.

It is well known that the capacity on networks is related to the networking architecture. For some type central controlled infrastructure system, e.g. a single cellular cell with FDMA CDMA or TDMA, the capacity upper bound is often assured. But the capacity on common wireless networks is still illegible, even including the multi-cell cellular system (T.M.Cover & J.A.Thomas 2006).

Without regard to the architecture and the access mode, the abstract capacity of a wireless system could be classified in two types:

- For the typical inference limited systems, the capacity of each node should be (Gupta & Kumar 2000; Kumar 2003) :

$$C_{node} = \theta(1/\sqrt{K}) \text{ or } C_{node} = \theta(1/\sqrt{K\log K}) \qquad (1)$$

- For a X networking , in which each node has useful information to all the other nodes, the capacity of each node should be (Cadambe & Jafar 2007; Cadambe & Jafar 2008; Cadambe & Jafar 2009) :

$$C_{node}(SNR) = \theta(1) \qquad (2)$$

Where $\theta(\bullet)$ indicates the relation of equivalence; K is the number of nodes. Formula (1) shows that the capacity of a node is inverse ratio to the $\sqrt{K}$ or $\sqrt{K\log K}$. In the other words, the capacity is decided by the SINR or SIR. Formula (2) shows the capacity could be

unattached to the number of the nodes in the system. In the other words, if all the signal power could be taken as useful mutual information other than interference, the capacity should be limited by the SNR other than used SINR or SIR. In fact, formula (2) assumed the networking as an ideal cooperative MIMO system.
For a X networking with S source nodes, D destination nodes and R relay nodes, say each nodes has full-duplex ability, the upper bound of capacity should be (Cadambe & Jafar 2007; Cadambe & Jafar 2008; Cadambe & Jafar 2009):

$$C_{node}(SNR) = \theta\left[SD / K(S + D - 1)\right] \quad (3)$$

This means the capacity on multi-hop systems should be less than the one hop system. However, Wireless mesh network (WMN) has been regarded as an alternative technology for last-mile broadband access, as in fig 1.

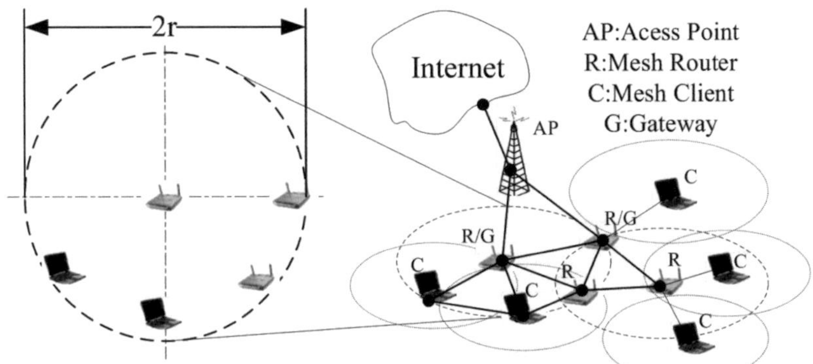

Fig. 1. A typical application of WMN. Typical nodes in WMN are Mesh Routers and Mesh Clients. Mesh clients form ad hoc sub-networks. Mesh routers form the mesh backbone for the mesh clients. Each node in WMN could act as a relay, forwarding traffic generated by other nodes.

Most industrial standards groups are actively specifying WMN, e.g. IEEE 802.11/802.15/802.16 and 3GPP LTE. For the combination of infrastructure and self-organized networking brings many advantages such as low up-front cost, robustness and reliable service coverage, etc. While WMN can be built upon existing technologies, spot test proved that the performance is still far below expectations. One of the most challenge problem is the avaliable capacity based practicable rule(Goldsmith 2005). Gennerally, similar capacity problems are slided over by simplier resource redundance(Akyildiz & Xudong 2005). In this paper, the Asymptotic Capacity on WMN will be talked about, mainly based on the former paper(Chen, Zhu et al. 2008).

2. Characteristic of multi-hops wireless mesh networking

2.1 The optimal architecture of multi-hop networking is still illegible

The shared channel leads to hidden terminals and exposed terminals(Gallager 1985). It is a series of handshake signals that could resolve these problems to a certain extent(Karn Sept.1990; Bharghavan, Demers et al. Aug. 1994). In balance, the capacity has to bound the successful throughput on collision-free transmissions as in fig 2.

Due to lack of any centralized controls and possible node mobility, it is hard to transplant the mature techniques from the central controlled or wired networking to the multi-hops wireless networking with high resource efficiency, which used to rely on the networking infrastructure (Basagni, Turgut et al. 2001) (Haartsen 2000) (Akyildiz & Xudong 2005; Nandiraju, Nandiraju et al. 2007). And the medium access scheme is also a challenge for the self-organized neworking(Gupta & Kumar 2000): Use of TDMA or dynamic assignment of frequency bands is complex since there is no centralized control; FDMA is inefficient in dense networks; CDMA is difficult to implement due to the inorganization networking . It is hard to keep track of the frequency-hopping patterns and/or spreading codes for all the nodes. the optimal architecture to the multi-hop systems is still illegible (Goldsmith 2005).

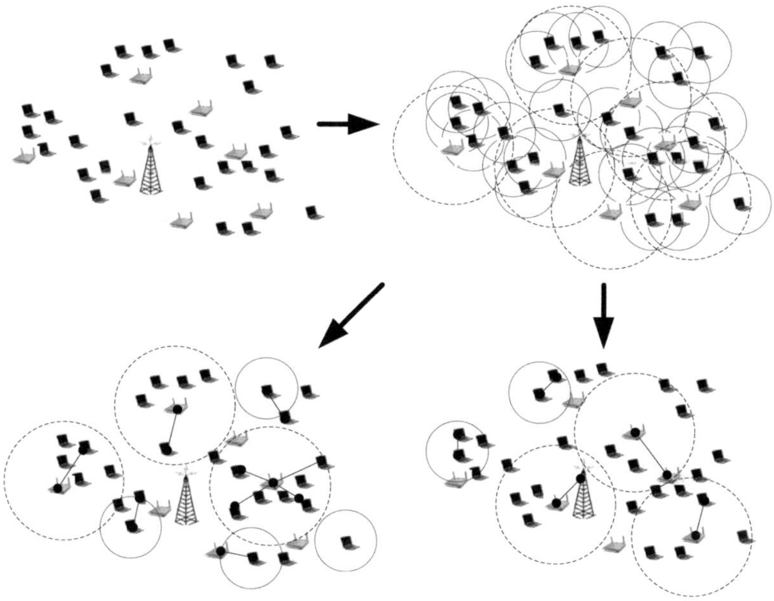

Fig. 2. Whether one hop networking or multiple hop netowrkig, practicable wireless communication system should be based on available resource reuse. The communication should be hop hy hop.

2.2 Power Gains of ideal multi-hop link

With an ideal linearity multi-hop chain, obviously the shorter propagating distance the more power gains. Say σ_n^2 is the noise variance, P is the transfer power of each node, $K \bullet d^{-\gamma}, \gamma \geq 2$ is the path loss, where K is constant, d is the whole distance and γ is path loss facter. Thus the end to end frequency normalized capacity is:

$$C = \log\left[1 + \frac{K \bullet P}{\sigma_n^2 d^{\gamma}}\right] \tag{4}$$

Say N_{hop} is the number of hops. d_i is the distance of the i-th hop, obviously $d \leq \sum_{i=1}^{N_{hop}} d_i$. Say $d_{max} = \max\{d_i\}$, thus:

$$C = \log\left[1+\frac{KP}{\sigma_n^2 d_i^\gamma}\right] \geq \log\left[1+\frac{KP}{\sigma_n^2 d_{\max}{}^\gamma}\right] \quad (5)$$

Since N_{hop} times relay, the SNR gain of N_{hop} systems is:

$$\left[\frac{1}{N_{\text{hop}}}\left(\frac{KP}{\sigma_n^2 d_{\max}{}^\gamma}\right) \div \left(\frac{KP}{\sigma_n^2 d^\gamma}\right)\right]_{\text{dB}} = 10\lg\left(\frac{1}{N_{\text{hop}}}\left(\frac{d}{d_{\max}}\right)^\gamma\right) \quad (6)$$

Whrere $N_{\text{hop}} \geq 1, \gamma \geq 2$. If $d / d_{\max} = N_{hop}$, the gain is $10(\gamma-1)\lg\left(N_{\text{hop}}\right)$ dB.

2.3 Constraints of multi-hop systems

Even if the multi-hop link is ideal, increasing with N_{hop}, the link need at least N_{hop} times transfer cost, e.g. the delay will be direct ratio with N_{hop}. Say the maximum capacity of each hop is constant 1. As a) in fig 3, despite of the hidden and exposed terminals problems, the last hop near the destination node is the bottleneck determining the capacity, with the fairness scheme. It is obviously that capacity per-node is $1/N_{\text{hop}}$. As b) in fig 3, with virtual circuit mode, each hop relay has the same payload, thus there is only one efficient payload from the source to the destination, capacity per-node also is $1/N_{\text{hop}}$. In balance either absolute fairness scheme or monopolization mode, the utmost throughput per-node is $1/N_{\text{hop}}$.

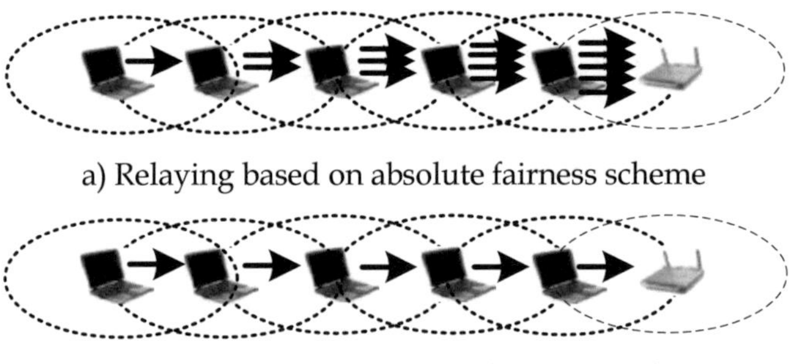

a) Relaying based on absolute fairness scheme

b) Relaying based on virtual circuit mode

Fig. 3. Constraints of multi-hop systems

Due to the shared channels, the hidden and exposed terminals problems are inevitable in multi-hop fashion communication. By using multiple channels/radios, or the other methods to decreases the delay, but the transfer do not truly enhance the resource utilization efficiency.

Considerate access competition, say each hop is independent and has probability p_c to success, if the transfer time is limited to 1, thus the access probability of a N_{hop} hops chain is:

$$p_S = p_c^{N_{\text{hop}}} \quad (7)$$

If without limitation of retransfer times, the access probability is 1:

$$1=\prod_{i=1}^{N_{hop}}\sum_{i=0}^{\infty}p_c(1-P_c)^i \tag{8}$$

Say the delay of each competiction time is T , the expection of total delay is:

$$\begin{aligned}E(T_D)&=T\bullet\sum_{j=1}^{N_{hop}}\sum_{i=0}^{\infty}(1+i)\bullet P_c\bullet(1-P_c)^i\\&=T\bullet N_P\,/\,P_c\end{aligned} \tag{9}$$

Take the average retransfer times regarded as:

$$N'_{hop}=N_{hop}\,/\,P_c \tag{10}$$

Thus the actual spectrum efficiency is:

$$1\,/\,N'_{\text{hop}}=P_c\,/\,N_{\text{hop}} \tag{11}$$

2.4 Mobility is dilemma

There are many research focus on mobility of mesh nodes (Gupta & Kumar 2000; Jangeun & Sichitiu 2003; Tavli 2006). It could proved that the mobility of nodes, either random or bounded, could improve the capacity of multi-hop wireless networks by deducing the hops between the source-destination chains, as in fig 4(Grossglauser & Tse 2002; Diggavi, Grossglauser et al. 2005). But Mobility is obviously a dilemma problem. Because too much mobility limited the capacity of multi-hop wireless networks, if considerate the cost (Jafar 2005)

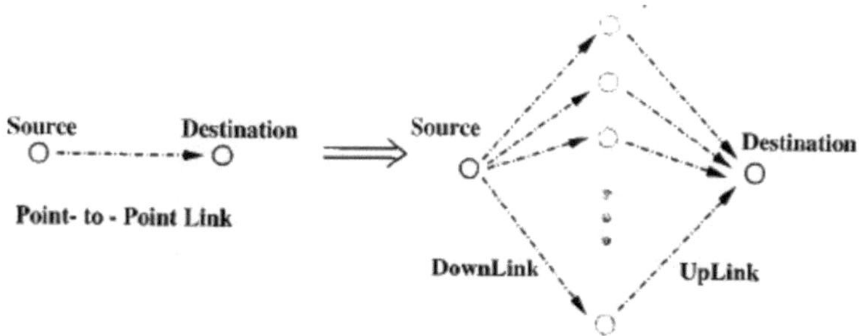

Fig. 4. Say the mobility is random, the mobile relay node has enough storage, the node as in a certain area or move along a fix path. The message could be transfered to the destination in probability with less hops.

3. Probability model on random multi-access multi-hop system

3.1 Assumption

- Say R is the radium of wireless network coverage, and N is the number of nodes on the area, thus the node density is $\rho_D=N\,/\,\pi R^2$;
- Considerate the path fading, Say each node has the same coverage, r is the radium;

- ω dentes the transfer capability during a transfer period.Say ω is the same for each node;
- Say the location of the nodes is symmetrical if the scale is lager than $2(1+2\Delta)r$, and the locations is random if the scale is smaller than $2(1+2\Delta)r$. Where Δ is the interference limitation facter. Thus the number of node in a node cell, n_{cell}, is random.
- Say each node learn the transfer direction and send the message to these direction, and there is ideal whole networking synchronization, thus if one node get the channel at a competition slot, the transfer will be success during the next slot. In the other words, if each node has the same sending probability and similar payload, each hop of the multi-hop chain could be model as independent.

3.2 Traffic model

The networks traffics could mainly be classified in three styles: unicast traffic (Gupta & Kumar 2000) , multicast traffic (Tavli 2006) and backhaul traffic(Jangeun & Sichitiu 2003). Note that the capacity of broadcast traffics and the backhaul traffics are equivalent in (Jangeun & Sichitiu 2003; Tavli 2006). The collision domain of backhaul traffics obviously happen to the nodes near the gateway, while the broadcast traffics are transferring the same payload. In any case, each transmission traffics must be hop-by-hop even if the node has possible mobility as in (Grossglauser & Tse 2002; Diggavi, Grossglauser et al. 2005). This means that the efficiency of a multi-hop chain is decide by the hops, at least partially. And each node in the chain(s) could carry no more than ω / N_{hop} efficient payload. For the different traffics there are different equivalent hops.

- For unicast traffics, Take N_{hop} as the sum hops in the multi-hop chain;
- For broadcast traffic, Take N_{hop} as the sum hops of all the broadcast source-termination pairs;
- For multicast traffic, Take N_{hop} as the sum hops of each multi-hop chain.

3.3 The connectivity model

The model is similar to the connectivity model in (Miorando & Granelli 2007). Model the spatial positions of each nodes as a Poisson distribution as in (Miorando & Granelli 2007) (Takagi & Kleinrock 1984). We have assumed each node could get the neighbors positions information, thus each node transmits its traffic directly to the very neighbor and the probability has k forward node is:

$$p(k;\lambda = n_f) = \frac{e^{-n_f} n_f^k}{k!} \tag{12}$$

For Omni-antenna, take $n_f = n_{cell} / 2$ as in [20]. For smart antenna technology, n_f could be a weighted n_{cell}. Denote E(.) as the mathematical expectation. In any case:

$$n_f = c_1 E(n_{cell}), \quad c_1 \in [0,1] \tag{13}$$

For simplify the analysis, normalized ρ as n_{cell} / N, thus

$$\rho = E(n_{cell}) / N = (\rho_D \pi r^2) / (\rho_D \pi R^2) = (r / R)^2 \tag{14}$$

(13) can be rewrite (15) as:

$$n_f = c_1 \rho N, \quad c_1 \in (0,1], \rho \in (0,1] \tag{15}$$

By the model, the probability a node has no avaliable next hop relay or terminal node is:

$$p_{isol} = P(k=0; n_f) = e^{-n_f} \tag{16}$$

$$E(p_{isol}) = e^{-E(n_f)} = e^{-c_1 \rho N} \tag{17}$$

3.4 The access model

Even if a node has available relay, it does not mean the node could always transmit the message successfully. With fading and shared wireless channels, a competitive access should be necessarily either in fully self-organized sytems or partially self-organized system. Therefore, a node with sending probability a does not mean has the accessable probability a. Assumed that the whole networking is synchronous as IEEE 802.11 DCF (Pham, Pham et al. 2005; Samhat, Samhat et al. 2006; Khayyat, Gebali et al. 2007), and the nodes have the same probability to send. Thus the collision of each-hop is independent and has the same probability distribution. In any case, assumed each node could send the message successfully with probability u, while the sending probability is a, with some backoff algorithm. Thus the successfully probability of a n hop chain is:

$$p_f = u^n \tag{18}$$

The mathematical expectation of p_f is:

$$E(p_f) = \sum_{k=1}^{\rho_D \pi (r+\Delta)^2} \frac{e^{-n_{cell}} {n_{cell}}^k}{k!} (u)^k \tag{19}$$

Where take $\lambda = n_{cell} = \rho_D \pi r^2 = \rho N$. Considerate the collision probability will increase rapidly with the density of the nodes, in this case $u \bullet n_{cell}$ will be smaller.

$$E(p_f) \approx e^{-\rho \bullet N} \left(e^{u \bullet \rho \bullet N} - 1 \right) \tag{20}$$

while $\rho_D \pi (r+\Delta)^2 \geq 5$.

4. Asymptotic capacity model on multi-hop systems

4.1The capacity model

Say the traffic over the j-th sub-channel has $h_{i,j}$ hops. Derived from the throughput definition in (Gupta & Kumar 2000), the average capacity of each node can be defined as:

$$C_{X(i)} = \sum_{j}^{N_{ch}(i)} \left\{ \prod_{hop=1}^{h_{i,j}} [(1-p_{isol,hop}) p_{f,hop}] \right\} \omega_{i,j} / h_{i,j} \tag{21}$$

Thus:

$$E\left(C_{X(i)}\right)=E\left\langle \sum_{j}^{N_{ch}(i)}\left\{\prod_{hop=1}^{h_{i,j}}[(1-p_{isol,hop})p_{f,hop}]\right\}\omega_{i,j}/h_{i,j}\right\rangle$$
$$=\sum_{j}^{N_{ch}(i)}\left\{\prod_{hop=1}^{h_{i,j}}\left\{\left[1-E\left(p_{isol,hop}\right)\right]E\left(p_{f,hop}\right)\right\}\right\}\omega_{i,j}/h_{i,j} \quad (22)$$
$$=\sum_{j}^{N_{ch}(i)}[(1-E(p_{isol}))E(p_f)]^{h_{i,j}}\omega_{i,j}/h_{i,j}$$

For multiple sub-channel just provide more QoS with more complexity without more avaiable capability, the capacity formula could be simplified as single channel:

$$E(C_{X(i)})=[(1-E(p_{isol}))E(p_f)]^{h_i}\omega/h_i \quad (23)$$

4.2 The upper bound on capacity for unicast traffics

Derived from "arbitrary networks" in (Gupta & Kumar 2000) and formula (23), the upper bound capacity on the ideal unicast traffics happens to be while each node just communicates to the one hop neighbors, $h_{ij}=1$, and has maximum $N/2$ communication pair, obtain:

$$E(C)=\sum E(C_{X(i)})=\frac{N}{2}(1-E(p_{isol}))E(p_f)\omega \quad (24)$$

And the normalized capacity is:

$$S=\frac{E(C)}{N\omega}=\frac{1}{2}(1-E(p_{isol}))E(p_f) \quad (25)$$

4.3 The upper bound on capacity for broadcast traffics

Case broadcast traffics, in a networks with N nodes, the N nodes received the same message from the same source, thus the average efficiency almost is ω/N when N is large enough. The upper bound on capacity for broadcast traffic is:

$$\arg\max[E(C)]=\arg\max\left[\sum_i E(C_{X(i)})\right]$$
$$=\frac{1}{N}\arg\max\left\{\sum_i\left[(1-E(p_{isol}))E(p_f)\right]^{h_{i,j}}\omega_i\right\} \quad (26)$$

Say D is the radius of the area covered WMN; define $M=\lceil D/r\rceil$. For simplify analysis, say D is divided exactly by r, thus M=D/r. As in fig 5, the nodes covering the k=0 circle just needs one hop to the AP; the nodes covering the k=1 ring needs at least two hops. Thus the nodes covering the k ring, k<=M, need at least k+1 hops. It is obviously that the number of nodes in the k ring is:

$$N_k = (2k+1)(r/D)^2 N = (2k+1)N/M^2 \tag{27}$$

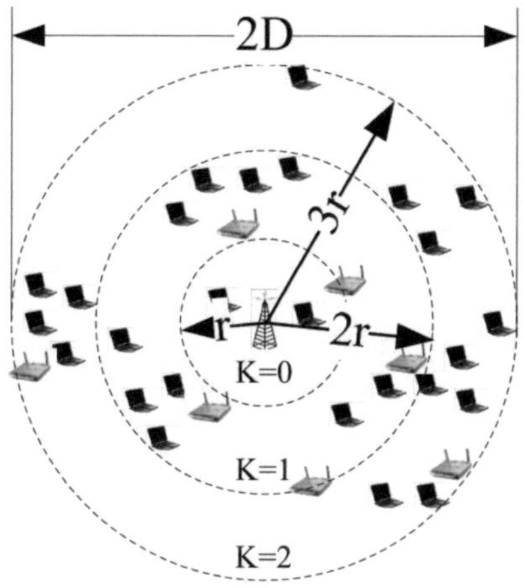

Fig. 5. A scenario for broadcast traffics, case M=3

If the number of AP is 1,

$$\begin{aligned}\max[E(C)] &= \frac{\omega}{N}\left\{\sum_{k=0}^{M}((2k+1)N/M^2)[(1-E(p_{isol}))E(p_f)]^{k+1}\right\} \\ &= \frac{\omega}{M^2}\left\{\sum_{k=0}^{M}(2k+1)[(1-E(p_{isol}))E(p_f)]^{k+1}\right\}\end{aligned} \tag{28}$$

And the normalized capacity is

$$S = \frac{\max[E(C)]}{N\omega} = \frac{1}{NM^2}\left\{\sum_{k=0}^{M}(2k+1)[(1-E(p_{isol}))E(p_f)]^{k+1}\right\} \tag{29}$$

If there are N_A APs, for each AP, similarly get

$$N_{k,R} = (2k+1)N/{M_A}^2 N_A \quad ,k = 0,1,2,...,M_A \tag{30}$$

$$\begin{aligned}\max[E(C)] &= \frac{\omega}{N}\max\left\{N_A\sum_{k=0}^{M_R}N_{k,R}[(1-E(p_{isol}))E(p_f)]^{k+1}\right\} \\ &= \frac{\omega}{{M_A}^2}\bullet\max\left\{\sum_{k=0}^{M_R}(2k+1)[(1-E(p_{isol}))E(p_f)]^{k+1}\right\}\end{aligned} \tag{31}$$

$$S = \frac{\max[E(C)]}{N\omega} = \frac{1}{N{M_A}^2}\bullet\max\left\{\sum_{k=0}^{M_R}(2k+1)[(1-E(p_{isol}))E(p_f)]^{k+1}\right\} \tag{32}$$

4.4 The upper bound on capacity of backhual traffics

For the backhaul traffics, each multi-hop chains has the same capacity ω / h ,thus:

$$\max[E(C)] = \arg\max\left\{\sum_i(\sum_j^{N_{ch}(i)}[(1-E(p_{isol}))E(p_f)]^{h_{i,j}}\omega_{i,j}/h_{i,j})\right\} \tag{33}$$

Similar say $M = D / r$ is constant, If there are 1 mesh routers obtaions:

$$\max[E(C)] = \frac{N\omega}{M^2}\left\{\sum_{k=0}^{M}(2k+1)[(1-E(p_{isol}))E(p_f)]^{k+1}/(k+1)\right\} \tag{33}$$

$$S = \frac{\max[E(C)]}{N\omega} = \frac{1}{M^2}\left\{\sum_{k=0}^{M}(2k+1)[(1-E(p_{isol}))E(p_f)]^{k+1}/(k+1)\right\} \tag{34}$$

If there are multiple routers:

$$\max[E(C)] = \frac{\omega N}{M_A{}^2}\bullet\left\{\sum_{k=0}^{M_R}(2k+1)[(1-E(p_{isol}))E(p_f)]^{k+1}/(k+1)\right\} \tag{35}$$

$$S = \frac{\max[E(C)]}{N\omega} = \frac{1}{M_A{}^2}\bullet\left\{\sum_{k=0}^{M_R}(2k+1)[(1-E(p_{isol}))E(p_f)]^{k+1}/(k+1)\right\} \tag{36}$$

5. Conclusion

Say $E(p_{isol})$ is constant,which the density of a networks is cnostant, the capacity on a network is decided by the access probability. With (20) , to get the extremum, obstain:

$$\frac{d\left[E(p_f)\right]}{dv} = -\rho \bullet N \bullet e^{v\bullet\rho\bullet N} \neq 0 \tag{37}$$

$$\frac{d\left[E(p_f)\right]}{d\rho} = -v \bullet N \bullet e^{-v\bullet\rho\bullet N} + N \bullet e^{-\rho\bullet N} \tag{38}$$

$$\frac{d\left[E(p_f)\right]}{dN} = -v \bullet \rho \bullet e^{-v\bullet\rho\bullet N} + \rho \bullet e^{-\rho\bullet N} \tag{39}$$

(38) and (39) leads the same conclusion:

$$\rho \bullet N = \frac{\ln v}{v-1} \tag{40}$$

While

$$E(p_f) = e^{-\rho\bullet N}\left(e^{u\bullet\rho\bullet N} - 1\right) = e^{-v\bullet\frac{\ln v}{v-1}} - e^{-\frac{\ln v}{v-1}} \tag{41}$$

The relationship of (40) is shown in fig 6

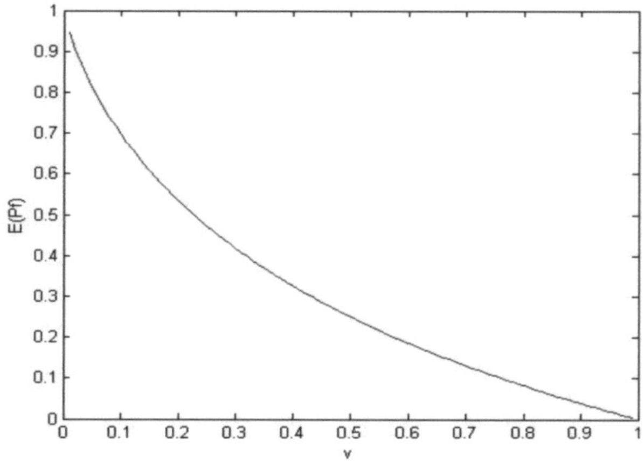

Fig. 6. $E(p_f) - v$ relationship

6. References

Akyildiz, I. F. and W. Xudong (2005). "A survey on wireless mesh networks." Communications Magazine, IEEE 43(9): S23-S30.

Basagni, S., D. Turgut, et al. (2001). Mobility-adaptive protocols for managing large ad hoc networks. Communications, 2001. ICC 2001. IEEE International Conference on.

Berrou, C., A. Glavieux, et al. (1993). Near Shannon limit error-correcting coding and decoding: Turbo-codes. . Communications, 1993. ICC 93. Geneva. Technical Program, Conference Record, IEEE International Conference on.

Bharghavan, V., A. Demers, et al. (Aug. 1994). MACAW: A media access protocol for wireless LANs. Proc. SIGCOMM'94 Conf. on Communications Architectures, Protocols and Applications.

Cadambe, V. R. and S. A. Jafar (2007). Degrees of Freedom of Wireless Networks - What a Difference Delay Makes. Signals, Systems and Computers, 2007. ACSSC 2007. Conference Record of the Forty-First Asilomar Conference on.

Cadambe, V. R. and S. A. Jafar (2008). Can feedback, cooperation, relays and full duplex operation increase the degrees of freedom of wireless networks? Information Theory, 2008. ISIT 2008. IEEE International Symposium on.

Cadambe, V. R. and S. A. Jafar (2009). "Degrees of Freedom of Wireless Networks With Relays, Feedback, Cooperation, and Full Duplex Operation." Information Theory, IEEE Transactions on 55(5): 2334-2344.

Chen, Y., G. Zhu, et al. (2008). On the Capacity and Scalability of Wireless Mesh Networks. Wireless Communications, Networking and Mobile Computing, 2008. WiCOM '08. 4th International Conference on.

Diggavi, S. N., M. Grossglauser, et al. (2005). "Even One-Dimensional Mobility Increases the Capacity of Wireless Networks." Information Theory, IEEE Transactions on 51(11): 3947-3954.

E.Telatar (1999). "Capacity of Multi-Antenna Gaussian Channels." European Transactio on Telecommunications 10(6): 585-595.

G.J.Foschini (1996). "Layered Space-Tie Architecture for Wirless Communicaton in a Fading Environment when Using Multi-Element Antennas." Bell Labs Technology Journal 1(2): 41-59.

Gallager, R. (1985). "A perspective on multiaccess channels." Information Theory, IEEE Transactions on 31(2): 124-142.

Goldsmith, A. (2005). wireless commnunications, Cambridge University Press.

Grossglauser, M. and D. N. C. Tse (2002). "Mobility increases the capacity of ad hoc wireless networks." Networking, IEEE/ACM Transactions on 10(4): 477-486.

Gupta, P. and P. R. Kumar (2000). "The capacity of wireless networks." Information Theory, IEEE Transactions on 46(2): 388-404.

Haartsen, J. C. (2000). "The Bluetooth radio system." Personal Communications, IEEE 7(1): 28-36.

Jafar, S. A. (2005). "Too much mobility limits the capacity of wireless ad hoc networks." Information Theory, IEEE Transactions on 51(11): 3954-3965.

Jangeun, J. and M. L. Sichitiu (2003). "The nominal capacity of wireless mesh networks." Wireless Communications, IEEE [see also IEEE Personal Communications] 10(5): 8-14.

Karn, P. (Sept.1990). MACA: A new channel access method for packet radio. Proc. 9th Computer Networking Conf.

Khayyat, K. M. J., F. Gebali, et al. (2007). Performance Analysis of the IEEE 802.11 DCF. Signal Processing and Information Technology, 2007 IEEE International Symposium on.

Kumar, P. R. (2003). "A correction to the proof of a lemma in "The capacity of wireless networks"." Information Theory, IEEE Transactions on 49(11): 3117.

Miorando, E. and F. Granelli (2007). On Connectivity and Capacity of Wireless Mesh Networks. Communications, 2007. ICC '07. IEEE International Conference on.

Nandiraju, N., D. Nandiraju, et al. (2007). "Wireless Mesh Networks: Current Challenges and Future Directions of Web-In-The-Sky." Wireless Communications, IEEE [see also IEEE Personal Communications] 14(4): 79-89.

Pham, P. P., P. P. Pham, et al. (2005). Performance Analysis of the IEEE 802.11 DCF. Communications, 2005 Asia-Pacific Conference on.

Samhat, A., A. Samhat, et al. (2006). Performance analysis of the IEEE 802.11 DCF with imperfect radio conditions. Wireless and Mobile Communications, 2006. ICWMC '06. International Conference on.

T.M.Cover and J.A.Thomas (2006). Elements of Information Theory (second edition), John Wiley & Sons.

Takagi, H. and L. Kleinrock (1984). "Optimal Transmission Ranges for Randomly Distributed Packet Radio Terminals." Communications, IEEE Transactions on [legacy, pre - 1988] 32(3): 246-257.

Tavli, B. (2006). "Broadcast capacity of wireless networks." Communications Letters, IEEE 10(2): 68-69.

8

The Performance of Wireless Mesh Networks with Apparent Link Failures

Geir Egeland[1], Paal E. Engelstad[2], and Frank Y. Li[3]

[1]*Department of Electrical and Computer Engineering, University of Stavanger*
[2]*Simula and Telenor Corporate Development, University of Oslo*
[3]*Department of Information and Communication Technology, University of Agder*
Norway

1. Introduction

A wireless multi-hop network is a network consisting of a group of nodes interconnected by the means of wireless links. The nodes in such a network, which are often self-configured and self-organized, communicate with each other over multiple hops through a routing protocol. Examples of such networks include Wireless Mesh Networks (WMNs) IEEE802.11s (2010), Mobile Ad Hoc Networks (MANETs) Chlamtac et al. (2003) and Wireless Sensor Networks (WSNs) Gharavi & Kumar (2003). The performance and the reliability of these networks depend heavily on the routing protocol's capability to detect link failures between neighboring nodes as well as its link-maintenance mechanism to recover a path from source to destination when a link-failure happens.

While MANETs generally appear more dynamic due to node mobility, the network topology for WMNs and WSNs remains comparatively stable. No matter which network form is concerned, however, these networks exhibit ad hoc features since wireless links are intrinsically unreliable. In the majority of cases, link failures are present in a multi-hop network regardless of the use of link-maintenance mechanisms. Sometimes link failures are unavoidable, such as when a mobile node deliberately leaves a network or is subject to the exhaustion of its battery power. In another case a link would cease to be operative when two nodes move outside each others' radio transmission range. In addition to these, a set of link failures which we refer to as *apparent link-failures* exist. They are primarily caused by radio links being vulnerable to radio induced interference, but also appear when a link-maintenance mechanism erroneously assumes a link to be inoperable due to *loss of beacons*. A beacon is a short packet transmitted periodically to a node's one-hop neighbors and its purpose is to detect neighbors and to keep links alive. Beacons are normally broadcast, and are thus not acknowledged, i.e. they are unreliable and vulnerable to overlapping transmissions from hidden nodes Tobagi & Kleinrock (1975). Moreover, common protection mechanisms against hidden nodes (such as RTS/CTS of the IEEE 802.11 MAC protocol IEEE802.11 (1997)) are not applicable, since unicast data transmission using RTS/CTS will only provide protection for packet reception at the node that issued the CTS.

1.1 Motivation and methods

Although a huge number of efforts have been made in the research community during the past decade on various facets of wireless multi-hop networks, little attention has been paid

to the reliability aspect of such networks. In this chapter, we propose an analytical model for apparent link-failures in static mesh networks where the location of each node is carefully planned (referred to hereafter as *planned mesh network*). A planned mesh network typically appears as a consequence of the high costs associated with interconnecting nodes in a network with wired links. For example, ad hoc technology can in a cost-efficient manner, extend the reach of a wired backbone through a wireless backhaul mesh network. Apparent link-failures are often a significant cause for performance degradation of mesh networks, and thus a model is needed in order to diminish their effect. For instance, with a model in place it is possible to detect and avoid undesirable topologies that might lead to a high frequency of such failures.
The proposed model makes use of the assumption that the probability of losing a beacon due to a packet collision with transmissions from hidden nodes (p_e), is much larger than the probability of losing beacons due to transmissions from one-hop neighbors (p_{coll}). The probability that a receiving node considers a link to be inoperative at the time a beacon is expected, is then estimated through analysis using a Markov model. Furthermore, an algorithm which is used for determining the number of hidden nodes and the associated traffic pattern is introduced so that the model can be applied to arbitrary topologies.

1.2 Significance of our results

By avoiding poorly planned topologies, not only the reliability of mesh networks can be increased, but also the general performance of such networks can be improved. Apparent link-failures are often a significant cause for performance degradation of ad hoc networks since erroneous routing information may be spread in the network when apparent link-failures happen. Also, it might lead to a disconnected topology or less optimal routes to a destination. Analysis of a real life network Li et al. (2010) has demonstrated that it takes a significant amount of time to restore failed links Egeland & Li (2007). An example of the effect of these failures is illustrated in Fig. 1. Using a well known network simulator ns2 (2010) we have measured the throughput from node $d_8 \rightarrow d_7$ in the topology shown in Fig. 1(a). As the load from the hidden nodes increases, the throughput from node $d_8 \rightarrow d_7$ is reduced, because the routing protocol forces the data packets to traverse longer paths in order to bypass the apparent link-failure or simply because node d_7 drops packets when buffers are filled as a result of having no operational route to node d_8. The throughput would remain relatively stable if the apparent link-failures were eliminated, as seen from the "No apparent link failure" graph in Fig. 1(b).
The model presented in this chapter allows a node to calculate the probability of losing connectivity to its one-hop neighbors caused by beacon loss. Utilizing the model, we demonstrate how a node in a mesh network operated on the *Optimized Link State Routing* (OLSR) Clausen & Jacquet (2003) routing protocol can apply the apparent link-failure probability as a criterion to decide when to unicast and when to broadcast beacons to surrounding neighbors, thus improving the packet delivery capability.

1.3 Related work

In Voorhaen & Blondia (2006) the performance of neighbour sensing in ad hoc networks is studied, however, only parameters such as the transmission frequency of the Hello-messages and the link-layer feedback are covered. In Ray et al. (2005) a model for packet collision and the effect of hidden and masked nodes are studied, but only for simple topologies, and the work is not directly applicable to the Hello-message problem. The work in Ng & Liew (2004) addresses link-failures in wireless ad hoc networks through the effect of routing instability.

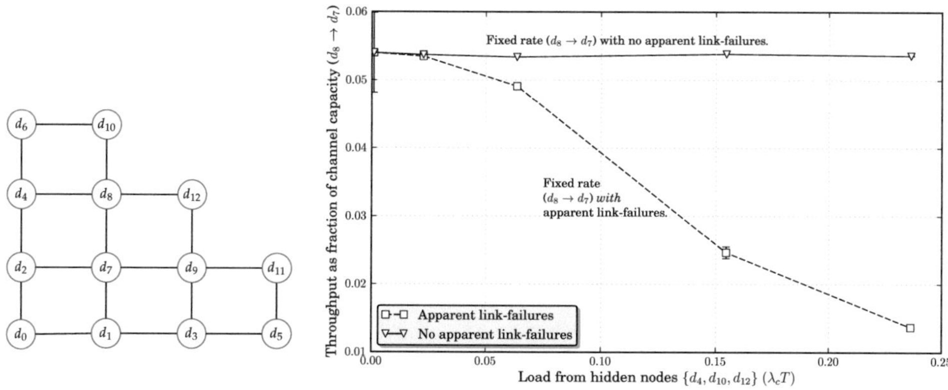

(a) Example topology (b) Throughput from node $d_8 \to d_7$

Fig. 1. Performance with and without apparent link-failures. The possibility of apparent link failures is artificially removed by not allowing the links to time out when beacons are lost.

Here the authors study the throughput of TCP/UDP in networks where the routing protocol falsely assumes a link is inoperable. However, what causes a link to become unavailable to the routing protocol is not studied. A model for packet collision and the effect of hidden and masked nodes are studied in Ray et al. (2004), but only for simple topologies, and the work is not directly applicable to loss of beacons. Not much published work relates directly to the modeling of apparent link-failures caused by loss of beacons. In Egeland & Engelstad (2009) the reliability and availability of a set of mesh topologies are studied using both a distance-dependent and a distance-independent link-existence model, but the effects of beacon-based link maintenance and hidden nodes are ignored. Here it is assumed that apparent link-failures are a result of radio-induced interference only. The work in Gerharz et al. (2002) studies the reliability of wireless multi-hop networks with the assumptions that link-failures are caused by radio interference.

2. Network model

2.1 Network terminology

This chapter reuses the terminology of wireless mesh networks in order to describe the architecture of a planned mesh network, more specifically of the IEEE 802.11s specification IEEE802.11s (2010) of mesh networks. In this terminology a node in a mesh network is referred to as a *Mesh Point* (MP). Furthermore, an MP is referred to as a *Mesh Access Point* (MAP) if it includes the functionality of an 802.11 access point, allowing regular 802.11 Stations (STAs) access to the mesh infrastructure. When an MP has additional functionality for connecting the mesh network to other network infrastructures, it is referred to as a *Mesh Portal* (MPP). A mesh network is illustrated in Fig. 2.

A mesh network can be described as a graph $G(V,E)$ where the nodes in the network serve as the vertices $v_j \in V(G)$. Any two distinct nodes v_j and v_i create an edge $\epsilon_{i,j} \in E(G)$ if there is a direct link between them. In order to provide an adequate measure of network reliability, the use of probabilistic reliability metrics and a probabilistic graph is necessary. This is an undirectional graph where each node has an associated probability of being in an operational state, and similarly for each edge, i.e. the random graph $G(V,E,p)$ where p is

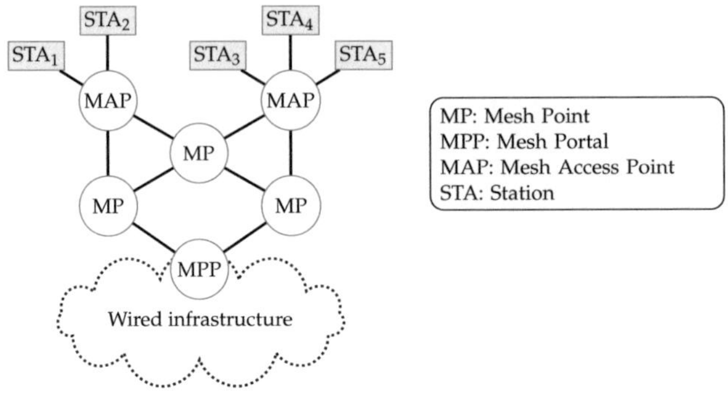

Fig. 2. A wireless mesh network connected to a fixed infrastructure.

the link-existence probability. An underlying assumption in the analysis is that the existence of a link is determined independently for each link. This means that the link $\epsilon_{s,d}$ may fail independently of the link $\epsilon_{i,j} \in E(G) \setminus \{\epsilon_{s,d}\}$. As the link failure probability in general is much higher than the node failure probability, it is natural to model the nodes $v_j \in V(G)$ in the topology as invulnerable to failures. Thus, a mesh network can be described and analyzed as a random graph.

2.2 Link maintenance using beacons

In a multi-hop network, links are usually established and maintained proactively by the use of one-hop beacons which are exchanged between neighboring nodes periodically. Beacons are broadcast in order to conserve bandwidth, as no acknowledge messages are expected from the receivers of these beacons. Thus, the link status of every link on which a beacon is received can be effectively obtained through beacon transmissions. Since broadcast packets are not acknowledged, beacons are inherently unreliable. A node anticipates to receive a beacon from a neighbor node within a defined time interval and can tolerate that beacons occasionally will be missing due to various error events like channel fading or packet collision. However, a node failed to receive a number of $(\theta+1)$ consecutive beacons will accredit that the node on the other side of the link is permanently unreachable and that the link is inoperable. The value of the configurable parameter θ is a tradeoff between providing the routing protocol with stable and reliability links (a large θ), and the ability to detect link-failures in a timely and fast manner (a small θ). Since beacons are broadcast, they are unable to take the advantage of the Request-To-Send/Clear-To-Send (RTS/CTS) signaling that protects the IEEE 802.11 MAC protocol's IEEE802.11 (1997) unicast data transmission against hidden nodes. Although some beacon loss is avoided using RTS/CTS for the *unicast data traffic* in the network, it will only affect the links of the node that issues the CTS. The consequence is that beacons will be susceptible to collisions with traffic from hidden nodes *even if* RTS/CTS is enabled. Thus, the utilization of a link may be prevented if the link is assumed to be inoperable due to beacon loss. Examples of routing protocols that make use of beacons are the proactive protocol OLSR Clausen & Jacquet (2003) and an optional mode of operation for the reactive *Ad hoc On-Demand Distance Vector* (AODV) routing protocol Perkins et al. (2003).

A major difference between various beacon-based schemes is how the routing protocol determines if a failed link is operational again. Stable links are desirable, and introducing a link too early can lead to a situation where a link oscillates between an operational and a non-operational state. A solution that avoids this situation is by measuring the Signal-to-Noise Ratio (SNR) of the failed link and define the link as operational only when

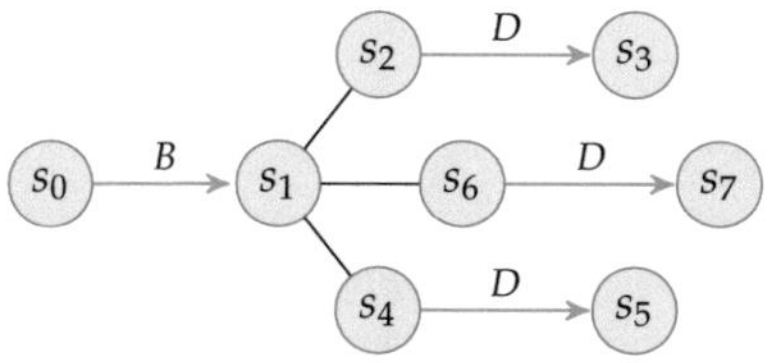

(a) Isolated hidden nodes

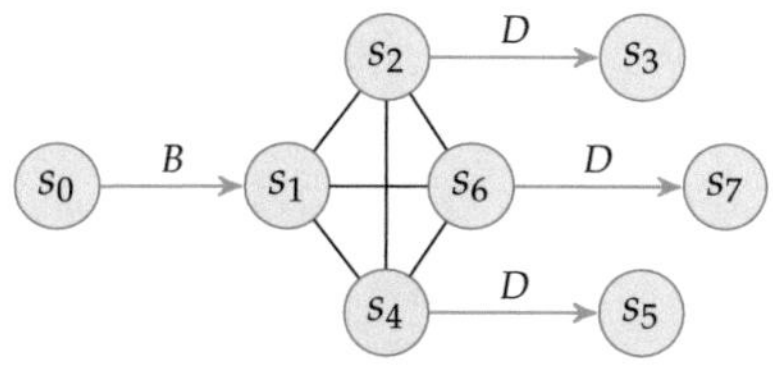

(b) Connected hidden nodes

Fig. 3. Sample topologies where the hidden nodes $\{s_2, s_4, s_6\}$ are isolated or connected. When the hidden nodes send data (D), this may collide with the beacons (B) sent by node s_0.

both beacons are being received and the received SNR is above a defined threshold Ali et al. (2009). However, if SNR measurement is not available or not practical, a simple solution is to introduce some kind of hysteresis by requiring a number of consecutive beacons to be received ($\theta_h + 1$) before the link is assumed to be operational. This is the solution chosen in this analysis.

3. Apparent link-failures due to beacon loss

3.1 Assumptions for the beacon-based link maintenance

Before we can determine the apparent link-failure probability, a model for identifying losing a single beacon caused by overlapping transmissions must be found. In order to simplify the analysis, the model is based upon three assumptions. First, it is assumed that a beacon sent by a node has a negligible probability of colliding with a beacon from any of the neighboring nodes. This is a fair assumption, since beacons are short packets that are transmitted periodically and at a random instant at a relatively low rate. Secondly, it is assumed that the probability of a beacon colliding with a data transmission from any of the (non-hidden) neighboring nodes also is negligible, i.e. $p_e \gg p_{coll}$. This assumption is also fair, since a MAC layer often has mechanisms that reduce such collisions to a minimum. Examples of such mechanisms are the collision avoidance scheme of the IEEE 802.11 MAC protocol with randomized access to the channel after a busy period, and the carrier- and virtual sense of the physical layer. Accordingly to the IEEE 802.11 standard, a beacon will be deferred at the transmitter if there is ongoing transmission on the channel. Therefore, the probability that beacons are lost, is a result of *overlapping data packet transmissions from hidden nodes only.* Thirdly, we make the assumption that the packet buffers of a node can be modeled as an $M/M/1$ queue Kleinrock (1975) and that the packet arrival rate is Poisson distributed with parameter λ_c and that the channel access and data packet transmission times are exponential distributed with parameter $1/\mu$.

These assumptions allow us to verify the model in a simple manner. Even though traffic in a real network may follow other distributions, the results presented later in the chapter suggest that the assumptions are fair. The bounds for beacon loss probability based on a large number of random independent traffic scenarios will be presented, and these capture more of the characteristics of the traffic in a real-life network.

3.2 Probability of losing a beacon p_e

Consider the topology in Fig. 3(a). We need to find firstly the probability (p_e) that the beacon from s_0 and a data packet from the hidden node s_2 collide. Let $q_{s_2}(0)$ denote the probability of

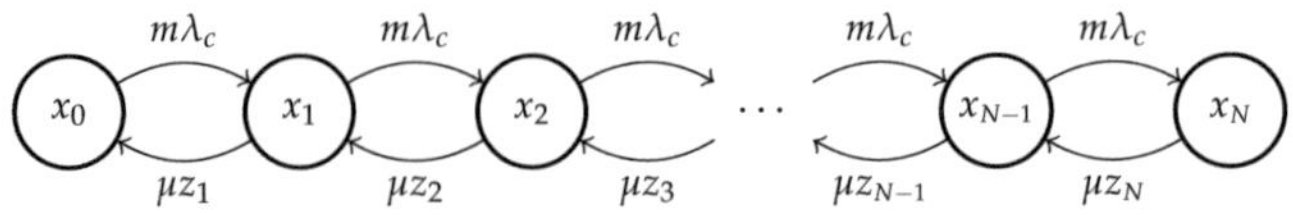

Fig. 4. A Markov model of the total number of packets waiting to be transmitted by the m hidden nodes, where λ_c is the packet arrival rate, $1/\mu$ is the service time and z_n is the average number of the m hidden nodes transmitting simultaneously.

node s_2 having zero packets awaiting in its buffer. p_e can be expressed as Dubey et al. (2008):

$$\begin{aligned} p_e &= \Pr\{\text{Collision}|q_{s_2}(0) > 0\} \cdot \Pr\{q_{s_2}(0) > 0\} \\ &+ \Pr\{\text{Collision}|q_{s_2}(0) = 0\} \cdot \Pr\{q_{s_2}(0) = 0\} \\ &= (1 - p_0) \cdot 1 + (1 - e^{-\lambda_c \omega_b / T_p}) \cdot p_0 \end{aligned} \tag{1}$$

where p_0 is the probability that the hidden node s_2 has zero packets awaiting to be transmitted. The parameters T_p and ω_b represent the average transmission time of the data packet and of the beacon packet, respectively. Both these transmission times are assumed to be exponentially distributed. The probability that a node has i data packets in its packet queue is given by $p_i = (1 - \rho)\rho^i$, where $\rho = \lambda_c/\mu$, thus $p_0 = 1 - \rho$ Kleinrock (1975).

3.2.1 Isolated hidden nodes

We will now evaluate the probability that a beacon collides with data transmissions from a set of hidden nodes using the topology illustrated in Fig. 3(a). In this sample topology, the hidden nodes are assumed to be *isolated*, i.e. outside the transmission range of each other. Individually, the probability that one of them sends a data packet which overlaps with a beacon from node s_0 is given by Eq. (1) (denoted p_e). The number of data packets from $\{s_2, s_4, s_6\}$ overlapping with a beacon from s_0 is binomially distributed $B(m, p_e)$ where m is the number of hidden nodes. The probability that a beacon is lost can then be expressed as:

$$p_e^I = \sum_{k=1}^{m} \binom{m}{k} p_e^k (1 - p_e)^{m-k}. \tag{2}$$

3.2.2 Connected hidden nodes

In Fig. 3(b) the hidden nodes are all within radio transmission range of each other. When all the hidden nodes are connected, the calculation of the beacon loss probability is not as straightforward, and we need to make further simplified assumptions. Firstly, it is assumed that the nodes access the common channel according to a *1-persistent* CSMA protocol Kleinrock & Tobagi (1975). This might seem like a contradiction, since it was stated earlier that we assumed a MAC protocol that reduces the collisions between non-hidden neighbours to a minimum. However, for the case where the hidden nodes are connected, there will be a parameter (z_n) in the model that can be set to control to which extent transmissions between the hidden nodes are permitted to collide with each other. Secondly, it is assumed that the arrival rates at the different hidden nodes are not coupled, hence a Markov model can be used for the analysis.

Consider the Markov chain illustrated in Fig. 4. Each state represents the sum of all packets queuing up in the m hidden nodes. Here z_n is the average number of hidden nodes transmitting when a total of n packets are distributed amongst the hidden nodes.

We are now able to find the probability of being in state x_0, which is the case for which none of the hidden nodes have packets awaiting transmission (p_0^C). Using standard queuing theory Kleinrock (1975), it can easily be shown that this probability is given by:

$$p_0^C = \left[1 + \sum_{i=1}^{N} (m\rho)^i \left(\prod_{n=1}^{i} z_{n,i}\right)^{-1}\right]^{-1}, \quad \rho = \frac{\lambda_c}{\mu} \tag{3}$$

where $z_{n,i}$ is the average number of the m nodes transmitting simultaneously and is calculated according to:

$$z_n = \begin{cases} \dfrac{\sum_{k=1}^{n} k\binom{m}{k}\binom{n-1}{k-1}(1-\rho^m)}{\sum_{k=1}^{n}\binom{m}{k}\binom{n-1}{k-1}} & n<m, \; \rho=\lambda_c/\mu \\ \dfrac{\sum_{k=1}^{m-1} k\binom{n-1}{k-1}(1-\rho^m)}{\sum_{k=1}^{m-1}\binom{n-1}{k-1}} + m\rho^m & n\geq m, \; \rho=\lambda_c/\mu. \end{cases} \tag{4}$$

The probability that one or more of the m nodes having zero packets in its buffer, given the sum of packets in the buffers is n, is given by the term $1-\rho^m$ in Eq. (4). The combinations of k of m buffers containing packets, constrained by a total sum of n packets is given by $\binom{n-1}{k-1}$. By substituting p_0 in Eq. (1) with p_0^C (Eq. (3)), the probability that transmissions from the connected hidden nodes overlap with a beacon can be calculated as:

$$p_e^C = 1 - p_0^C \cdot e^{-\lambda_c \omega_b / T_p}. \tag{5}$$

Before attempting to model more complex traffic patterns, i.e. arbitrary packet flows between different nodes, we must ensure that the basic model is capturing all possible transmission configurations. In fact, the initial model did not take into account the possibility that a neighbouring node receiving the beacon could be transmitting any data packets. Therefore, an approximate model will be provided, where the channel access time of the neighbouring node receiving the beacon is also taken into account. This model will be used in the next sub-section when random traffic patterns is analysed.

Again, consider the sample topology illustrated in Fig. 3(a). Let us assume that node s_1 has a traffic load with the rate λ_c and the probability that it gains access to the channel in order to transmit a packet is p_{s_1}. If the nodes $\{s_1, s_2, s_4, s_6\}$ are modelled as M/M/1 queues, the probability that e.g. node s_2 has no packets in its buffer can be expressed as:

$$q_{s_2}(0) = \left[1 + \sum_{k=1}^{N} \left(\frac{\rho}{1-\rho p_{s_1}}\right)^k\right]^{-1}, \rho = \lambda_c/\mu. \tag{6}$$

An approximate expression for p_{s_1} is the probability that none of the neighbour nodes of s_1 have a packet in its buffer. The probability p_{s_1} is then given by $\prod_{i\in\{2,4,6\}} q_{s_i}(0)$ and can now be written as:

$$p_{s_1} \approx \left[1 + \sum_{k=1}^{N} \left(\frac{\rho}{1-\rho p_{s_1}}\right)^k\right]^{-m} \tag{7}$$

where solutions for p_{s_1} can be found numerically and $m = |\{s_2, s_4, s_6\}|$. For the case of isolated hidden nodes in Fig. 3(a), the parameter p_0 in Eq. (1) can now be expressed as $q_{s_i}(0)$ in Eq. (6).

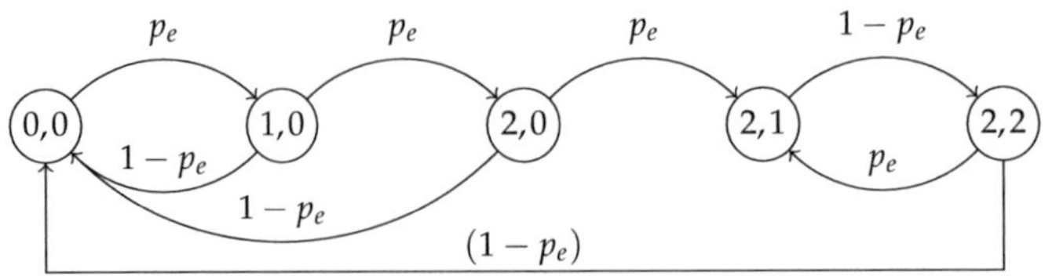

Fig. 5. A Markov model of a link-sensing mechanism with θ=2 and θ_h=1. The probability of losing a single beacon (p_e) is random and independent.

For the connected hidden nodes in Fig. 3(b), the probability p_{s_1} is equal to $1/(m+1)$, since each of the $m+1$ nodes gets an equal share of the common channel. Thus, p_0^C is rewritten as:

$$p_0^C = \left[1 + \sum_{i=1}^{N} (m\rho)^i \left(\prod_{n=1}^{i} z_{n,i} \left[1 - \frac{1}{m+1}\right]^i \right)^{-1} \right]^{-1}. \tag{8}$$

When the hidden nodes are connected, i.e. within each others transmission range, a packet arriving at one of the hidden nodes might have to wait until an ongoing transmission is finished before it is transmitted. When all the buffers are filled, the m hidden nodes will transmit simultaneously after an ongoing transmission is finished, thus emptying the buffers at a rate of $m{\cdot}\mu$. If we however change the model for the connected case, and enforce that the hidden nodes access the channel once at a time, the rate of emptying the buffers of the hidden nodes is reduced to μ, and can be calculated using Eq. (8) with z_n=1$\forall n$. The model will now resemble the IEEE 802.11 MAC protocol, which has mechanisms that aim to reduce collisions on the channel to a minimum. This will represent an *upper bound* for the beacon loss probability. We can now use the beacon loss probabilities in Eqs. (1)–(8) to calculate the link-failure probability p_f.

3.3 A model for apparent link-failures

If we assume that the event of losing a beacon is random and independent, apparent link-failures can be analyzed using a Markov model as shown in Fig. 5 where the state variable $s_{i,j}$ describes the number of $i\in[0,\theta]$ beacons lost and $j\in[0,\theta_h]$ the number of beacons received in the hysteresis state. Solving the state equation in the model, it is easy to show that the probability of apparent link-failure (p_f) is the sum of the state probabilities $\sum_{j=1}^{\theta_h} p_{i,j}$. Thus, p_f can be expressed as:

$$p_f = \frac{(2-p_e)p_e^3}{(p_e^3 - p_e + 1)} \tag{9}$$

where p_e is the probability of losing a single beacon.

3.4 Analysis of the model's performance

In order to test the model's accuracy, a discrete-event simulation model was used. The simulator can model a two-dimensional network where every node transmits with the same power on the same channel. The sensing range (r_{cp}) of the physical layer is equal to the transmission range (r_{rx}). Even though this is not the case in a real-life network, it simplifies our analysis and provides to certain extent of topology control. Every node experiences the same path loss versus distance and has the same antenna gain and receiver sensitivity. A node receives a packet correctly only if the packet does not overlap with any other packet

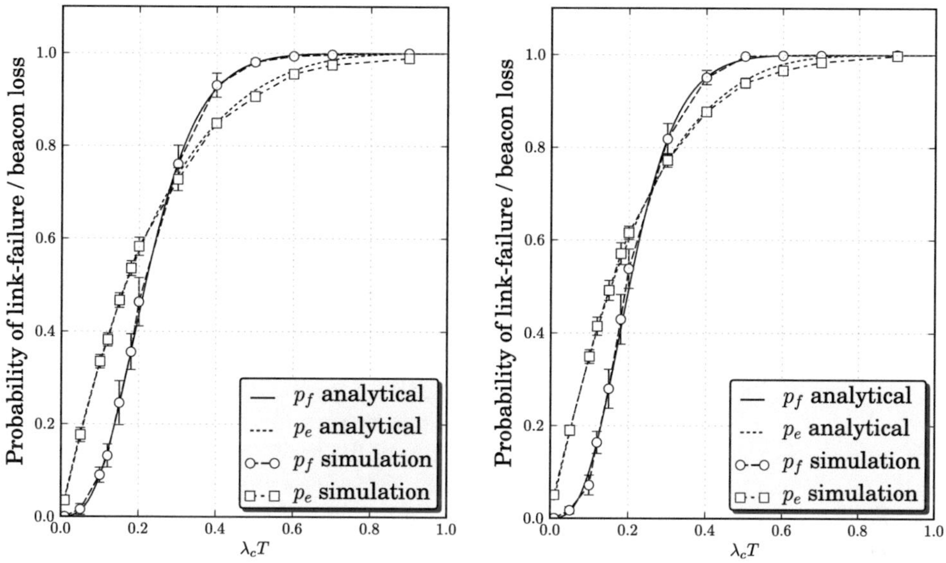

(a) Results for Fig. 3(a)

(b) Results for Fig. 3(b)

Fig. 6. The probability of losing a beacon (p_e) and the probability of link-failure (p_f) for the topologies in Fig. 3. The simulation results are shown with a 95% confidence interval.

IP/MAC layer	Values	Physical layer	Values	Simulation	Values
Beacon/ Data	30/ 100 bytes	Propagation model	Free Space	Simulation/ transient time	900s/25s
MAC protocol	CSMA/CA	Data rate	11Mbps	Traffic/ Distribution	Poisson
Queue Length	50	Turn time	10 μs	Replications	50 times

Table 1. Simulation parameters.

transmitted by a node within its range. The propagation delay is assumed to be negligible and the nodes are static. The beacon-loss probability (Eqs. (1)–(8)) was verified in Egeland & Engelstad (2010), using both the simulation model and the widely used *ns2* network simulator ns2 (2010).

The results in Fig. 6 show the beacon loss probability (p_e) and the link-failure (p_f) probability for the topologies in Fig. 3. Both analytical and simulated results are shown. The simulation parameters are listed in Tab. 1. As can be verified from the figure, the results from our simulation model match well with the analytical results. The results confirm that the model provides sufficient accuracy, even though the model assumes that the length of the data packets are exponential distributed while a fixed packet length is used in the simulations.

4. Apparent link-failures in arbitrary mesh topologies

4.1 Link-failure probability for complex traffic patterns

The apparent link-failure probability in Eq. (9) is only applicable for a topology with a specific connectivity between the nodes. In order to apply the apparent link-failure model on links in an arbitrary mesh topology with a given traffic pattern, an algorithm is needed to determine the number of hidden nodes and the associated traffic pattern that have impact on the rate of which the hidden nodes empty their buffers.

A wireless mesh topology can also be described as a *directed graph* $G=(V,E)$, where the nodes in the network serve as the vertices $v_j \in V(G)$ and any pair of nodes $v_j \rightarrow v_i$ creates an edge $\epsilon_{i,j} \in E(G)$ if there is a direct link between them. A random traffic pattern where a set of nodes transmit data over a link $\epsilon_{i,j} \in E(G)$ with the probability p_{tx} will also form a directed graph $S(V,E,p_{tx})$ that is a subset of G. It is assumed that every node $v_j \in S$ generates data packets at the same rate. Algorithm (1) calculates the number of neighbor nodes (h_u) of the vertice n that are hidden from a vertice $i \in V(G) : \epsilon_{i,n} \in E(G)$ where $h_u = |\{j, \forall j : j \in V(G) \wedge \epsilon_{n,j} \in E(G) \wedge \exists \epsilon_{j \rightarrow k \in V(S)} \in E(S)\}|$. In addition, it returns a flag (0|1) that indicates whether or not vertice n transmits data traffic. Applying Eq. (9) on these parameters will give the upper bound link-failure probability p_f for the link $\epsilon_{n \rightarrow i}$.

For the calculation of the lower bound, an average value for the number of hidden nodes is used, which is denoted h_l in Alg. (1). The rationale behind this is that for a set of nodes $R \subseteq V(S)$ hidden from node i, the carrier sense nature of the MAC protocol will in the case of two nodes $\{k,z\} \in R$ where $\exists z \neq k : \epsilon_{z,k} \in E(G)$ result in that only a subset of the nodes in R can transmit data at any given time. The parameter h_l is the average number of nodes in R that transmit data at a given time. For the calculation of the lower bound this will give a more accurate estimate than using h_u as the number of hidden nodes in Eq. (2).

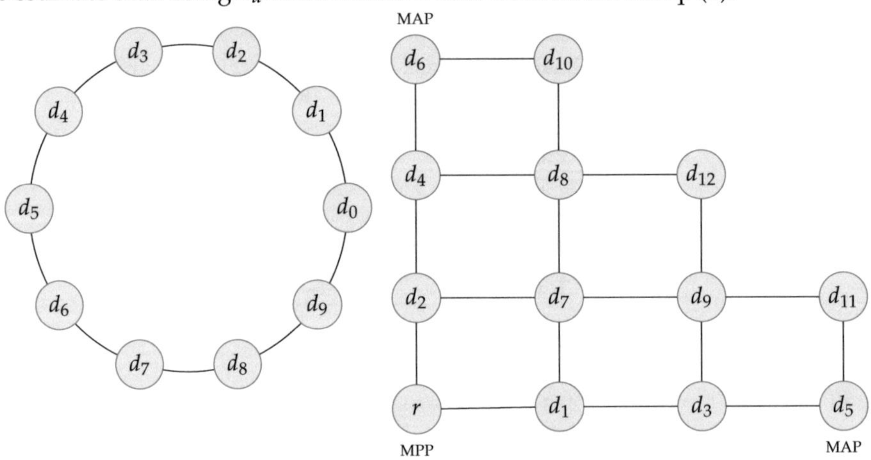

(a) Topology: Ring with 10 nodes (b) Topology: Connected MAPs with redundant MPs

Fig. 7. The distribution of nodes in two example mesh topologies.

4.2 Random pattern of bursty traffic

In this section we investigate how the analyzes of the topologies in Fig. 3 can be applied to more complex mesh topologies. Without loss of generality, we now focus on the two topologies in Fig. 7 as examples, observing that the analysis can easily be generalized for any arbitrary mesh topology. The topologies in Fig. 7 do not resemble the topologies in Fig. 3, but equations Eqs. (1)–(9) will together with Alg. (1) be able provide an upper and lower bound for the apparent link-failure probability p_f.

The simplest approach to analyzing a bursty traffic pattern is to generate a snapshot of the traffic in the topology. We assume that the time between each snapshot is sufficiently long

Algorithm 1 $\vec{H}(G,S)$

```
Require: An undirected graph G(V,E), a directed graph S ⊆ G.
H ← ∅
for i ∈ V(G) do
    J ← {j,∀j : ϵ_{i,j} ∈ E(G)}
    for n ∈ J do
        R ← {r,∀r ≠ i : ϵ_{n,r} ∈ E(G)}
        for k ∈ R do
            if |{j,∀j : ϵ_{k,j} ∈ E(S)}| > 0 ∧ k ∉ G_i then
                h_u ← h_u + 1
            end if
        end for
        N ← ∅
        for k = 0 to 2^{|R|} do
            n_i ← 0; ca ← ∅
            for p = 0 to |R| do
                if k —rshift→ p&1 ∧ ϵ_{n,R_p} ∈ E(S) then
                    ca ← ca ∪ ϵ_{n,R_p}
                    n_i ← n_i + 1
                end if
            end for
            if not [∃z:ϵ_{n,z}∈ca ∧ ∃w≠z:ϵ_{n,w}∈ca:ϵ_{z,w}∈E(G)] then
                N ← N ∪ n_i
            end if
        end for
        h_l ← ⟦ (1/|N|) Σ_{k=0}^{|N|} N_k ⟧
        L⃗ ← (i,n)
        H ← {H} ∪ {(L⃗, h_u, h_l, |{j,∀j : ϵ_{n,j} ∈ E(S)}|?0 : 1))}
    end for
end for
return H
```

for the traffic patterns of each snapshot to be considered independent and that for each link in the topologies in Fig. 7, a burst of data packets is transmitted with the probability p_{tx}. Each node generates data packets within a burst according to a Poisson process with the rate parameter λ_c. If the topology is described as a graph $G(V,E)$, the traffic pattern given by the graph $S(V,E,p_{tx}) \subseteq G$ is a snapshot that will represent a possible data transmission pattern. By generating a large number of random snapshots for a given p_{tx} $\left(S_{i\in\{0,M\}}\right)$, the overall average apparent link-failure probability for a given λ_c can be found.

Fig. 8 shows the average upper and lower bound for the apparent link-failure probability for λ_c=0.2. The apparent link-failure probability for the topologies in Fig. 7 is calculated using Alg. (1) and Eqs. (1)–(9) on the randomly generated traffic patterns. The figure also shows simulation results for the average apparent link-failure. As the simulation results demonstrate, the analytical upper and lower bounds provide a good indicator of the average link-failure probability even though it can be seen that the gap between the upper and lower bound increases as $p_{tx} \rightarrow 1$. This is a result of a complex traffic pattern and interaction between the nodes that the simple model does not incorporate. At low values for p_{tx}, the model's upper and lower bound is as expected, more accurate.

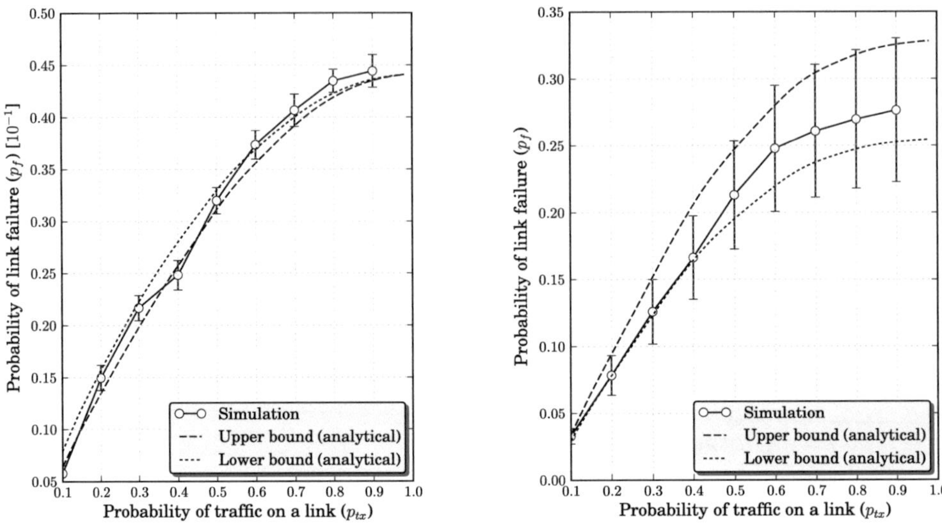

(a) Results for topology in Fig. 7(a) (b) Results for topology in Fig. 7(b)

Fig. 8. Apparent link-failure probability for Fig. 7 ($\lambda_c = 0.2$). Simulation results are shown with a 95% confidence interval.

In Fig. 9 the upper and lower bound link-failure probability for different values of λ_c is shown. As can be seen from the figure, for small and large values of λ_c, the gap between lower and upper bound is negligible. The reason for this is that when $\lambda_c \gtrsim 0$, the sum of the packets awaiting transmission in the buffers of the hidden nodes is almost zero in both the isolated and the connected cases. Therefore, the apparent link-failure probabilities are almost identical. For the case when $\lambda_c \lesssim 1$, the sum of packets awaiting transmission in the buffers of the hidden nodes is always greater that zero, i.e. there is always a packets waiting to be transmitted. Hence, the difference in apparent link-failure probability is almost negligible. For $0.2 < \lambda_c < 0.6$, there exist various combinations of empty and non-empty buffers for the isolated and the connected cases, thus it is expected that there will be a difference in the upper and lower bound.

5. Network availability

If a network operates successfully at time t_0, the network reliability yields the probability that there were no failures in the interval $[0,t]$ Shooman (2002). The analysis of network reliability assumes for simplicity that there are no link repairs in the network. This is not exactly true for mesh networks, since a link-maintenance mechanism will ensure that a failed link is restored. The metric used to describe repairable networks is *availability*. The network availability is defined as the probability that at any instant of time t, the network is up and available, i.e. the portion of the time the network is operational Shooman (2002). This section focuses on the availability at the steady-state, found as $t \to \infty$, i.e. when the transient effects from the initial conditions are no longer affecting the network.

A typical availability measure is the k-terminal availability, namely the probability that a given subset k of K nodes are connected. For a graph $G(V,E,p)$, the k-terminal availability for the k nodes $\subseteq V(G)$ can be found as:

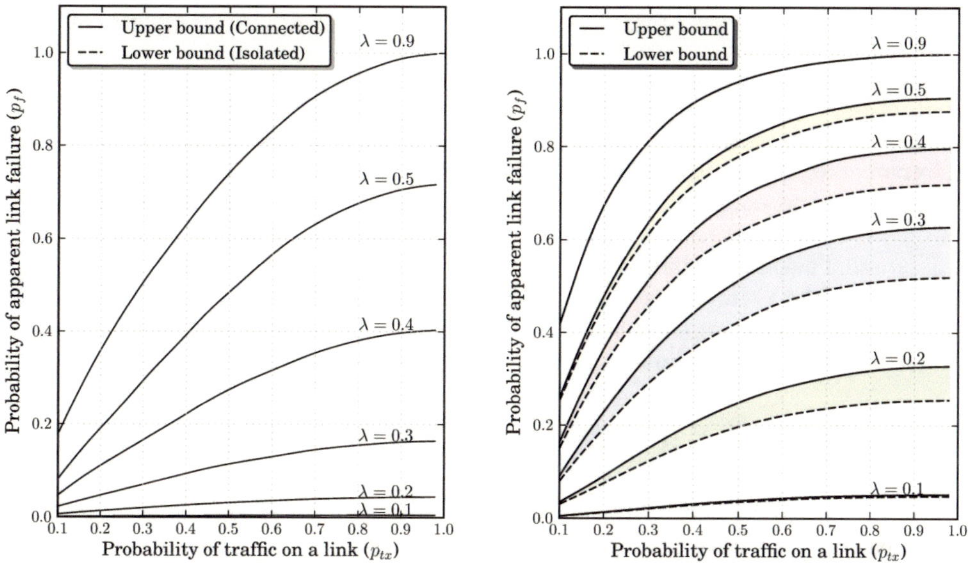

(a) Results for topology in Fig. 7(a) (b) Results for topology in Fig. 7(b)

Fig. 9. Analytical results for the upper/lower bound of the apparent link-failure probability for the topologies in Fig. 7.

$$P_A(K=k) = \sum_{i=w_k(G)}^{|E(G)|} T_i^k(G)(1-p)^i p^{|E(G)|-i} \tag{10}$$

$$= 1 - \sum_{i=\beta(G)}^{|E(G)|} C_i^k(G) p^i (1-p)^{|E(G)|-i} \tag{11}$$

where $T_i^k(G)$ in Eq. (10) denotes the tieset with cardinality i, i.e. the number of subgraphs connecting k nodes with i edges. Furthermore, $w_k(G)$ is the size of the minimum tieset connecting the k nodes. In Eq. (11), $C_i^k(G)$ denotes the number of edge cutsets of cardinality i and $\beta(G)$ denotes the cohesion.

5.1 *k*-terminal availability with apparent link-failures

The network availability (Eq. (11)) is a measure of the robustness of a wireless mesh network and is determined by the structure and the link-failure probability of the links, provided the node-failure probability is negligible.

For a topology described as a graph G, which includes $k-1$ different distribution nodes $d_i \in V(G)$ and a set of root nodes $r_i \in V(G)$ (normally one root node serves a set of distribution nodes), a distribution node corresponds to a MAP while the root node corresponds to an MPP, according to the terminology of IEEE 802.11s. For normal network operation, the transit traffic in an IEEE802.11s network is directed along the shortest path between a root node r and each distribution node, $d_i \in G(V)$. The network is not operating as expected if a distribution node is disconnected from the root node, i.e. the network has failed. Thus, the network is fully operational only if there is an operational path between the root node and each of the distribution nodes. This is true if, and only if, the root node r and the $k-1$ distribution nodes

are all connected. Thus, the reliability of the network may be analyzed using the k-terminal reliability.
The expression for the network availability in Eq. (11) assumes a fixed and identical link-failure probability for all the links in a topology. However, the apparent link-failure model can provide exact probabilities for every link in a topology. In the following we compare the availability using an average apparent link-failure probability with the availability using an exact and a simulated-based apparent link-failure probability.

5.1.1 k-terminal availability based on an average p_F (P_A^a)

As in Section 4, the average apparent link-failure probability is calculated according to Eqs. (1)–(9) and Alg. (1). For a number of $|S|=|\{S_0,\ldots,S_{M-1}\}|=5000$ random patterns of bursty traffic, the average apparent link-failure probability is expressed as:

$$\overline{p}_F = \frac{1}{|S|\times|E(G)|}\sum_{s\in S}\sum_{\epsilon_{i,j}\in E(G)} p_f(i,j)_s \times p_f(j,i)_s \tag{12}$$

where p_f is calculated according to Eq. (9). The k-terminal availability based on an undirectional average link-failure probability is given by:

$$P_A^a\left[G(V,E,\overline{p}_F)\right] = 1 - \sum_{i=\beta}^{|E(G)|} C_i\,(\overline{p}_F)^i\,(1-\overline{p}_F)^{|E(G)|-i} \tag{13}$$

5.1.2 k-terminal availability using simulation (P_A^m)

Using a Monte Carlo simulation, the availability of each topology is calculated where the existence of a link $\epsilon_{i,j}\in E(G)$ depends on the probability $1-p_F(i,j)$. An estimate for the k-terminal availability can then be calculated for $s\in S$ ($|S|=5000$) random bursty traffic patterns as:

$$P_A^m\left[G(V,E,p_F)\right] = \frac{1}{|S|}\times\left[\begin{array}{l}\text{Number of graphs where}\\ k\text{ nodes are connected}\end{array}\right] \tag{14}$$

5.1.3 k-terminal availability using exact calculation (P_A^e)

Since we can calculate the apparent link-failure probability of every link, it is also possible to calculate an exact value for the k-terminal availability. Let us define $L\subseteq E(G)$ as a set of links that are removed from the graph $G(V,E)$. For a traffic pattern $s\in S$, we define:

$$T(L)_s = \prod_{\forall\epsilon_{i,j}\in L} p_F(i,j)_s \times \prod_{\substack{\forall\epsilon_{q,r}\in\\ E(G)\setminus L}} \left[1-p_F(q,r)_s\right]. \tag{15}$$

An exact calculation of the k-terminal availability for $|S|$ bursty traffic patterns is then given by:

$$P_A^e\left[G(V,E,p_F)\right] = 1 - \frac{1}{|S|}\sum_{s\in S}\sum_{\substack{\forall L:V_k(G)\subseteq V(G)\\ \text{is not connected}}} T(L)_s. \tag{16}$$

5.1.4 Availability of example topologies

In this section we apply Eqs. (13)–(16) on the topologies in Fig. 7 using a scenario where the network is configured to allow the STAs to access the MAPs at one frequency band (e.g. using 802.11b or 802.11g) and use another frequency band for the communication between the MPs.

Since the extra equipment cost of such a configuration often is minimal compared with the costs associated with site-acquisition, it is anticipated that many commercial mesh networks will implement a MAP at each MP in the network. For such a configuration, the *all-terminal* availability ($P_A(K{=}k)$) of the network is of interest, which is shown in Fig. 10 (upper and lower bound). The figure shows that the all-terminal availability based on an average p_F (P_A^a) differs slightly from the exact calculations (P_A^e) for the topology in Fig. 7(b). This is caused by the fact that nodes at the border of the topology have fewer neighbors than the nodes in the center area of the topology. For larger 2D-grid topologies, this effect will be reduced and we will have $P_A^a \approx P_A^e$. This is easy to deduce, since the average number of neighbors in an $N \times N$ grid network is $4-4/N$. As N increases, the nodes in the network experience comparable one-hop neighbor/hidden node conditions, due to the topology's regular structure. This is also illustrated in Fig.11(b), where the availability is calculated without the border nodes, i.e. $P_A(\{c_0,\ldots,c_8\})$.

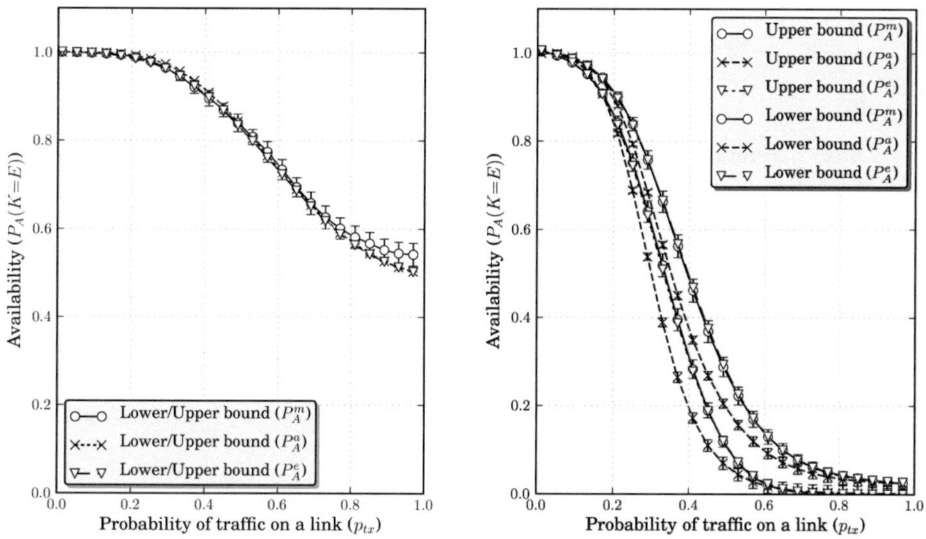

(a) Results for topology Fig.7(a) (b) Results for topology in Fig.7(b)

Fig. 10. The upper/lower bound all-terminal availability, $P_A(K{=}k)$ for the topologies in Fig. 7 ($\lambda_c{=}0.4$).

6. A random geometric graph model approach to apparent link-failures

The main drawback in the previous sections is that it does not take into account correlations between different links. For example, if two ad hoc nodes s_a and s_b are physically very close to each other, and another ad hoc node s_c is farther away, the existence of the links $\epsilon_{a,c}$ and $\epsilon_{b,c}$ is expected to be correlated in reality. So far Eqs. (1)–(9) do not model this correlation.
In this section, we further extend the apparent link-failure model to encompass random geometric graphs Haenggi et al. (2009). A *random geometric graph* $G(V,E,r)$ is a geometric graph in which the $n = |V(G)|$ nodes are independently and uniformly distributed in a metric space. In other words, it is a random graph for which a link between two nodes s_a and s_b exists if, and only if, their Euclidean distance is such that $\| s_a - s_b \| \leq r_0$, where r_0 is the transmission range of the nodes.

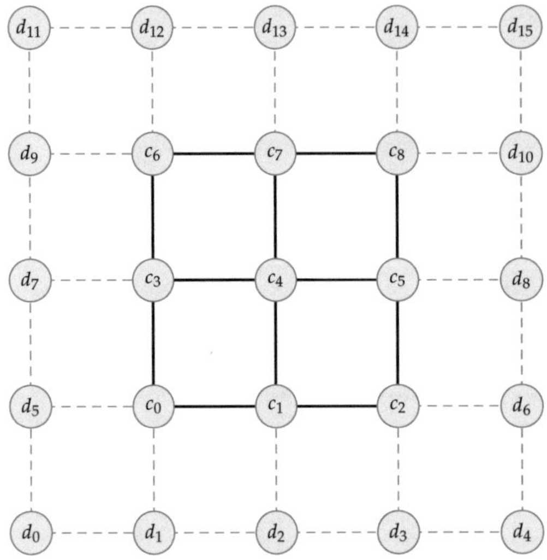

(a) Every node transmits with probability $p_{tx} = 1$ and at rate $\lambda_c = 0.4$

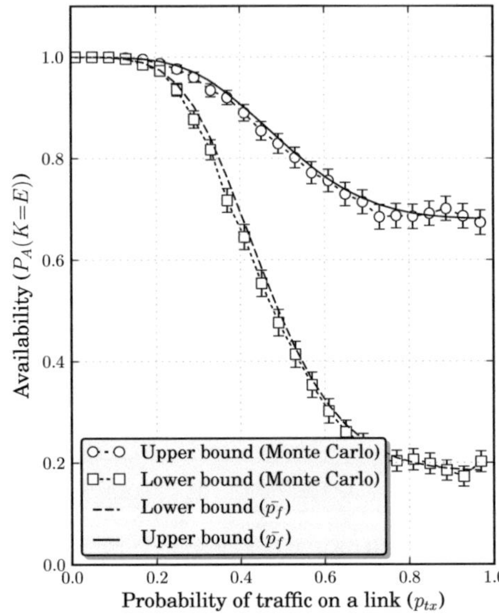

(b) Upper and lower bound all-terminal reliability for the nodes $\{c_0, \ldots, c_8\}$.

Fig. 11. Illustration of the border effect. Since every node in $\{c_0, \ldots, c_8\}$ experiences equal amount of hidden nodes, using the average apparent link-failure probability ($\overline{p}_f$) gives the same all-terminal reliability measure as the Monte Carlo simulation.

6.1 The node degree

We first establish an expression for the probability that n_0 of all n nodes are within a certain area A_0 in the system plane Ω. The expected number of nodes per unit area is then $\rho = n/\Omega$. This probability is in Bettstetter (2002) shown to be:

$$P(d = n_0) = \frac{\left(\frac{A_0}{\Omega} n\right)^{n_0}}{n_0!} \cdot e^{-\frac{A_0}{\Omega} n} = \frac{(\rho A_0)^{n_0}}{n_0!} \cdot e^{-\rho A_0} \tag{17}$$

for large n and large Ω. If a node's radio transmission range r_0 covers an area $A_0 = \pi r_0^2$, the probability that a randomly chosen node has n_0 neighbors is:

$$P(d = n_0) = \frac{\left(\rho \pi r_0^2\right)^{n_0}}{n_0!} \cdot e^{-\rho \pi r_0^2}. \tag{18}$$

A probabilistic bound for the minimum node degree of a homogenous ad hoc network is shown to be Bettstetter (2002):

$$P(d_{min} \geq n_0) = \left(1 - \sum_{i=0}^{n_0 - 1} \frac{(\rho \pi r_0^2)^i}{i!} \cdot e^{-\rho \pi r_0^2}\right)^n. \tag{19}$$

6.2 Average number of hidden nodes of an area A_0

For a given node density and transmission range, we now find the average number of hidden nodes for any given node. Consider the intersecting circles in Fig. 12. Let us assume that the

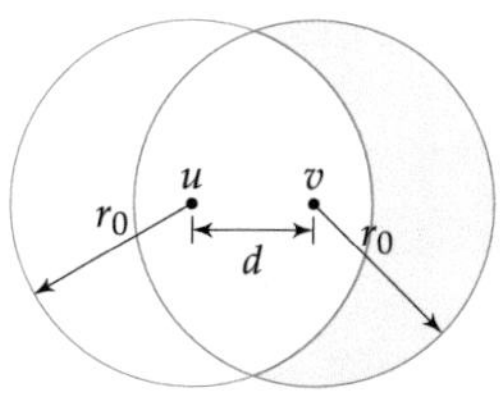

Fig. 12. Analysis of area containing hidden nodes when node u sends a beacon to node v.

points u and v represent to nodes separated by a distance d, each with a transmission range of r_0. If each node covers region S_u and S_v when they transmit a packet, we are interested in finding the area S_{v-u}, since if any node is located in this area, this node will appear hidden from node u. From Tseng et al. (2002) this area is given by $|S_{v-u}| = |S_v| - |S_{u\cap v}| = \pi r^2 - \text{INTC}(d)$, where INTC$(d)$ is the intersection area of the two circles:

$$\text{INTC}(d) = \int_{d/2}^{r_0} \sqrt{r_0^2 - x^2} dx. \tag{20}$$

When the node v is randomly located within u's transmission range, the average area of S_{v-u} is:

$$S_{v-u} = B_0 = \frac{3\sqrt{3}}{4\pi} \pi r_0^2 \tag{21}$$

If n nodes are randomly and uniformly distributed on an area Ω following a homogenous Poisson point process, the probability of finding b_0 nodes in the area B_0 is given by Eq. (17), substituting A_0 with B_0. Thus,

$$P(d = b_0) = \frac{(\rho B_0)^{b_0}}{b_0!} \cdot e^{-\rho B_0} = \frac{\left(\rho \frac{3\sqrt{3}}{4\pi} \pi r_0^2\right)^{b_0}}{b_0!} \cdot e^{-\rho \frac{3\sqrt{3}}{4\pi} \pi r_0^2}. \tag{22}$$

6.3 Connectivity

A topology is said to be *k-connected* ($k = 1,2,3,\ldots$) if for each node pair there exist at least k mutually independent paths connecting them. For a topology described as a graph $G(V,E)$ where $|V(G)| = n$, the probability that G, with $n \gg 1$ where each node has a transmission range r_0 and a homogenous node density ρ is k-connected is Bettstetter (2002):

$$P(G \text{ is } k\text{-connected} \approxeq P(\text{node } i \text{ has } d_{min} \geq k),\ \forall i \in V(G). \tag{23}$$

A beacon from node u to v in Fig.12 will fail to be received if any nodes in the area B_0 (S_{v-u}) transmit a data packet. The apparent link-failure probability with m hidden nodes ($p_f(m)$) is given by Eq. (9). From Eq.(22), we can easily find the probability that node u in Fig. 12 has zero hidden nodes to be $e^{-\rho B_0}$. The probability that the link between node u and v is operational if k nodes are located within node u's transmission range can be calculated as:

$$p_{ok}(k) = e^{-\rho B_0} + \sum_{m=1}^{n-k} \frac{(\rho B_0)^m}{m!} \cdot e^{-\rho B_0} \left(1 - \left[p_f(m)\right]^2\right). \tag{24}$$

If we make the assumption that the one-hop links of node u fail independently, the probability that k of the links are operational is:

$$P(\text{node } u \text{ is } k\text{-connected}) = P(d_{min} \geq k) = \sum_{i=k}^{n} (1 - [1 - p_{ok}(k)]^i) \cdot \frac{(\rho A_0)^i}{i!} \cdot e^{-\rho A_0}. \tag{25}$$

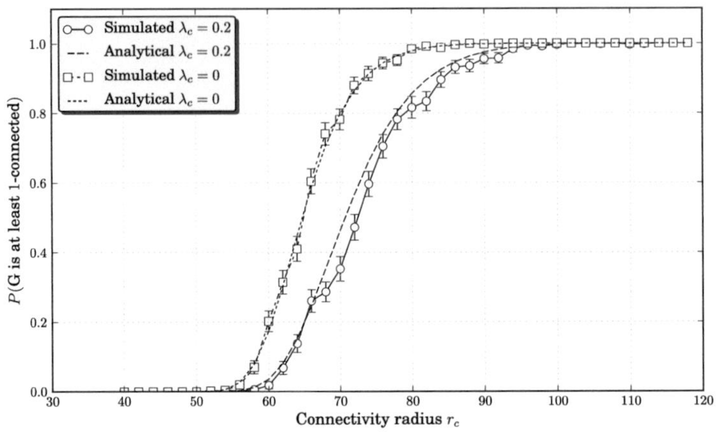

Fig. 13. P(k-connected) with usual Euclidian distance metric. $A = 1000 \times 1000$ ($\rho = 5 \cdot 10^{-4}$)

The probability that a graph $G(V,E)$ where $|V(G)|$ is k-connected is given by:

$$P(G \text{ is } k\text{-connected}) \approxeq \left(\sum_{i=k}^{n} (1-[1-p_{ok}(k)]^i) \cdot \frac{(\rho A_0)^i}{i!} \cdot e^{-\rho A_0} \right)^n . \tag{26}$$

Fig. 13 shows the probability of a topology with $n = 500$ nodes being at least 1-connected, i.e. $P(d_{min} \geq 1)$. The apparent link-failure probability (p_f) is calculated using the lower bound. Every node in the topology transmits data packets with probability $p_{tx} = 1$ which are Poisson distributed with parameter λ_c. Both analytical and Monte Carlo simulation results are shown. The simulation results are based on 1000 randomly generated topologies from which links are removed based on traffic load and the number of hidden nodes of a link. As the figure shows, the simulation results match well with the analytical model. Also, the probability that the topology is connected increases as the transmission range of the nodes is gradually increased, which is as expected. The figure also demonstrates that more neighbors are needed in order to have a connected topology as λ_c, i.e. the traffic rate of the hidden nodes is increased.

7. Using unicast beacon in the presence of apparent link-failures

Having studied the probability of apparent link-failures and its effect on network availability, it is also of interest to explore the measures for diminishing the influence of apparent link-failures. There are several methods for this purpose, such as:

- **Increasing beacon loss parameter (θ):** This method will require more consecutive beacon loss before a node determines that the link is inoperable. However, in cases where a node or a link becomes permanently unavailable due to other reasons than apparent link-failures, this will result in a longer time interval before a new route is calculated; and
- **Reducing hysteresis (θ_h):** This method will bring a link back to operational much faster, however it can result in oscillation between an operational and non-operational state of the link.

Another simple yet effective solution to apparent link-failures is to introduce *unicast beacon transmissions*. This method has the advantage that the MAC layer will retransmit the beacon

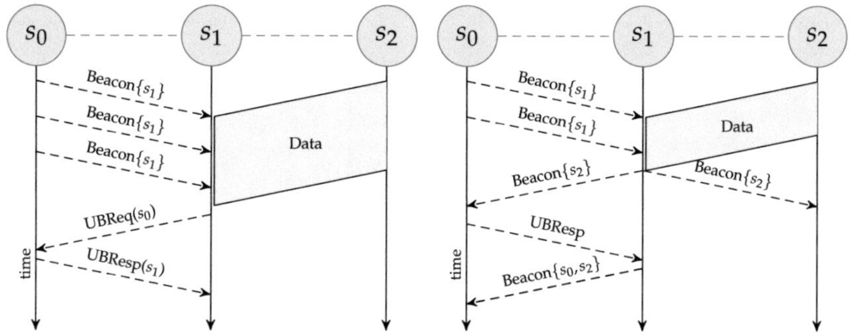

(a) Explicit unicast beacon (b) Implicit unicast beacon

Fig. 14. Handshake of UBReq and UBResp messages.

a defined number of times if an acknowledge is not received. In addition, it is possible to protect the beacon using the RTS/CTS signalling of the MAC layer.

A request for a beacon is called a *Unicast Beacon Request* (UBReq) message and a response to this is called a *Unicast Beacon Response* (UBResp) message. Both these messages can be the same packet format as normal beacon, with the difference that they use a unicast destination address instead of a broadcast destination address. A unicast beacon can be triggered in either end of a link. Consider the topology in Fig. 14(a). Let us assume that node s_0 has discovered s_1 as a neighbor and vice versa. Then, at some point node s_2 transmits data such that node s_1 fails to receive the beacons from node s_0. Node s_1 can then send a UBReq message to node s_0 which answers with a UBResp message. This prevents node s_1 from defining the one-hop link to node s_0 as inoperable. The UBReq and UBResp messages will also be vulnerable to overlapping transmissions from hidden nodes. To overcome this, the link sensing mechanism can protect the UBReq and UBResp messages using RTS/CTS at the MAC layer.

Now consider Fig. 14(b). A UBResp could also be triggered implicitly if node s_0 receives broadcast beacons from node s_1 but fails to find its address in the beacon message. This indicates that s_1 has not received broadcast beacons from node s_0. Node s_0 could therefore send a UBResp message to node s_1, indicating that it can hear node s_1, whereupon node s_1 will include s_0 in its next beacon.

We implemented the unicast beacon scheme in ns2 by modifying the OLSR routing protocol, allowing unicast beacons to be protected by RTS/CTS signalling. Using *No Route To Host* packets drop as an indicator, we can calculate the availability of a network. The simulation parameters are shown in Table 2 and the topologies are shown in Fig.7.

IP/MAC layer	**Values**	**Physical layer**	**Values**	**Simulation**	**Values**
Data	1500 bytes	Propagation model	Free Space	Simulation/ transient time	500s/25s
MAC protocol	CSMA/CA	Data rate	11Mbps	Traffic/ Distribution	Poisson
Queue Length	10			Replications	50 times

Table 2. ns2 simulation parameters.

Fig.15 illustrates the all-terminal availability. The analytical and simulated results are shown for normal OLSR beacon scheme. As can be observed from the figure, the simulated average availability for OLSR is much lower than the analytical one. This is as expected, since our

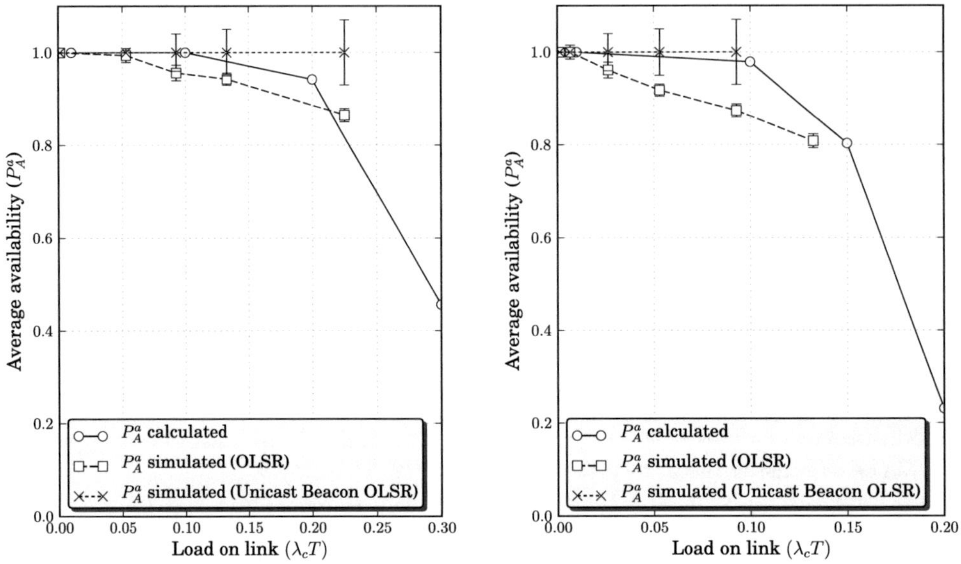

(a) Availability for the topology in Fig.7(a) (b) Availability for the topology in Fig.7(b)

Fig. 15. Average availability for the topologies in Fig.7. Results for standard beacon transmission and unicast beacon transmission protected by RTS/CTS are shown.

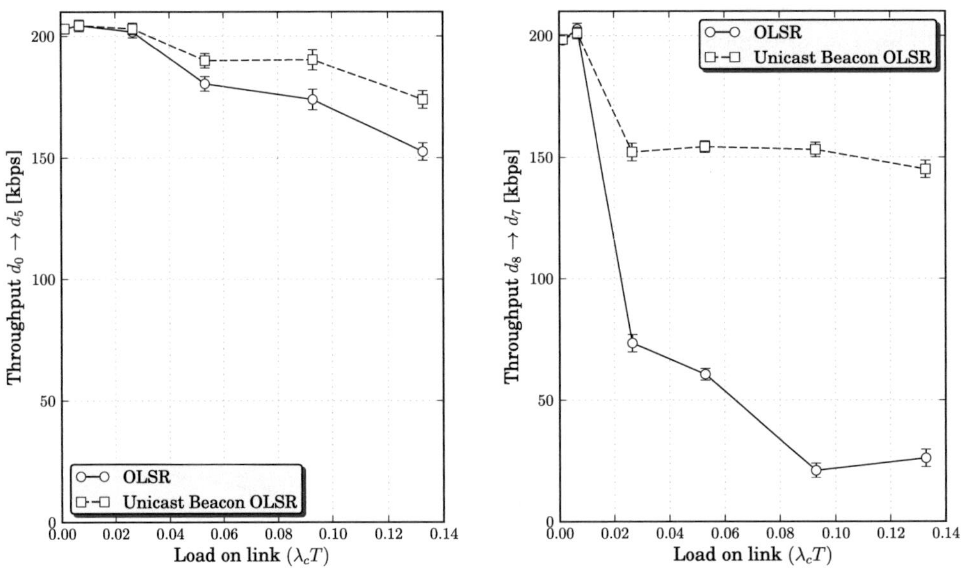

(a) Throughput node $d_0 \to d_5$ in Fig.7(a) (b) Throughput node $d_8 \to d_7$ in Fig.7(b)

Fig. 16. Average throughput for the topologies in Fig.7. Results for standard beacon transmission and unicast beacon transmission protected by RTS/CTS are shown. The source nodes transmit at a fixed rate of 200 kbps while the load on all links is gradually increased.

simple model does not take MAC retransmissions into account. MAC retransmissions will increase the average load (λ_c) on the channel, thus increasing the probability of beacon loss.The figure also shows that the simulated results from unicast beacon scheme provide much higher availability as the load on the channel increases.

8. Conclusions

This chapter introduces an approximate model for the probability of apparent link-failures in beacon-based link maintenance schemes. The model is extended to provide a rough upper and lower bound for arbitrary topologies. Through extensive simulations, it has been confirmed that the model provides acceptable accuracy for simple topologies. Furthermore, more advanced topologies with random traffic patterns and bursty traffic have been studied, where the model can provide an average upper and lower bound for the link-failure probability with satisfactory accuracy. In addition, the work has demonstrated how the apparent link-failure model can be used to investigate the availability of mesh topologies and that using an average apparent link-failure probability can serve as a good indicator for the availability of a given topology. However, the k-terminal reliability problem is known to belong to a class of NP-complete problems Valiant (1979), which has similar complexity as calculating the exact network availability. Applying approximate methods to the k-terminal probability is possible, but this is a topic for future work. In order to provide intuition about the effects of apparent link-failures in large network with randomly distributed nodes, random geometric graph analysis has been applied. Based on existing work on random geometric graphs, we have extended our link-failure model so that connectivity calculations can be performed for topologies where apparent link-failures are present.

Last but not least, a simple remedy for apparent link-failures has been introduced where unicast beacons are used to mitigate beacon loss caused by overlapping transmissions. This solution has been implemented for the OLSR routing protocol and the performance improvements have been verified using the *ns2* simulation tool.

9. References

Ali, H. M., Naimi, A. M., Busson, A. & Vèque, V. (2009). Signal strength based link sensing for mobile ad-hoc networks, *Telecommunication Systems* 42(3-4): 201–212.

Bettstetter, C. (2002). On the minimum node degree and connectivity of a wireless multihop network, *MobiHoc '02: Proceedings of the 3rd ACM international symposium on Mobile ad hoc networking & computing*, ACM, New York, NY, USA, pp. 80–91.

Chlamtac, I., Conti, M. & Liu, J. J.-N. (2003). Mobile ad hoc networking: Imperatives and challenges, *Ad Hoc Networks, Elsevier* 1(1): 13–64.

Clausen, T. & Jacquet, P. (2003). Optimized link state routing protocol (olsr), ietf rfc 3626.

Dubey, A., Jain, A., Upadhyay, R. & Charhate, S. (2008). Performance evaluation of wireless network in presence of hidden node: A queuing theory approach, *Modeling and Simulation, 2008. AICMS 08. Second Asia International Conference on*, pp. 225–229.

Egeland, G. & Engelstad, P. E. (2009). The availability and reliability of wireless multi-hop networks with stochastic link failures, *IEEE J.Sel. A. Commun.* 27(7): 1132–1146.

Egeland, G. & Engelstad, P. E. (2010). A model for the loss of Hello-Messages in a wireless mesh network, *IEEE ICC 2010 - Ad-hoc, Sensor and Mesh Networking Symposium*, Cape Town, South Africa.

Egeland, G. & Li, Y, F. (2007). Prompt route recovery via link break detection for proactive routing in wireless ad hoc networks, *10th International Symposium Wireless Personal Multimedia Communications (WPMC)*, Jaipur, India.

Gerharz, M., Waal, C. D., Frank, M. & Martini, P. (2002). Link stability in mobile wireless ad hoc networks, *In Proceedings of the 27th Annual IEEE Conference on Local Computer Networks (LCN'02)*.

Gharavi, H. & Kumar, S. (2003). Special issue on sensor networks and applications, *Proceedings of the IEEE* 91(8).

Haenggi, M., Andrews, J., Baccelli, F., Dousse, O., Franceschetti, M. & Towsley, D. (2009). Guest editorial: geometry and random graphs for the analysis and design of wireless networks, *Selected Areas in Communications, IEEE Journal on* 27(7): 1025 –1028.

IEEE802.11 (1997). Wireless LAN medium access control (MAC) and physical layer (PHY) specification.

IEEE802.11s (2010). Lan/man specific requirements - part 11: Wireless medium access control (mac) and physical layer (phy) specifications: Amendment: Ess mesh networking.

Kleinrock, L. (1975). *Theory, Volume 1, Queueing Systems*, Wiley-Interscience.

Kleinrock, L. & Tobagi, F. (1975). Packet switching in radio channels: Part i–carrier sense multiple-access modes and their throughput-delay characteristics, *Communications, IEEE Transactions on* 23(12): 1400–1416.

Li, F., Bucciol, P., Vandoni, L., Fragoulis, N., Zanoli, S., Leschiutta, L. & Lázaro, O. (2010). Broadband internet access via multi-hop wireless mesh networks: Design, protocol and experiments, *Wireless Personal Communications* .
URL: *http://dx.doi.org/10.1007/s11277-009-9907-9*

Ng, P. C. & Liew, S. C. (2004). Re-routing instability in ieee 802.11 multi-hop ad-hoc networks, *Local Computer Networks, 2004. 29th Annual IEEE International Conference on*, pp. 602–609.

ns2 (2010). The Network Simulator NS-2, `http://www.isi.edu/nsnam/ns/`.

Perkins, C., Belding-Royer, E. & Das, S. (2003). Ad hoc on-demand distance vector (aodv) routing, ietf rfc 3561.

Ray, S., Carruthers, J. B. & Starobinski, D. (2004). Evaluation of the masked node problem in ad-hoc wireless lans, *IEEE Transactions on Mobile Computing* 4: 430–442.

Ray, S., Starobinski, D. & Carruthers, J. B. (2005). Performance of wireless networks with hidden nodes: a queuing-theoretic analysis, *Comput. Commun.* 28(10): 1179–1192.

Shooman, M. L. (2002). *Reliability of Computer Systems and Networks: Fault Tolerance, Analysis, and Design*, John Wiley and Sons, Inc.

Tobagi, F. & Kleinrock, L. (1975). Packet switching in radio channels: Part ii–the hidden terminal problem in carrier sense multiple-access and the busy-tone solution, *Communications, IEEE Transactions on* 23(12): 1417–1433.

Tseng, Y.-C., Ni, S.-Y., Chen, Y.-S. & Sheu, J.-P. (2002). The broadcast storm problem in a mobile ad hoc network, *Wirel. Netw.* 8(2/3): 153–167.

Valiant, L. G. (1979). The complexity of computing the permanent, *Theor. Comput. Sci.* 8: 189–201.

Voorhaen, M. & Blondia, C. (2006). Analyzing the impact of neighbor sensing on the performance of the olsr protocol, *Modeling and Optimization in Mobile, Ad Hoc and Wireless Networks, 2006 4th International Symposium on*, pp. 1–6.

9

Pursuing Credibility in Performance Evaluation of VoIP Over Wireless Mesh Networks

Edjair Mota[1], Edjard Mota[1], Leandro Carvalho[1], Andréa Nascimento[1] and Christian Hoene[2]
[1]*Federal University of Amazonas*
[2]*Tübingen University*
[1]*Brazil*
[2]*Germany*

1. Introduction

There has been an increasingly interest in real-time multimedia services over wireless networks in the last few years, for the most part due to the proliferation of powerful mobile devices, and the potential ubiquity of wireless networks. Nevertheless, there are some constraints that make their deployment over Wireless Mesh Networks (WMNs) somewhat difficult. Due to the dynamics of WMNs, there are significant challenges in the design and optimization of such services. Impairments like packet loss, delay and jitter affects the end-to-end speech quality (Carvalho, 2004). Experimenters have been proposing solutions to the challenges found so far, and comparing them before implementation is a mandatory task. There exists a necessity of designing efficient tools for enhancing the computational effort of the performance modeling and analysis of VoIP over WMNs. Structural complexity of such highly dynamic systems causes that in many situations computer simulation is the only way of investigating their behavior in a controllable manner, allowing the experimenter to conduct independent and repeatable experiments, in addition to the comparison of a large number of system alternatives. Stochastic simulation is a flexible, yet powerful tool for scientifically getting insight into the characteristics of a system being investigated. However, to ensure reproducible results, stochastic simulation imposes its own set of rules. The credibility of a performance evaluation study is greatly affected by the problem formulation, model validation, experimental design, and proper analysis of simulation outcomes.
Therefore, a fine-tuning of the parameters within a simulator is indispensable, so that it closely tracks the behavior of a real network. However, the lack of rigor in following the simulation methodology threatens the credibility of the published research (Pawlikowski et al., 2002; Andel & Yasinac, 2006; Kurkowski et al., 2005).
The aim of this chapter is to provide a detailed discussion of these aspects. To do so, we used as a starting point the observation that the optimized use of the bandwidth of wireless networks definitely affects the quality of VoIP calls over WMN. Since the payload size of VoIP packets is usually smaller than the header size, much network resource is spent for conveying control information instead of data information. Hence, VoIP header compression is an alternative to reduce the use of the bandwidth needed to transmit control information, thereby increasing the percentage of bandwidth used to carry payload information.

However, this mechanism can make the VoIP system less tolerant to packet loss, which can be harmful in WMN, due to its high rate of packet loss. Additionally, in a multi-hop wireless environment, simple schemes of header compression may not be enough to increase or maintain the speech quality. An interesting alternative approach in this context is the use of header compression in conjunction with packet aggregation (Nascimento, 2009), aiming to eliminate the intolerance to packet loss without reducing the compression gain.
Although these issues are not unique to simulation of multimedia transmission over wireless mesh networks, we focus on issues affecting the WMN research community interested in VoIP transmissions over WMN. After modeling thoroughly the issues of VoIP over WMN, we built a simulation model of a real scenario at the Federal University of Amazonas, where we have been measuring the speech quality of VoIP transmissions by means of a tool developed by our groups. Then, we modeled a bidirectional VoIP traffic, and proposed a carefully selected set of experiments and simulation details such as the sources of randomness and analysis of the output data, closely following sound methodology for each phase of the experimentation with simulation.

2. Background

2.1 Wireless mesh networks

Wireless mesh network (WMN) is a promising communication technology that has been successfully tested in academic trials and is a mature candidate to implement metropolitan area networks. Compared to fiber and copper based access networks, it can be easily deployed, maintained and expanded on demand. It offers network robustness, and reliable service coverage, besides its low costs of installation and operation. In many cities, such as Berlin and Bern, WMN has been used to provide Internet access for many users.
In its more general form, a WMN consists of a set of wireless mesh routers (WMRs) that interconnect with each other via wireless medium to form a wireless backbone. These WMRs are usually stationary and work as access points or gateways to the Internet to wireless mesh clients (WMCs). High fault tolerance can be achieved in the presence of network failures, improper operation of WMRs, or wireless link inherent variabilities. Based on graph theory, (Lili et al., 2009) suggested a method to analyze the fault-tolerant and communication delay in a wireless mesh network, while (Queiroz, 2009) investigated the routing table maintenance issue, by proposing and evaluating the feasibility of applying the Bounded Incremental Computation model (Ramalingam & Reps, 1996) to satisfy scalability issues. Such kind of improvement is essential to real time multimedia application in order to reduce the end-to-end delay.
Even being accepted as a good solution to provide access to the telephone service, the WMN technology poses some problems being currently investigated such as routing algorithms, self-management strategies, interference, to say a few. To understand how to achieve the same level of quality of multimedia applications in wired networks, it is imperative to grasp the nature of real-time multimedia traffic, and then to compare it against the problems related to the quality of multimedia applications.

2.2 Voice over IP

A VoIP call placed between two participants requires three basic types of protocol: signaling, media transmission, and media control. The signaling protocols (e.g. H.323, SIP) establish, maintain and terminate a connection, which should be understood as an association between applications, with no physical channel or network resources associated with it. The media transmission protocols (e.g. RTP) are responsible for carrying out the actual content of the call – the speaker's voice – encoded in bits. Finally, the media control protocols (e.g. RTCP) convey voice packet transmission parameters and statistics, ensuring better end-to-end packet delivery. In this work, our attention is focused upon the voice stream. So, lets briefly introduce the main logical VoIP components of the media transmission path.

As illustrated in Figure 1, the sender's voice is captured by a microphone and digitalized by an A/D conversor. The resulting discrete signal is then encoded and compressed by some codec into voice frames. One or more frames can be encapsulated into a voice packet by adding RTP, UDP and IP headers. Next, the voice packets are dispatched to the IP network, where they can get lost due to congestion or transmission errors. The transmission delay of packets – i.e, the time needed to deliver a packet from the sender to the receiver – is variable and depends on the current network condition and the routing path (Hoene et al., 2006).

At the receiver, the arriving packets are inserted into a dejitter buffer, also known as playout buffer, where they are temporarily stored to be isochronously played out. If packets are too late to be played out in time, they are discarded and considered as lost by the application. After the dejitter buffer the speech frames are decoded. If a frame is missing, the decoder fills the gap by applying some Packet Loss Concealment (PLC) algorithm. Finally, the digital signal is transformed into an acoustic signal.

Since IP networks were not designed to transport real-time traffic, an important aspect in VoIP communications is the assessment of speech quality. It is imperative that new voice services undergo a significant amount of testing to evaluate their performance. Speech quality is a complex psychoacoustic outcome of the perception process of the human auditory system (Grancharov & Kleijn, 2008). Its measurement can be carried out using either subjective or objective methods.

Subjective methods, specified in ITU-T Rec. P.800 (ITU-T, 1996), require that a pool of listeners rates a series of audio files using a five-level scale (1 – bad, 2 – poor, 3 – fair, 4 – good, 5 – excellent). The average of all scores thus obtained for speech produced by a particular system represents its Mean Opinion Score (MOS). The main reason for the popularity of this test is its simplicity (Grancharov & Kleijn, 2008). However, the involvement of human listeners makes them expensive and time consuming. Moreover, subjective tests are not suitable to monitor the QoS of a network on a daily basis. This has made objective methods very attractive for meeting the demands for voice quality measurement in communications networks.

Among the objective (or instrumental) methods, the E-model, defined in the ITU-T Rec. G.107 (ITU-T, 1998), is one of the most used. It computes, in a psychoacoustic scale, the contribution of all impairment factors that affect voice quality. This does not imply that the factors are uncorrelated, but only that their contributions to the estimated impairments are independent and each impairment factor can be computed separately (Myakotnykh & Thompson, 2009). Although initially designed for transmission planning of telecommunication systems (Raake, 2006; ITU-T, 1998), the E-model was modified by (Clark, 2003; Carvalho et al., 2005) to be used for VoIP network monitoring.

The output of the E-model is the R factor, which ranges from 0 (worst) to 100 (excellent) and

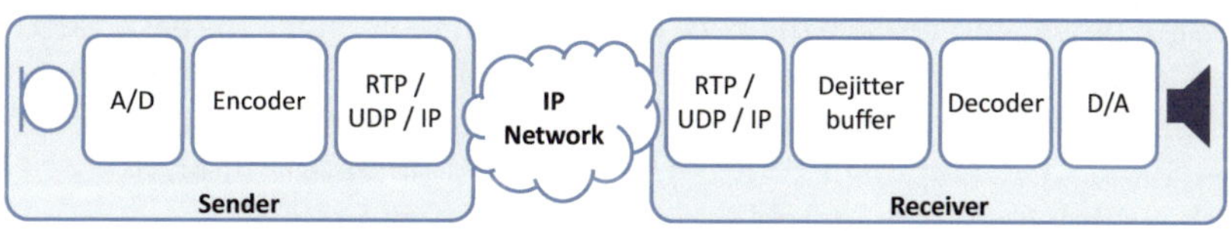

Fig. 1. VoIP components of the media transmission path.

can be converted to the MOS scale. Voice calls whose R factor value is below 60 (or 3.6 in MOS scale) are not recommended (ITU-T, 1998). For VoIP systems, the R factor can be obtained by the following simplified expression (Carvalho et al., 2005):

$$R = 93.2 - Id(codec, delay) - Ie{,}eff(codec, loss, PLB) \quad (1)$$

where Id represents the impairments associated with end-to-end delay, and $Ie{,}eff$ represents the impairments associated with codec compression, packet loss rate and packet loss behavior (PLB) during the call.

The measurement tool proposed in (Carvalho et al., 2005) for speech quality evaluation based on the E-model was adapted as a patch to the Network Simulator code (McCanne & Floyd, 2000). It was used for validating our simulation models concerning the transmission of VoIP over WMNs.

2.3 Performance evaluation

Measurement and stochastic simulation are the main tools commonly used to assess the performance of multimedia transmission over WMNs. Success in the development of complex wireless networks is partially related to the ability of predicting their performance already in the design phase and subsequent phases of the project as well. Dynamic increasing of the complexity of such networks and the growth of the number of users require efficient tools for analyzing and improving their performance. Analytical methods of analysis are neither general nor detailed enough, and in order to get tractability, they need sometimes to make assumptions that require experimental validation. On the other hand, simulation, formerly known as a last resort method, is a flexible and powerful tool adequate for prototyping such complex systems. To the factors that have additionally stimulated the use of simulation, one could include: faster processors, larger-memory machines and trends in hardware developments (e.g. massively parallel processors, and clusters of distributed workstations).

Straightforward simulation of complex systems, such as WMNs, takes frequently a prohibitively amount of computer time to obtain statistically valid estimates, despite increasing processing speed of modern computers. It is not rare the simulation take some hours to estimate a performance metric corresponding to a few seconds of real time. (Mota et al., 2000) investigated the influence of jitter on the quality of service offered by a wireless link, and reported a simulation time as long as 180 hours using just 9 wireless terminals, though simulation has been executed in a fast workstation dedicated to that purpose.

Such phenomenon results from the statistical nature of the simulation experiments. Most simulation models contain stochastic input variables, and, thereby, stochastic output variables, the last ones being used for estimating the characteristics of the performance parameters of the simulated system. In order to obtain an accurate estimate with known (small) statistical error, it is necessary to collect and analyze sequentially a substantial amount of simulation output data, and this can require a long simulation run.

Efficient statistical tools can be used to impact the running time of an algorithm by choosing an estimator with substantially lower computational demand. It would be a mistake to think that more processing power can replace the necessity for such tools, since the associated pitfalls can be magnified as well (Glynn & Heidelberger, 1992). The need for effective statistical methods to analyze output data from discrete event simulation has concerned simulation users as early as (Conway, 1963). Development of accurate methods of statistical analysis of simulation output data has attracted a considerable scientific interest and effort.

Even though, credibility of stochastic simulation has been questioned when applied to practical problems, mainly due to the application of not robust methodology for simulation projects, which should comprise at least the following:

- The correct definition of the problem.
- An accurate design of the conceptual model.
- The formulation of inputs, assumptions, and processes definition.
- Build of a valid and verified model.
- Design of experiments.
- Proper analysis of the simulation output data.

3. Model credibility

3.1 Problem definition

To formulate a problem is so important as to solve it. There is a claim credited to Einstein that states: "The formulation of a problem is often more essential than its solution, which may be merely a matter of mathematical or experimental skill". The comprehension of how the system works and what are the main specific questions the experimenter wants to investigate, will drive the decisions of which performance measures are of real interest.
Experts are of the opinion that the experimenter should write a list of the specific questions the model will address, otherwise it will be difficult to determine the appropriate level of details the simulation model will have. As simulation's detail increases, development time and simulation execution time also increase. Omitting details, on the other hand, can lead to erroneous results. (Balci & Nance, 1985) formally stated that the verification of the problem definition is an explicit requirement of model credibility, and proposed high-level procedure for problem formulation, and a questionnaire with 38 indicators for evaluating a formulated problem.

3.2 Sources of randomness

The state of a WMN can be described by a stochastic or random process, that is nothing but a collection of random variables observed along a time window. So, input variables of a WMN simulation model, such as the transmission range of each WMC, the size of each packet transmitted, the packet arrival rate, the duration of periods ON an OFF of a VoIP source, etc, are random variables that need to be:

1. Precisely defined by means of measurements or well-established assumptions.
2. Generated with its specific probability distribution, inside the simulation model during execution time.

The generation of a random variate - a particular value of a random variable - is based on uniformly distributed random numbers over the interval [0, 1), the elementary sources of randomness in stochastic simulation. In fact, they are not really random, since digital computers use recursive mathematical relations to produce such numbers. Therefore, it is more appropriate to call then pseudo-random numbers (PRNs).
Pseudo-random numbers generators (PRNGs) lie in the heart of any stochastic simulation methodology, and one must be sure that its cycle is long enough in order to avoid any kind of correlation among the input random variables. This problem is accentuated when there is

a large number of random variables in the simulation model. Care must be taken concerning PRNGs with small periods, since with the growth of CPU frequencies, a large amount of random numbers can be generated in a few seconds (Pawlikowski et al., 2002). In this case, by exhausting the period, the sequence of PRNs will be soon repeated, yielding then correlated random variables, and compromising the quality of the results.

As the communication systems become even more sophisticated, their simulations require more and more pseudo-random numbers which are sensitive to the quality of the underlying generators (L'Ecuyer, 2001). One of the most popular simulation packages for modeling WMN is the so called ns-2 (Network Simulator) (McCanne & Floyd, 2000). In 2002, (Weigle, 2006) added an implementation of the MRG32k3, a combined multiple recursive generator (L'Ecuyer, 1999), since it has a longer period, and provides a larger number of independent PRNs substreams, which can be assigned to different random variables. This is a very important issue, and could be verified before using a simulation package. We have been encouraging our students to test additional robust PRNGs, such as Mersenne Twister (Matsumoto & Nishimura, 1998) and Quantum Random Bit Generator – QRBG (Stevanović et al., 2008).

3.3 Valid model

Model validation is the process of establishing whether a simulation model possesses a satisfactory range of accuracy consistent with the real system being investigated, while model verification is the process of ensuring that the computer program describing the simulations is implemented correctly. Being designed to answer a variety of questions, the validity of the model needs to be determined with respect to each question, that is, a simulation model is not a universal representation of a system, but instead it should be an accurate representation for a set of experimental conditions. So, a model may be valid for one set of experimental conditions and invalid for another.

Although it is a mandatory task, it is often time consuming to determine that a simulation model of a WMN is valid over the complete domain of its intended applicability. According to (Law & McComas, 1991), this phase can take about 30%–40% of the study time. Tests and evaluations should be conducted until sufficient confidence is obtained and a model can be considered valid for its intended application (Sargent, 2008).

A valid simulation model for a WMN is a set of parameters, assumptions, limitations and features of a real system. This model must also address the occurrence of errors and failures inherent, or not, to the system. This process must be carefully conducted to not introduce modeling errors. It should be a very good practice to present the validation of the used model, and the corresponding deployed methodology so independent experimenters can replicate the results. Validation against a real-world implementation, as advocated by (Andel & Yasinac, 2006), it is not always possible, since the system might not even exist. Moreover, high fidelity, as said previously, is often time consuming, and not flexible enough. Therefore, (Sargent, 2008) suggests a number of pragmatic validation techniques, which includes:

- Comparison to other models that have already been validated.
- Comparison to known results of analytical models, if available.
- Comparison of the similarity among corresponding events of the real system.
- Comparison of the behavior under extreme conditions.
- Trace the behavior of different entities in the model.

– Sensitivity analysis, that is, the investigation of potential changes and errors due changes in the simulation model inputs.

For the sake of example, Ivanov and colleagues (Ivanov et al., 2007) presented a practical example of experimental results validation of a wireless model written with the Network Simulator (McCanne & Floyd, 2000) package for different network performance metrics. They followed the approach from (Naylor et al., 1967), to validate the simulation model of a static ad-hoc networks with 16 stations by using the NS-2. The objective of the simulation was to send a MPEG4 video stream from a sender node to a receiving node, with a maximum of six hops. The validation methodology is composed of three phases:

Face validity This phase is based on the aid of experienced persons in the field, together with the observation of real system, aiming to achieve high degree of realism. They chose the more adequate propagation model and MAC parameters, and by means of measurements on the real wireless network, they found the values to set up those parameters.

Validation of Model Assumption In this phase, they validated the assumptions of the shadowing propagation model by comparing model-generated and measured signal power values.

Validation of input-output transformation In this phase, they compared the outputs collected from the model and the real system.

3.4 Design of experiments

To achieve full credibility of a WMN simulation study, besides developing a valid simulation model, one needs exercise it in valid experiments in order to observe its behavior and draw conclusions on the real network. Careful planning of what to do with the model can save time and efforts during the investigation, making the study efficient. Documentation of the following issues can be regarded as a robust practice.

Purpose of the simulation study The simple statement of this issue will drive the overall planning. Certainly, as the study advances and we get deeper understanding of the system, the ultimate goals can be improved.

Relevant performance measures By default, most simulation packages deliver a set of responses that could be avoided if they are not of interest, since the corresponding time frame could be used to expand the understanding of the subtleties of WMN configurations.

Type of simulation Sometimes, the problem definition constraints our choices to the deployment of terminating simulation. For example, by evaluating the speech quality of a VoIP transmission over a WMN, we can choose a typical conversation duration of 60 seconds. So, there is no question about starting or stopping the simulation. A common practice is to define the number of times the simulation will be repeated, write down the intermediate results, and average them at the end of the overall executions. We have been adopting a different approach based on steady-state simulation approach. To mitigate the problem of initialization bias, we rely on Akaroa 2.28 (Ewing et al., 1999) to determine the length of the warm-up period, during which data collected during are not representative of the actual average values of the parameters being simulated, and cannot be used to produce good estimates of steady-state parameters. To rely on arbitrary choices for the run length of the simulation is an unacceptable practice, which compromises the credibility of the entire study.

Experimental Design The goal of a proper experimental design is to obtain the maximum information with the minimum number of experiments. A factor of an experiment is a controlled independent variable, whose levels are set by the experimenter. The factors can range from categorical factors such as routing protocols to quantitative factors such as network size, channel capacity, or transmission range (Totaro & Perkins, 2005). It is important to understand the relationship between the factors since they impact strongly the performance metrics. Proper analysis requires that the effects of each factor be isolated from those of others so that meaningful statements can be made about different levels of the factor.

As a simple checklist for this analysis, we can enumerate:

1. Define the factors and their respective levels, or values, they can take on;
2. Define the variables that will be measured to describe the outcome of the experimental runs (response variables), and examine their precision.
3. Plan the experiments. Among the available standard designs, choose one that is compatible with the study objective, number of design variables and precision of measurements, and has a reasonable cost. Factorial designs are very simple, though useful in preliminary investigation, especially for deciding which factors are of great impact on the system response (the performance metric). The advantages of factorial designs over one-factor-at-a-time experiments is that they are more efficient and they allow interactions to be detected. To thoroughly know the interaction among the factors, a more sophisticated design must be used. The approach adopted in (C.L.Barrett et al., 2002) is enough in our problem of interest. The authors setup a factorial experimental design to characterize the interaction between the factors of a mobile ad-hoc networks such as MAC, routing protocols, and nodes' speed. To characterize the interaction between the factors, they used ANOVA (analysis of variance), a well-known statistical procedure.

3.5 Output analysis

A satisfactory level of credibility of the final results cannot be obtained without assessing their statistical errors. Neglecting the proper statistical analysis of simulation output data cannot be justified by the fact that some stochastic simulation studies might require sophisticated statistical techniques.

A difficult issue is the nature of the output observations of a simulation model. Observations collected during typical stochastic simulations are usually strongly correlated, and the classical settings for assessing the sample variance cannot be applied directly. Neglecting the existence of statistical correlation can result in excessively optimistic confidence intervals. For a thorough treatment of this and related questions, please refer to (Pawlikowski, 1990). The ultimate objective of run length control is to terminate the simulation as soon as the desired precision of relative width of confidence interval is achieved. There is a trade-off since one needs a reasonable amount of data to get the desired accuracy, but on the other hand this can lengthen the completion time. Considering that early stopping leads to inaccurate results, it is mandatory to decrease the computational demand of simulating steady-state parameters (Mota, 2002).

Typically, the run length of a stochastic simulation experiment is determined either by assigning the amount of simulation time before initiating the experiment or by letting the simulation run until a prescribed condition occurs. The latter approach, known as sequential

procedure, gather observations at the output of the simulation model to investigate the performance metrics of interest, and a decision has to be taken to stop the sampling. It is evident that the number of observations required to terminate the experiment is a random variable since it depends on the outcome of the observations.
According to this thought, carefully-designed sequential procedures can be economical in the sense that we may reach a decision earlier compared to fixed-sample-sized experiments. Additionally, to decrease computational demands of intensive stochastic simulation one can dedicate more resources to the simulation experiment by means of parallel computing. Efficient tools for automatically analyzing simulation output data should be based on secure and robust methods that can be broadly and safely applied to a wide range of models without requiring from simulation practitioners highly specialized knowledge. To improve the credibility of our simulation to investigate the proposal of using bandwidth efficiently for carrying VoIP over WMN, we used a combination of these approaches, namely, we applied a sequential procedure based on spectral analysis (Heidelberger & Welch, 1981) under Akaroa-2, an environment of Multiple Replications in Parallel (MRIP) (Ewing et al., 1999).
Akaroa-2 enables the same sequential simulation model be executed in different processors in parallel, aiming to produce independent an identically distributed observations by initiating each replication with strictly non-overlapping streams of pseudo-random numbers. It controls the run length and the accuracy of final results.
This environment solve automatically some critical problems of stochastic simulation of complex systems:

1. Minimization of bias of steady-state estimates caused by initial conditions. Except for regenerative simulations, data collected during *transient phase* are not representative of the actual average values of the parameters being simulated, and cannot be used to produce good estimates of steady-state parameters. The determination of its length is a challenging task carried out by a sequential procedure based on spectral analysis. Underestimation of the length of the transient phase leads to bias in the final estimate. Overestimation, on the other hand, throws away information on the steady state and this can increase the variance of the estimator.
2. Estimation of the sample variance of a performance measure and its confidence interval in the case of correlated observations in equilibrium state;
3. Stopping the simulation within a desired precision selected by the experimenter.

Akaroa-2 was designed for full automatic parallelization of common sequential simulation models, and full automated control of run length for accuracy of the final results Ewing et al. (1999). An instance of a sequential simulation model is launched on a number of workstations (operating as simulation engines) connected via a network, and a central process takes care of collecting asynchronously intermediate estimates from each processor and calculates conveniently an overall estimate.
The only things synchronized in Akaroa-2 are substreams of pseudo-random numbers to avoid overlapping among them, and the load of the same simulation model into the memory of different processors, but in general this time can be considered negligible and imposes no obstacle.
Akaroa-2 enables the same simulation model be executed in different processors in parallel, aiming to produce IID observations by initiating each replication with strictly non-overlapping streams of pseudo-random numbers provided by a combined multiple recursive generator (CMRG) (L'Ecuyer, 1999).

Essentially, a master process (*Akmaster*) is started on a processor, which acts as a manager, while one or more slave processes (*akslave*) are started on each processor that takes part in the simulation experiment, forming a pool of simulation engines (see Figure 2). Akaroa-2 takes care of the fundamental tasks of launching the same simulation model on the processors belonging to that pool, controlling the whole experiment and offering an automated control of the accuracy of the simulation output.

At the beginning, the stationary Schruben test (Schruben et al., 1983) is applied locally within each replication, to determine the onset of steady state conditions in each time-stream separately and the sequential version of a confidence interval procedure is used to estimate the variance of local estimators at consecutive checkpoints, each simulation engine following its own sequence of checkpoints.

Each simulation engine keeps on generating output observations, and when the amount of collected observations is sufficient to yield a reasonable estimate, we say that a checkpoint is achieved, and it is time the local analyzer to submit an estimate to the global analyzer, located in the processor running akmaster.

The global analyzer calculates a global estimate, based on local estimates delivered by individual engines, and verifies if the required precision was reached, in which case the overall simulation is finished. Otherwise, more local observations are required, so simulation engines continue their activities.

Whenever a checkpoint is achieved, the current local estimate and its variance are sent to the global analyzer that computes the current value of the global estimate and its precision.

NS-2 does not provide support for statistical analysis of the simulation results, but in order to control the simulation run length, ns-2 and Akaroa-2 can be integrated. Another advantage of this integration is the control of achievable speed-up by adding more processors to be run

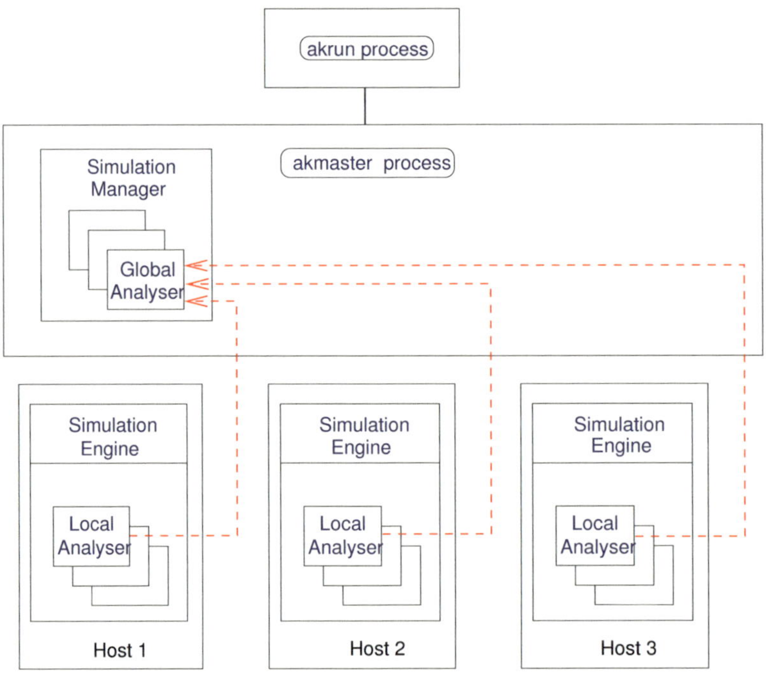

Fig. 2. Schematic diagram of Akaroa.

in parallel. A detailed description of this integration can be found in (*The ns-2akaroa-2 Project*, 2001).

4. Case study: header compression

4.1 Problem definition

One of the major challenges for wireless communication is the capacity of wireless channels, which is especially limited when a small delay bound is imposed, for example, for voice service. VoIP signaling packets are typically large, which in turn could cause a long signaling and media transport delay when transmitted over wireless networks (Yang & Wang, 2009). Moreover, VoIP performance in multi-hop wireless networks degrades with the increasing number of hops (Dragor et al., 2006).

VoIP packets are divided into two parts, headers and payload, that travel on RTP protocol over UDP. The headers are control information added by the underlying protocols, while the payload is the actual content carried out by the packet, that is, the voice encoded by some codec. As Table 1 shows, most of the commonly used codecs generates packets whose payload is smaller than IP/UDP/RTP headers (40 bytes).

In order to use the wireless channel capacity efficiently and make VoIP services economically feasible, it is necessary to apply compression techniques to reduce the overheads in the VoIP bearer and signaling packets. The extra bandwidth spared from control information traffic can be used to carry more calls in the same wireless channel or to allow the use of better quality codec to encode the voice flow.

Header compression in WMNs can be implemented in the mesh routers. Every packet received by a router from a mesh client should be compressed before being forwarded to the mesh backbone, and each packet forwarded to a mesh client should be decompressed before being forwarded out of the backbone. This guarantees that only packets with compressed headers would be transported among mesh backbone routers.

Header compression is implemented by eliminating redundant header information among packets of the same flow. The eliminated information is stored into data structures on the compressor and the decompressor, named context. When compressor and decompressor are under synchronization, it means that both compressor context and decompressor context are updated with the header information of the last sent/received packet of the flow. Figure 3 shows the scheme of header compression.

Codec	Bit rate (kbps)	Packet duration (ms)	Payload size (bytes)
G.711	64.0	20	160
G.726	32.0	20	80
G.728	16.0	20	40
G.729a	8.0	20	20
G.723.1 (MP-MLQ)	6.3	30	24
G.723.1 (ACELP)	5.3	30	20
GSM-FR	13.2	20	33
iLBC	13.33	30	50
iLBC	15.2	20	38

Table 1. Payload size generated by the most used codecs.

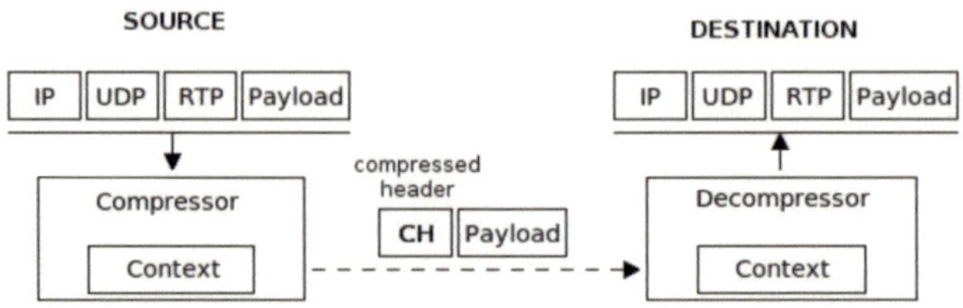

Fig. 3. General header compression scheme.

When a single packet is lost, the compressor context will be updated but the decompressor context will not. This may lead the decompressor to perform an erroneous decompression, causing the loss of synchronization between the edges and lead to the discard of all following packets at the decompressor until synchronization is restored. This problem may be crucial to the quality of communication on highly congested environments.

WMNs offer a high error rate in the channel due to the characteristics of the transmission media. Since only a device can transmit at a time, when more than one element transmits at the same time a collision occurs, as in the problem of the hidden node, which can result in loss of information in both transmitters. Moreover, many other things can interfere with communication, as obstacles in the environment, and receiving the same information through different paths in the propagation medium (multi-path fading). With these characteristics, the loss propagation problem may worsen, and the mechanisms of failure recovery by the algorithms may not be sufficient, especially in the case of bursty loss. Furthermore, the bandwidth in wireless networks is limited, making the allowed number of simultaneous users also limited. The optimal use of available bandwidth can maximize the number of users on the network.

4.2 Robust header compression – RoHC

The Compressed RTP (CRTP) was the first proposed header compression algorithm for VoIP, defined in the Request for Comments (RFC) 2508 (Casner & Jacobson, 1999). It was originally developed for low-speed serial links, where real-time voice and video traffic is potentially problematic. The algorithm compresses IP/UDP/RTP headers, reducing their

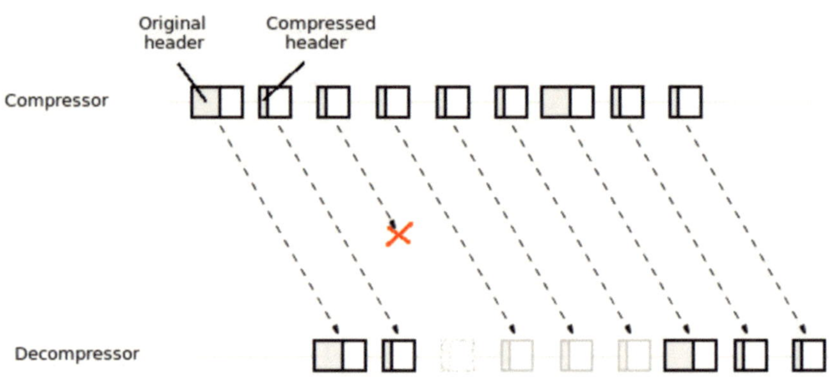

Fig. 4. Loss propagation problem.

size to approximately 2 bytes when the UDP checksum header is not present, and 4 bytes otherwise.

CRTP was designed based on the unique header compression algorithm available until that date, the Compressed TCP (CTCP) Jacobson (1990), which defines a compression algorithm for IP and TCP headers in low-speed links. The main feature of CRTP is the simplicity of its mechanism.

The operation of CRTP defines sending a first message with all the original headers information (FULL_HEADER), used to establish the context in the compressor and decompressor. Then, the headers of following packets are compressed and sent, carrying only the delta information of dynamic headers. FULL_HEADER packets are also periodically sent to the decompressor, in order to maintain synchronization between the contexts, or when requested by the decompressor through a feedback channel, if the decompressor detects that there was a context synchronization loss.

CRTP does not present a good performance over wireless networks, since it was originally developed for reliable connections (Koren et al., 2003), and characteristic of wireless networks present high packet loss rates. This is because the CRTP does not offer any mechanism to recover the system from a synchronization loss, presenting the loss propagation problem. The fact that wireless networks do not necessarily offers a feedback channel available to request for context recovery also influences the poor performance of CRTP.

The Robust Header Compression (RoHC) algorithm (Bormann et al., 2001) and (Jonsson et al., 2007) was developed by the Internet Engineering Task Force (IETF) to offer a more robust mechanism in comparison to the CRTP. RoHC offers three operating modes: unidirectional mode (U-mode), bidirectional optimistic mode (O-mode) and bidirectional reliable mode (R-mode). Bidirectional modes make use of a feedback channel, as well as the CRTP, but the U-mode defines communication from the compressor to the decompressor only. This introduces the possibility of using the algorithm over links with no feedback channel or where it is not desirable to be used.

The U-mode works with periodic context updates through messages with full headers sent to the decompressor. The B-mode and R-mode work with request for context updates made by the decompressor, if a loss of synchronization is detected. The work presented in (Fukumoto & Yamada, 2007) showed that the U-mode is most advantageous for wireless asymmetrical links, because the context update does not depend on the request from the decompressor through a channel that may not be available (by the fact that it is asymmetric link).

The RoHC algorithm uses a method of encoding for the values of dynamic headers that are transmitted in compressed headers, called Window-Least Significant Bits (W-LSB). This encoding method is used for headers that present small changes. It encodes and sends only the least significant bits, which the decompressor uses to calculate the original value of the header together with stored reference values (last values successfully decompressed). This mechanism, by using a window of reference values, provides a certain tolerance to packet loss, but if there is a burst loss that exceeds the window width, the synchronization loss is unavoidable.

To check whether there is a context synchronization loss, the RoHC implements a check on the headers, called Cyclic Redundancy Check (CRC). Each compressed header has a header field that carries a CRC value calculated over the original headers before the compression process. After receiving the packet, the decompressor retrieves the headers values with the information from the compressed header and from its context, and executes again the calculation of the CRC. If the value equals the value of the CRC header field, then the compression is considered

successful, otherwise it is found a synchronization loss.
The RoHC offers a high compression degree, and high robustness, but its implementation is quite complex compared to other algorithms. Furthermore, RoHC has been implemented for cellular networks, which typically have one single wireless link, and it considers that the network delivers packets in order.

4.3 Static compression + aggregation

A header compression algorithm that does not need synchronization of contexts could eliminate any possibility of discarding packets at the decompressor due to packet loss, and eliminate all necessary process for updating and context re-synchronization. However, the cost to implement such an algorithm may be reflected in the compression gain, which may be lower with respect to algorithms that require synchronization.
If it is not possible to maintain the synchronization, the decompressor cannot decompress the headers of received packets. Whereas usually the decompressor is based on the information of previously received packets of the same stream to update its context, the loss of a single packet can result in the context synchronization loss, and then the decompressor may not decompress the following packets successfully, even if they arrive on time and without errors at the decompressor, and it is obliged to discard them. In this case we say that the loss was propagated as the loss of a single packet leads to the decompressor to discard all the following packets (Figure 4).
To alleviate the loss propagation problem, some algorithms use context update messages. Those messages are sent periodically, containing all the complete information of the headers. When the decompressor receives an update message, it replaces the entire contents of its current context for the content of the update message. If it is unsynchronized, it will use the information received to update its reference values, and thus restore the synchronization.
One way to solve the problem of discarding packets at the decompressor due to context desynchronization was proposed in (Nascimento, 2009), by completely eliminating the need of keeping synchronization between compressor and decompressor. The loss propagation problem can be eliminated through the implementation of a compression algorithm whose contexts store only the static headers, and not the dynamic ones. If the contexts store static information only, there is no need for synchronization. This type of compression is called static compression.
The static compression has the advantage of no need of updating the context of compressor and decompressor. It only stores the static information, i.e., those that do not change during a session. This means that no packet loss will cause following packets to be discarded at the decompressor, thus eliminating the loss propagation problem. Another advantage presented by the static compression is the decrease in the amount of information to be stored in points where compression and decompression occur, as the context stores only the static information. However, the cost of maintaining contexts without the need for synchronization is reflected in the compression gain, since the dynamic information is sent in the channel and is not stored in context, as in conventional algorithms (Westphal & Koodli, 2005). This causes the compressed header size increase in comparison to conventional compression algorithms headers size, reducing the compression gain achieved.
The static compression can reduce the headers size to up to 35% of its original size. Some conventional algorithms, which require synchronization, can reduce the headers size to less than 10%. Experiments with static compression in this work showed that even though this algorithm does not present the loss propagation problem, its compression gain is not large

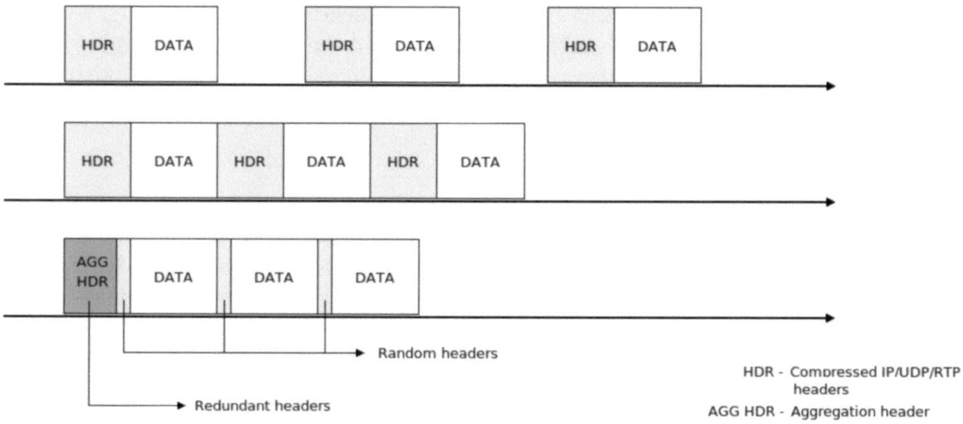

Fig. 5. Cooperative solution: compression + aggregation.

enough to offer significant gains in comparison to more robust algorithms. Therefore, it is suggested the use of technical aids to increase the compression gain achieved while using the static compression mechanism.

The static header compression use headers whose values do not change between packets of the same voice stream. However, some dynamic headers most of the time of a session also presents some redundancy between consecutive packets, because they follow a pre-established behavior pattern. One way to provide greater compression gain for the static header compression can take advantage of that redundancy often present. To use the dynamic information redundancy without returning to the problem of contexts synchronization and loss propagation, after the static compression process we can use a simple aggregation packet mechanism. The packet aggregation is a technique also used to optimize the bandwidth usage in wireless networks. Its main goal is, through the aggregation of several packets, to reduce the overhead of time imposed by the 802.11 link layer wireless networks MAC, reduce the number of packet loss in contention for the link layer, and decrease the number of retransmissions (Kim et al., 2006). In addition, aggregation also helps to save bandwidth consumption for control information traffic by decreasing the amount of MAC headers sent to the network.

An effective cooperation between the packet aggregation and packet header compression techniques requires thatonly packets of the same flow can be aggregated. The packet aggregation introduces a delay of the queuing process, since the compressor needs to expect the arriving of k packets to form an aggregation packet, where k is called aggregation degree. This additional delay reflects on the quality of the call, and that means that this type of mechanism is not the best option in environments with few wireless hops, or low traffic load. It is therefore important to use a low aggregation degree, since this value is directly proportional to the delay to be imposed on the traffic.

After the aggregation, the dynamic redundant information among the packets headers of the aggregated packets are taken from the compressed headers and kept into a single external header called aggregation header (Figure 5). By redundant information we mean that ones assuming sequential values or the same value among the aggregated packets.

The aggregation header contains the IP/UDP/RTP headers which value is equal for all aggregated packets. So when the aggregation packet reaches the destination, the decompressor will be able to rebuild the compressed header of each aggregated packet, from the aggregation header, and thus may continue with the process of deaggregation and subsequent static decompression. The experiments conducted in this study showed that the mechanism of compression and aggregation can increase the compression gain from about 60% (static compression only) to more than 80%.

4.4 Objective of the study

The main objective of this study is to evaluate the performance of the proposed approach based on the combination of static header compression and packet aggregation. We also aim to assess the performance of the algorithm RoHC U-mode, since it is an algorithm standardized by IETF, presenting a high compression gain, and presenting the loss propagation problem.

The objective of this chapter is to suggest a sound simulation methodology aiming to get reliable results of simulations of VoIP over WMN. To achieve this goal, we started by selecting an experimental environment based on two well-known simulation tools: ns-2 and Akaroa-2. The first one was selected due to its widely use in the scientific community, which enables the repeatability of the experiments. Moreover, ns-2 receives steadily support from active forums of developers and researchers. We used the version 2.29, which received a patch with improvements on physical and link layers modeling capabilities.

Akaroa-2 was deployed to guarantee the statistical quality of the results. We are interested in measures of the steady-state period, and Akaroa-2 is in charge of detecting the end of the transient period. Observations of that period are discarded by Akaroa-2, mitigating the bias effects that should appear in the final results otherwise. The carefully design of Akaroa-2 for detecting the end of the transient period is based on a formal method proposed in (Schruben et al., 1983), as opposed to simple heuristics. By integrating ns-2 and Akaroa-2, sources of randomness in our simulation model make use of the pseudo-random number of the latter, which we analyzed and accepted as adequate to our purposes.

4.5 Experimental design

For this study we opted for the use of end-to-end header compression, for no extra cost in the intermediate nodes between a source-destination pair. To make the use of end-to-end header compression over a WMN, it is necessary that the routers of the network are able to route packets with compressed headers. Since the header compression is applied also to the IP header, that means that the routers must implement the packets routing without extracting information from IP headers.

We decided to use routing labels, implemented with the Multi-protocol Label Switching (MPLS) (Rosen et al., 2001). The MPLS is known to perform routing between the network and link layers, thus performing routing on layer 2.5 (Figure 6). The MPLS works primarily with the addition of a label in the packets (and it is indifferent to the type of data transported, so it can be IP traffic or any other) in the first router of the backbone (edge router) and then the whole route through the backbone will be made by using labels, which are removed when the packets leave the backbone.

We used the implementation of MPLS for NS-2.26 available in (Petersson, 2004), it is called MPLS Network Simulator (MNS) version 2.0. It required a small adjustment on the module for use in version 2.29, and the structure of the wireless node of NS-2, because the original module only applies to wired networks.

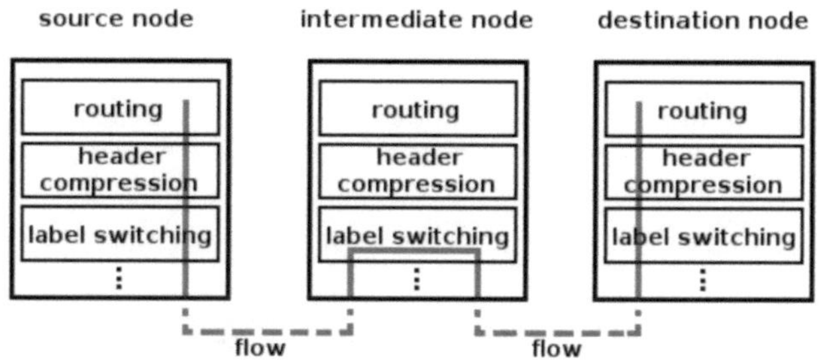

Fig. 6. Label routing performed by MPLS.

4.5.1 Factors

We compared the proposed scheme based on the combination pf static header compression and packet aggregation (SHC+AG) against ROHC, and the static header compression approach (SHC). Decisive values for state transition on compressor and decompressor state machines, like number of sent packets before changing to a higher state, or number of decompress failures before changing to a lower state are not discussed in the RoHC Request for Comments. In our experiments, those values were established in according to (Seeling et al., 2006; Fukumoto & Yamada, 2007).

4.5.2 Performance measures

Packet loss A factor that influences the quality of real-time applications is the packet loss in the network. VoIP applications offer some tolerance to packet loss, since small losses are imperceptible to the human ear. However, this tolerance is very limited, and high rates of packet loss could impose a negative impact on the speech quality and harm the understanding of the interlocutors.

Network delay It is a primary factor of influence on the speech quality. The time limit for the human ear does not perceive delay on the speech reproduction is 150 ms. Therefore, if the network imposes very large delays, the impact of this factor in the quality of the call will be noticeable.

MOS The MOS is intended to quantitatively describe the speech quality, taking into account several factors, including packet loss, delay, codec, compression, etc. Therefore, the MOS, presented in our work together with the metric of loss and delay, will give an idea of how those metrics affect the quality of the call as a whole. It also make it possible to check if the average quality of calls can be considered acceptable.

Compression gain This measure indicates how effective is the compression mechanism with respect to its ability on decreasing the headers size. The higher the algorithm compression gain, the greater is its ability on compressing the headers.

Bandwidth efficiency This measure indicates how much of bandwidth was used for payload transmission, thus quantifying the contribution of each header compression algorithm to a more optimal usage of available bandwidth. It is obtained through the ratio between the total number of transmitted bytes of payload and the total bytes effectively used for the transmission, including payload and headers.

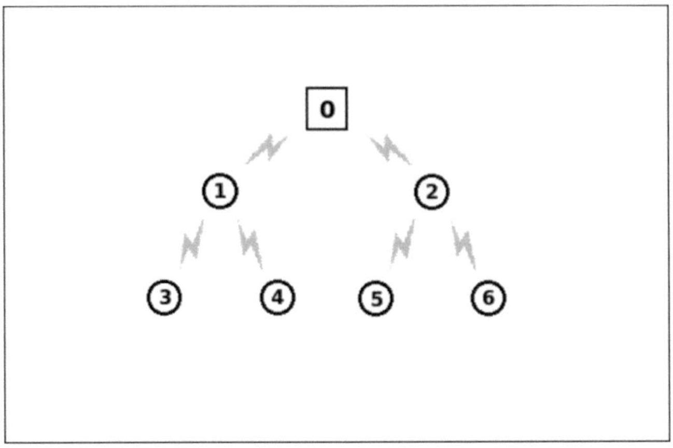

Fig. 7. Tree scenario used in the WMN simulation.

4.5.3 Scenario

In our experiments we simulated the IEEE 802.11b channel (IEEE, 2004), and we used the log-normal shadowing propagation model, with the changes suggested in (Schmidt-Eisenlohr et al., 2006), and the wireless channel was set based on (Xiuchao, 2004), customized according to measurements. The shadowing model was set to use pathloss exponent of 3.5 and standard deviation of 4.0 (Rappaport, 2001).

The selected scenario represents a mesh backbone with 7 routers positioned on tree form (Figure 7). In this scenario, the main idea is to evaluate the algorithms on a network whose routers need to handle traffic from different sources. On mesh networks such behavior is common. The VoIP calls were generated from the leaf nodes (nodes 3, 4, 5, and 6) destined to the node 0. This traffic behavior is usual in many mesh networks which have a gateway, a device that gives access to external networks or to the Internet.

4.5.4 Traffic model

Bidirectional VoIP calls were modeled as CBR ON/OFF traffic with 60 seconds of duration, configured to represent a voice stream coded by G.729a codec with 20ms of frame duration, 8 kbps of bit rate and with static dejitter buffer of 50ms. The codec G.729a was used, because it offers good quality in low transmission rate conditions.

4.5.5 Statistical plan

The mean value of the performance measures and the corresponding confidence interval were obtained by applying the sequential version of the Spectral Analysis method of estimation implemented by Akaroa-2. (Pawlikowski et al., 1998) concluded that this method of analysis under MRIP is very accurate. A method is said to be accurate when final confidence interval is quite close to the theoretical confidence interval of a simulation model whose analytical solution is known in advance.

Given the desired confidence level, Akaroa-2 collects a number of stead-state observations (samples) after deleting observations of the transient period. At predefined checkpoints determined by the Spectral Analysis method, Akaroa-2 calculates the estimative of the performance measure of interest, computes the confidence interval and then check the relative precision. If the relative precision is less than the maximum relative precision set by the experimenter, the simulation is finished, otherwise the simulation keeps generating samples

Algorithm	Compression gain
Robust Header Compression (RoHC)	0.8645
Static Header Compression (SHC)	0.6384
Static Header Compression + Aggregation (SHC+AG)	0.8274

Table 2. Compression gain of the header compression algorithms used in the simulation.

until the next checkpoint. For the experiments in this study, we set a confidence level of 95%, and a maximum relative precision of 5%. The run length control is automatically done by Akaroa-2.

4.6 Results analysis

We are going to depict only the main results, but the interested readers can access `http://grcm.dcc.ufam.edu.br` to get more details and, of course, the source code.

Table 2 shows the values obtained for the compression gain of the evaluated algorithms. The high compression gain presented by Robust Header Compression (RoHC) algorithm is due to the fact that header compression eliminates the static and dynamic information, leaving only the context identification information and the dynamic information when there is a change in its values.

The RoHC algorithm is able to decrease the headers size up to 2 bytes, which could provide an even greater compression gain. However, since it eliminates the dynamic information from the headers, the RoHC U-mode algorithm periodically needs to send context update messages with the objective of recovering the decompressor from a possible loss of synchronization. Those update messages have a larger size than the compressed headers, reaching almost the size of the original headers.

The frequency on which update messages are sent is a trade-off for header compression algorithms that need to update the context. The shorter this frequency is, the lower is the possibility of the decompressor context being outdated, however, the lower the compression gain. Therefore, the act of sending update messages and the frequency on which they are sent directly influence the RoHC compression gain. In our experiments, we sent messages to update the headers on every 10 packets of compressed headers according to the work presented in Seeling et al. (2006).

The static compression algorithm showed the smallest compression gain. Static compression eliminates from IP/UDP/RTP headers only the information classified as static and inferred, maintaining the dynamic information in the headers. Therefore, as expected, the static compression algorithm does not offer a compression gain so high as the RoHC algorithm, that also compresses the dynamic headers. The impact of this difference in compression gain on voice traffic behavior will be evaluated with the analysis of packet loss, delay and MOS.

The static compression and packet aggregation approach showed a compression gain almost as high as the RoHC algorithm. It means that the aggregation process has fulfilled the task of increasing the compression gain of the static compression. Although the compression gain did not exceed that obtained by the RoHC, the value displayed is approaching and that means that the static header compression and packet aggregation approach has generated headers of size almost as small as the headers compressed by RoHC.

The static compression showed a compression gain of 0.6384. The mechanism of packet aggregation offered an extra compression gain due to elimination of redundant dynamic header information of the aggregated packets. In this case, we can say that the compression gain of this approach is also influenced by the aggregation degree, which in our experiments

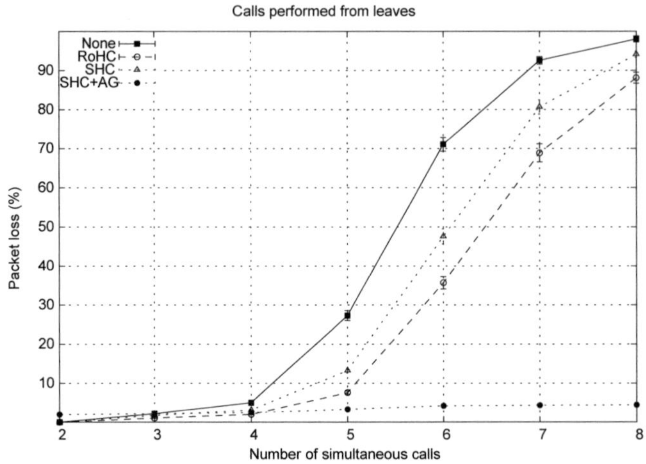

Fig. 8. Packet loss of calls performed over the tree scenario, using different header compression settings.

was two packets per aggregation packet. The aggregation degree poses a trade-off for overall speech quality, because the greater it is, the greater is the extra compression gain, but the packetization delay will be greater.

Figure 8 shows the values of packet loss obtained in the tree scenario. The number of simultaneous calls shown in the graph is the amount of simultaneous calls for each source-destination pair. Experiments were carried out with the number of simultaneous calls ranging from 2 to 8. For five or more simultaneous calls, the majority of the settings showed high packet loss rates.

The calls without header compression (*none*) showed the highest packet loss rate values. The packet loss for the SHC algorithm was higher than for the other compression algorithms.

The SHC+AG approach showed lower packet loss rate values if compared to the other algorithms. Although the aggregation increases the size of the packets, which could also increase the loss, this procedure also reduces the amount of packets sent to the network in a such a way proportional to the aggregation degree used. The packet aggregation on every two packets, as used in our experiments, resulted in the creation of slightly larger packets but not large enough to negatively impact the packet loss.

Then, aggregation reduces the amount of packets sent to the network, reducing the effort of the link layer. In addition, it provides a decrease in the amount of bytes sent to the network, key feature of the header compression. Therefore, besides the high compression gain offered by SHC+AG approach, the positive impact on the packet loss was also due to aggregation itself, which was primarily responsible for maintaining the packet loss rate below than those provided by other algorithms.

In this experiment, VoIP calls were performed to node 0, from all network nodes, and from the leaf nodes, at different times. Figure 9 shows the MOS values calculated on those calls.

The RoHC and SHC algorithms showed higher values of MOS, but showed no significant difference between them, which can be explained by the no significant difference between their packet loss rate. The SHC+AG approach showed the lowest MOS values for 2, 3, and 4 simultaneous calls, and the highest values with the increase of the simultaneous calls. Again, the MOS showed a more stable behavior with the increase on the number of calls, compared

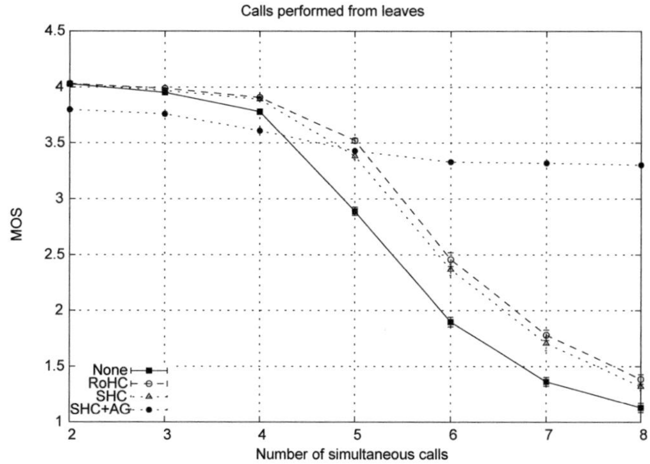

Fig. 9. MOS of calls performed over the tree scenario, using different header compression settings.

to other algorithms. This is justified by the behavior also more stable presented by metrics of delay and packet loss.

5. Conclusion

In this chapter we have considered a set of necessary conditions that should be fulfilled to give credibility to performance evaluation studies of VoIP transmission over WMN based on stochastic simulation. Since we have followed a sound methodology formed by the carefully choices in every stage of the simulation methodology, we can be sure that our results are reliable, no matter which results we have obtained. Certainly, the proposed compression scheme deserves additional fine tuning, but we are sure that future versions of it can be compared in an unbiased manner.

6. References

Andel, T. R. & Yasinac, A. (2006). On the Credibility of Manet Simulations, *Computer* 39(7): 48–54.

Balci, O. & Nance, R. (1985). Formulated Problem Verification as an Explicit Requirement of Model Credibility, *Simulation* 45(2): 76–86.

Bormann, C., Burmeister, C., Degermark, M., Fukushima, H., Hannu, H., Jonsson, L.-E., Hakenberg, R., Koren, T., Le, K., Liu, Z., Martensson, A., Miyazaki, A., Svanbro, K., Wiebke, T., Yoshimura, T. & Zheng, H. (2001). Robust Header Compression: Framework and four profiles, Request for Comments 3095.

Carvalho, L. S. G. (2004). *An e-model implementation for objective speech quality evaluation of voip communication networks*, Master's thesis, Federal University of Amazonas.

Carvalho, L. S. G., Mota, E. S., Aguiar, R., Lima, A. F., de Souza, J. N. & Barreto, A. (2005). An e-model implementation for speech quality evaluation in voip systems, *IEEE Symposium on Computers and Communications*, Cartagena, Spain, pp. 933–938.

Casner, S. & Jacobson, V. (1999). Compressing IP/UDP/RTP Headers for Low-Speed Serial Links, Request for Comments 2508.

Clark, A. D. (2003). Modeling the Effects of Burst Packet Loss and Recency on Subjective Voice

Quality, *IP Telephony Workshop*, Columbia University.

C.L.Barrett, Marathe, A., Marathe, M. & Drozda, M. (2002). Characterizing the interaction between routing and mac protocols in ad-hoc networks, *MobiHoc '02: Proceedings of the 3rd ACM international symposium on Mobile ad hoc networking & computing*, ACM, New York, NY, USA, pp. 92–103.

Conway, R. (1963). Some tactical problems in digital simulation, *Management Science* 10, 1: 47–61.

Dragor, N., Samrat, G., Kyungtae, K. & Rauf, I. (2006). Performance of voip in a 802.11 wireless mesh network, *Proceedings of the IEEE INFOCOM*, Barcelona, Spain, pp. 49–52.

Ewing, G. C., Pawlikowski, K. & Mcnickle, D. (1999). Akaroa2: Exploiting network computing by distributing stochastic simulation, *Proceedings of the 13th European Simulation Multi-Conference*, Warsaw, Poland, pp. 175–181.

Fukumoto, N. & Yamada, H. (2007). Performance Enhancement of Header Compression over Asymmetric Wireless Links Based on the Objective Speech Quality Measurement, *SAINT '07: Proceedings of the 2007 International Symposium on Applications and the Internet*, IEEE Computer Society, Washington, DC, USA, p. 16.

Glynn, P. & Heidelberger, P. (1992). Experiments with initial transient deletion for parallel replicated steady-state simulations, *Management Science* 38(3): 400–418.

Grancharov, V. & Kleijn, W. B. (2008). *Speech Quality Assessment*, Springer, chapter 5, pp. 83–99.

Heidelberger, P. & Welch, P. D. (1981). A spectral method for confidence interval generation and run length control in simulations, *Communications of the ACM* 24(4): 233–245.

Hoene, C., Karl, H. & Wolisz, A. (2006). A perceptual quality model intended for adaptive VoIP applications, *Int. J. Commun. Syst.* 19(3): 299–316.

IEEE (2004). IEEE 802.11TM Wireless Local Area Networks.
URL: *http://grouper.ieee.org/groups/802/11/*

Ivanov, S., Herms, A. & Lukas, G. (2007). Experimental validation of the ns-2 wireless model using simulation, emulation, and real network, *In 4th Workshop on Mobile Ad-Hoc Networks (WMAN07*, pp. 433–444.

Jacobson, V. (1990). Compressing TCP/IP Headers for Low-Speed Serial Links, Request for Comments 1144.

Jonsson, L.-E., Sandlund, K., Pelletier, G. & Kremer, P. (2007). Robust Header Compression: Corrections and Clarifications to RFC 3095, Request for Comments 4815.

Kim, K., Ganguly, S., Izmailov, R. & Hong, S. (2006). On Packet Aggregation Mechanisms for Improving VoIP Quality in Mesh Networks, *Proceedings of the Vehicular Technology Conference, VTC'06*, IEEE, pp. 891–895.

Koren, T., Casner, S., Geevarghese, J., Thompson, B. & Ruddy, P. (2003). Enhanced Compressed RTP (CRTP) for Links with High Delay, Packet Loss and Reordering, Request for Comments 3545.

Kurkowski, S., Camp, T. & Colagrosso, M. (2005). Manet simulation studies: the incredibles, *SIGMOBILE Mob. Comput. Commun. Rev.* 9(4): 50–61.

Law, A. M. & McComas, M. G. (1991). Secrets of successful simulation studies, *Proceedings of the 23rd conference on Winter simulation*, IEEE Computer Society, Washington, DC, USA, pp. 21–27.

L'Ecuyer, P. (1999). Good parameters and implementations for combined multiple recursive random number generators, *Operations Research* 47(1): 159–164.

L'Ecuyer, P. (2001). Software for uniform random number generation: Distinguishing the good and the bad, *Proceedings of the 33rd Conference on Winter Simulation*, IEEE Computer

Society, Virginia, USA, pp. 95–105.

Lili, Z., Huibin, W., Lizhong, X., Zhuoming, X. & Chenming, L. (2009). Fault tolerance and transmission delay in wireless mesh networks, *NSWCTC '09: Proceedings of the 2009 International Conference on Networks Security, Wireless Communications and Trusted Computing*, IEEE Computer Society, Washington, DC, USA, pp. 193–196.

Matsumoto, M. & Nishimura, T. (1998). Mersenne twister: a 623-dimensionally equidistributed uniform pseudo-random number generator, *ACM Trans. Model. Comput. Simul.* 8(1): 3–30.

McCanne, S. & Floyd, S. (2000). The Network Simulator.
URL: *http://www.isi.edu/nsnam/ns/*

Mota, E., Fitzek, F., Pawlikowski, K. & Wolisz, A. (2000). Towards Credible and Efficient Network Simulation Experiments, *Proceedings of the High Performance Computing Symposium, HCPS'2000*, Washington, DC, USA, pp. 116–121.

Mota, E. S. (2002). *Performance of Sequential Batching-based Methods of Output Data Analysis in Distributed Steady-state Stochastic Simulation*, PhD thesis, Technical University of Berlin.

Myakotnykh, E. S. & Thompson, R. A. (2009). Adaptive Rate Voice over IP Quality Management Algorithm, *International Journal on Advances in Telecommunications* 2(2): 98–110.

Nascimento, A. G. (2009). *Header compression to achieve speech quality in voip over wireless mesh*, Master's thesis, Federal University of Amazonas.

Naylor, T. H., Finger, J. M., McKenney, J. L., Schrank, W. E. & Holt, C. C. (1967). Management Science, *Management Science* 14(2): B92–B106. Last access at 23 Feb. 2009.
URL: *http://www.jstor.org/stable/2628207*

Pawlikowski, K. (1990). Steady-state simulation of queueing processes: survey of problems and solutions, *ACM Comput. Surv.* 22(2): 123–170.

Pawlikowski, K., Ewing, G. C. & Mcnickle, D. (1998). Coverage of Confidence Intervals in Sequential Steady-State Simulation, *Journal of Simulation Practise and Theory* 6(3): 255–267.

Pawlikowski, K., Jeong, H.-D. & Lee, J.-S. R. (2002). On Credibility of Simulation Studies of Telecommunication Networks, *IEEE Communications Magazine* 40(1): 132–139.

Petersson, M. (2004). MPLS Network Simulator verso 2.0 para NS-2 2.26.
URL: *http://heim.ifi.uio.no/ johanmp/ns-2/mns-for-2.26.tar.gz*

Queiroz, S. (2009). *Evaluation of incremental routing in 802.11 wireless mesh networks*, Master's thesis, Federal University of Amazonas.

Raake, A. (2006). *Speech Quality of VoIP: Assessment and Prediction*, John Wiley & Sons.

Ramalingam, G. & Reps, T. (1996). An incremental algorithm for a generalization of the shortest-path problem, *J. Algorithms* 21(2): 267–305.

Rappaport, T. S. (2001). *Wireless Communication Principles and Practice*, Prentice Hall PTR.

Rosen, E., Viswanathan, A. & Callon, R. (2001). Multiprotocol Label Switching Architecture, Request for Comments 3031.

Sargent, R. G. (2008). Verification and validation of simulation models, *Proceedings of the 40th Conference on Winter Simulation*, pp. 157–169.

Schmidt-Eisenlohr, F., Letamendia-Murua, J., Torrent-Moreno, M. & Hartenstein, H. (2006). Bug Fixes on the IEEE 802.11 DCF module of the Network Simulator Ns-2.28, Technical Report.
URL: *http://www.telematica.polito.it/fiore/index.html*

Schruben, L., Singh, H. & Tierney, L. (1983). Optimal tests for initialization bias in simulation output, *Operations Research* 31, 6: 1167–1178.

Seeling, P., Reisslein, M., Madsen, T. K. & Fitzek, F. H. (2006). Performance Analysis of Header Compression Schemes in Heterogeneous Wireless Multi—Hop Networks, *Wirel. Pers. Commun.* 38(2): 203–232.

Stevanović, R., Topić, G., Skala, K., Stipčević, M. & MedvedRogina, B. (2008). Quantum random bit generator service for monte carlo and other stochastic simulations, pp. 508–515.

The ns-2akaroa-2 Project (2001). Last access at 24 Aug. 2010. URL: *http://www.tkn.tuberlin.de/research/ns-2_akaroa-2/ns.html*

Totaro, M. W. & Perkins, D. D. (2005). Using statistical design of experiments for analyzing mobile ad hoc networks, *MSWiM '05: Proceedings of the 8th ACM international symposium on Modeling, analysis and simulation of wireless and mobile systems*, ACM, New York, NY, USA, pp. 159–168.

Union, I. T. (1996). Methods for subjective determination of transmission quality, *Recommendation P.800*, Telecommunication Standardization Sector of ITU, Geneva, Switzerland.

Union, I. T. (1998). The E-model, a computational model for use in transmission planning, *Recommendation G.107*, Telecommunication Standardization Sector of ITU, Geneva, Switzerland.

Weigle, M. C. (2006). Improving confidence in network simulations, *Proceedings of the Winter Simulation Conference*, Monterey, CA, pp. 2188–2194.

Westphal, C. & Koodli, R. (2005). Stateless IP Header Compression, *Proceedings of IEEE International Conference on Communication*, Mountain View, CA, USA, pp. 3236 – 3241.

Xiuchao, W. (2004). Simulate 802.11b Channel within Ns-2. URL: *http://www.comp.nus.edu.sg/ wuxiucha/research/reactive/report /80211ChannelinNS_2new.pdf*

Yang, Y. & Wang, X. (2009). *Compression Techniques for VoIP Transport over Wireless Interfaces*, CRC Press, chapter 5, pp. 83–99.

10

Virtual Home Region Multi-hash Location Management Service (VIMLOC) for Large-Scale Wireless Mesh Networks[1]

J. Mangues-Bafalluy, M. Requena-Esteso, J. Núñez-Martínez and A. Krendzel
Centre Tecnològic de Telecomunicacions de Catalunya (CTTC)
Av. Carl Friedrich Gauss, 7 – 08860 Castelldefels – Barcelona
Spain

1. Introduction

Wireless mesh networks (WMNs) have recently received much attention not only from the research community, but also from municipalities or non-tech-savvy user communities willing to build their own all-wireless network. One of the factors that has helped in making WMNs become popular is the widespread availability of low-cost wireless equipment, and particularly, IEEE 802.11 WLAN equipment. However, making these WMNs operationally efficient is a challenging task. In this direction, there has been a lot of work on the research issues highlighted in (Akyildiz & Wang, 2005). Nevertheless, such research topic as mobility management did not receive as much attention as others (e.g., channel assignment or routing).
In general, mobility management is split into two main functions, namely handoff management and location management. The former deals with maintaining the communication of the mobile node (MN) while (re-)attaching to a new attachment point, whilst the latter deals with locating the MN in the network when a new communication needs to be established.
Related to mobility, and at an architectural level, a common belief in the research community is that, unlike in an IP context, node identifiers and addresses (i.e., the current location in the network of those nodes) should not be integrated into a single identifier. The main purpose of this is to enable designing efficient mobility management schemes, and as part of them, efficient location management schemes (location services). This is particularly challenging in large-scale WMNs, due to the state information that must be stored in the nodes and the associated control overhead sent through the network. Related to this, position-based (geographic) routing algorithms are expected to improve scalability of large

---

[1] Based on "VIMLOC location management in wireless meshes: Experimental performance evaluation and comparison", by Mangues-Bafalluy et al., which appeared in Proc. ICC-2010, South Africa. © [2010] IEEE; "VIMLOC: Virtual Home Region Multi-Hash Location Service in Wireless Mesh Networks", by Krendzel et al., which appeared in Proc. Wireless Days-2008, United Arab Emirates. © [2008] IEEE; "Wireless Mesh Networking Framework for Fast Development and Testing of Network-level Protocols", by Requena-Esteso et al., which appeared in Proc. of the ICT-Mobile Summit-2009, Spain © [2009].

WMNs. In fact, by exploiting position information of nodes in the network both state information and control overhead can be substantially reduced when compared to more traditional flooding-based approaches.

Two building blocks are required for deploying an operational position-based routing scheme, namely a location management service and a position-based routing/forwarding algorithm (Mauve et al., 2001), (Camp, 2006). The location management service/scheme is needed to map between the identifier of a node (node_ID) and its current position in the network (i.e., location address (LA)) so that an underlying position-based routing/forwarding algorithm could take forwarding decisions based on the location information included in the packet header. A location management scheme is transparent/orthogonal from the viewpoint of the main underlying packet forwarding strategies, such as greedy forwarding (Camp, 2006), GPSR (Karp & Kung, 2000), restricted directional flooding (e.g. LAR (Ko & Vaidya, 2000)), etc.

In this chapter, we focus on a scalable distributed location management (DLM) scheme for large WMNs. Scalability is determined by the efficiency of a scheme in terms of overhead introduced in the network and state volume in the nodes to achieve two main goals: 1) a certain level of robustness, understood as the ability to make the location of a given node accessible even in the presence of impairments in the network, and 2) as accurate as possible location information, i.e., as up-to-date as possible.

Although a large number of location management schemes/services are available for mobile ad hoc networks (MANETs), up to our knowledge, there has not been a DLM scheme specifically designed for WMNs taking advantage of the availability of a highly static and non-power-constrained network backbone. Besides, location management schemes, even for MANETs, have only been simulated and there is no previous experimental evaluation over a real testbed implementation.

This chapter presents, up to our knowledge, the first DLM scheme, called *Vi*rtual Home Region *M*ulti-Hash *Loc*ation Service (VIMLOC), specifically designed to provide high robustness and accuracy in large-scale WMNs.

It also presents an experimental performance evaluation of VIMLOC under various network load conditions. Furthermore, it presents what is, up to our knowledge, the first experimental performance comparison over a WMN testbed of three different location management schemes, namely proactive, reactive, and VIMLOC. The interest of proactive and reactive schemes resides in that they represent the two main philosophies of operation in location management (Camp, 2006), and for this reason, they are taken as reference for the comparison with VIMLOC. All three schemes have been implemented in the Click modular router framework (Kohler et al., 1999). An extensive measurement campaign has been carried out to determine the efficiency, robustness, and accuracy each of these schemes.

This chapter is structured as follows. First, the most representative location services found in the literature for WMNs and MANETs are analyzed to define which ideas better match the requirements of large-scale WMNs. Second, these ideas are adapted to design a new robust and accurate DLM location service (VIMLOC) for WMNs, by introducing the new functional entities, components, and procedures. Third, the operation of VIMLOC in combination with a geographic routing scheme is explained. Then, the main building blocks of the implementation of VIMLOC using the Click modular router framework as well as the testbed developed to test the DLM scheme are described. After that, the experimental evaluation of VIMLOC is presented and discussed and its performance is compared over a WMN testbed with two different flooding-based philosophies, namely reactive and proactive.

2. Related work

Up to our knowledge, no location management scheme specially designed to take into account the requirements of a large-scale WMN (scalability, robustness, accuracy, benefits of stable backbone, etc.) can be found in the literature.

The traditional region-based location management scheme used in typical cellular networks and its improvement, called cluster-based location management scheme, have been theoretically analyzed in (Hu et al., 2007), (Hu et al., 2009) in the context of a mesh network based on the WiMAX technology. However, their idea of WMN is not exactly the same as the one we are considering in this chapter. The WiMAX-based mesh network consists of a base station, subscriber stations that act as client-side terminals through which mobile users can access the network, and mobile terminals. It is assumed that packets are forwarded to/from the base station, which serves as a gateway between the external network and the WiMAX mesh network and subscriber stations act as relays of the root base station, hence forming a tree. Therefore, this WMN is not really a fully distributed mesh network. Thus, these management schemes have no direct application to our scenarios.

In general, previous work on distributed location schemes/services may be found mostly for MANETs. As a basis for the development of a location service scheme for WMNs, some features of location services developed earlier for MANETs have to be revisited when taking into account the specificity of WMNs. For this reason, the main location schemes used in MANETs are analyzed below from the viewpoint of its possible applicability to WMNs.

In accordance with Mauve's classification (Mauve et al., 2001), existing location services for MANETs can be defined depending on what nodes actively participate in the location process, i.e., what nodes are servers storing location information. This can be either *all* nodes in the networks or *some* specific nodes. Besides, each server can store location information about positions of *all* nodes in the network or positions of *some* specific nodes.

On the other hand, in accordance with Camp's classification (Camp, 2006), location services can be divided into three types: proactive location database schemes, proactive location dissemination schemes, and reactive location schemes. In *proactive* location schemes nodes exchange location information periodically. Correspondingly, in a *reactive* location scheme location information requested when needed. In a proactive location *dissemination* scheme *all* nodes have location databases for *all* other nodes in the network. Therefore, a node can find in its local location table information about the position of any destination node of the network. On the other hand, in a proactive location *database* scheme, typically *all* nodes in the network maintain location databases for *some* other nodes. Thus, when a node needs position information about a destination node, it first requests the location database servers storing the destination node location.

The DREAM location service (Basagni et al., 1998) is an *all-for-all* proactive location dissemination scheme. From the viewpoint of large scale WMNs, it is not reasonable that each node is considered a server database for all other nodes given the state information required. Besides, it uses flooding to spread location information throughout the network. In other words, the number of one-hop transmissions of a location update procedure is very high and scales with $O(n)$ (Mauve et al., 2001). As a consequence, DREAM has low scalability and does not seem to be appropriate for large-scale WMNs.

The Reactive Location Service (RLS) (Kaseman et al, 2002) is classified as an *all-for-some* reactive location scheme. This scheme also uses flooding, but in its request procedure. Thus, the number of one-hop transmissions of a lookup procedure is very high (Kies, 2003), (Kies et al, 2004). Therefore, this scheme has low scalability as well, and thus, it does not seem to be efficient enough for a large-scale WMN.

Other location services are proactive location database schemes. They do not require flooding since specific nodes in the network serve as location databases for other specific nodes in the network (Camp, 2006).

The Row/Column location service (Stojmenovic, 1999) is a proactive location database scheme that uses the *all-for-some* approach. Spatial orientation in a certain direction (north/south, east/west) for location update and location request procedures is used in the scheme. However, an intersection between the north/south and east/west directions does not always occur, and as a result, the location reply may often contain out-of-date location information. Some improvements (Camp, 2006) to solve this problem lead to high implementation complexity of the mechanism.

The Hierarchical location service (Kies, 2003), (Kies et al., 2004) is another *all-for-some* proactive location database scheme that is characterized by very high implementation complexity, since it deals with several hierarchical levels. Besides, the approach followed to define the appropriate number of levels in the hierarchy is not specified in (Kies, 2003), (Kies et al., 2004). The main idea of the scheme is to select geographical regions (responsible cells) that contain a location server. However, the scheme is not quite robust, since there is just one location server in each of the defined geographic regions, which may lead to loss of location databases if the server fails (Kies et al., 2004).

The Uniform Quorum System (UQS) location service (Haas & Liang, 1999) is a proactive location database scheme that uses a non-position-based routing protocol for the virtual backbone consisting of a fixed number of nodes (a quorum). Location updates are sent to a subset (a write quorum) of available nodes and location requests are referred to a potentially different subset of nodes (a read forum) (Mauve et al., 2001). This feature increases implementation complexity and limits scalability of the service. Besides, the management of the virtual backbone is not described. The services can be configured as *all-for-all*, *all-for-some*, or *some-for-some* depending on how the size of the backbone and the quorum is selected (Mauve et al., 2001). However, it is mostly configured as a *some-for-some* approach.

Two other proactive location database services have been proposed to eliminate drawbacks of the UQS (Mauve et al., 2001). These are the Grid Location Service (GLS) (Li et al., 2000), (Grid project, 2003) and the Virtual Home Region (VHR) location service (Blazevic et al., 2001), (Wu, 2005), sometimes called the Homezone location service.

They are similar to each other in the sense that each node selects a subset of all available nodes as location servers, i.e. the all-for-some approach is used (Mauve et al., 2001). These services are similar as well from the viewpoint of communication complexity (the average number of one-hop transmissions to make a location update/look up and time complexity (the average time to perform a location update/look up) (Mauve et al., 2001).

However, the main drawback of the GLS is that location update/request procedures require that a chain of nodes based on node_IDs is found and traversed to reach the location server for a given node (Kies et al., 2004). Traversing the chain of arbitrary nodes may lead to significant update and request failures if the corresponding nodes in the chain cannot be reached (Kies et al., 2004). Furthermore, controlling node failures is quite difficult (Kies et al., 2004). Besides, if nodes are uniformly distributed throughout the network, the number of entries about positions of other nodes in the location database of a node (the state volume) increases logarithmically with the number of nodes, while in the VHR the state volume is constant (Mauve et al., 2001). Furthermore, the implementation complexity of GLS is higher than that of the previous schemes, except the UQS (Mauve et al., 2001).

As for the VHR, the position of the geographic (home) region that contains the location servers storing the location information of a certain node is found by applying a hash function to the node_ID. The main disadvantage of the service is the *single* home region (Mauve et al., 2001). As a consequence, if a node is far from its home region, update packets have to travel a long way to reach the home region. If an update packet is lost along this path, the location information stored in the home region for this node may become outdated. Moreover, since in MANETs all nodes can potentially move, it may be usual to have empty home regions, especially if node density is low.

Other schemes like GrLS and FSLS (Derhab & Badache, 2008), (Cheng et al., 2007), and some other similar schemes, are variations of previous location schemes developed to solve specific problems. However, some of the improvements are attained by introducing additional implementation complexity.

In conclusion, all the location schemes described above have some shortcomings when applied to large-scale WMNs. This is mainly because they were designed and tested with MANETs in mind, i.e., all nodes were supposed to have more or less the same characteristics, be mobile, and given their power constraints, they just mounted one radio, and thus, when applied to WMNs, they would not fully exploit the advantages of WMNs. Moreover, all these proposals give performance evaluation via simulation and/or asymptotical quantitative models. Thus, up to our knowledge, there has not been any experimental evaluation or comparison of such schemes neither for ad hoc nor for mesh networks.

The above analysis motivates our work on a DLM scheme for large-scale WMNs, called VIMLOC, which is described in the following section.

3. Overview of location management schemes: VIMLOC vs. legacy schemes

This section introduces the rationale and the main design principles behind our location management scheme (VIMLOC). It also explains the entities and procedures involved in its operation. Furthermore, we also briefly explain the operation of legacy proactive and reactive schemes, as in other sections of this chapter we are quantitatively comparing the performance of VIMLOC with that of such schemes.

3.1 VIMLOC

3.1.1 Motivation

As mentioned in the previous section, none of the location services developed earlier for MANETs can satisfy the requirements to large-scale WMNs. However, by analyzing such services thoroughly, it was found that some features of the VHR location service may be considered as the basis for the development of a location service scheme for WMNs. There are some reasons for this. First, this location service is scalable, i.e., the average number of one-hop transmission required to look up or update the position of a node scales with $O(n^{1/2})$ (Mauve et al., 2001). Second, the service has low implementation complexity compared to, for instance, the UQS or GLS (Mauve et al., 2001). Third, with appropriate modifications, it can take advantage of a mesh network backbone consisting of stable mesh routers that can help to avoid the problem of empty home regions. Fourth, the limitations of having a single home region can be avoided by increasing the number of home regions storing information for each node. Further additions described in the following subsections may as well help to improve the reliability and accuracy of the location service for WMNs.

In these subsections, the detailed description of a location management scheme called Virtual Home Region Multi-Hash Location Service (VIMLOC) is presented. It is based on the VHR concept, but it contains some distinguishing features conceived to increase its robustness and accuracy in the large-scale mesh networking environment for which it is designed.

3.1.2 Main ideas

VIMLOC is a *proactive* location database scheme. Conceived as a *some*-for-*some* approach, it is designed by taking into account the specific characteristics of the WMN architecture, namely the WMN backbone. In particular, and as opposed to the VHR scheme, not *all* nodes are considered as location servers. Since the mesh backbone consists of wireless mesh routers (WMRs) that are more stable (in terms of movement and power constraints) than mobile nodes (MNs), just these WMRs are considered as *some* nodes storing location information in a distributed way.

MNs do not act as location servers and just cache location information related to their flows. Furthermore, location databases do not contain the locations of all nodes in the network, but just of *some* selected ones in the network with their node_ID-to-location mapping. This globally saves location information state, thus, improving scalability.

VIMLOC is mainly conceived to increase robustness and accuracy. These are critical requirements for the successful delivery of packets in a position-based routing environment. Additionally, mechanisms to control the overhead generated by VIMLOC are also considered. In particular, the distinguishing features of the VIMLOC scheme follow:

- Multiple hash functions to increase robustness, i.e., one node has more than one virtual home region called *HomeGeoCluster* (HGC). This also allows load balancing of location servers
- Visited geographic region called *VisitedGeoCluster* (VGC) around a given node, in addition to its HGCs, for accuracy. It supports fine-grained mobility by diverting packets to the appropriate location as they approach the destination, i.e., arriving packets will follow the trail of the node
- "Lazy location updates" of HGCs to reduce update overhead throughout the network towards multiple HGCs. The VGC is updated more often than HGCs, thus localizing part of the overhead in a small region around the node and also attaining higher location accuracy
- Soft-state entries in location databases to avoid maintenance of stale entries. That is, the timer of the entries in the location table of those WMRs that are not anymore in the VGC allows removing stale entries.

Overall, this results in a scalable mechanism, as the average number of one-hop transmissions required to look up or update the position of a node scales with $O(n^{1/2})$.

Given that VIMLOC was designed to operate in a position-based routing environment, it is assumed that there is a coordinate space that allows assigning addresses to node identifiers, e.g., through a GPS system or by means of virtual coordinate spaces. It is also assumed that each node knows its location (i.e., address). Furthermore, the multiple hash functions used in the network are pre-defined and well-known by all nodes in network. For instance, they could be transferred to them by neighbors when joining the network.

3.1.3 VIMLOC functional entities and components

Each node n, $n = 1...N$ (where N is the number of nodes in the network, i.e., both WMRs and MNs) has a permanent node_ID. The current position of a node is defined by a temporary location address (LA).

The i-th *HomeGeoCluster* (HGC_i) of node n is the subset of WMRs inside the geographic region whose central location is obtained by applying the i-th hash function to the node_ID of node n, for i=1...k, where k is the total number of the hash functions used in the network.
The *VisitedGeoCluster* (VGC) of node n is the subset of WMRs forming the geographic/physical neighborhood (cluster) around node n.
Thus, each node has its own GeoClusters, in particular, some HGCs and one VGC, as it is shown in Fig. 1, in which k=2.
All WMRs inside a cluster (HGC or VGC) of node n have an entry in their location database for node n. In particular, the *location database* of an arbitrary WMR r contains an entry for a node m if $r \in VGC(m)$ or $r \in HGC_i(m)$, for any i=1…k. WMRs have location tables that are used to answer location queries. Location tables store soft state information to avoid maintenance of stale entries. MNs do not have location tables and just have caches that are used when they send packets to keep location information about nodes involved in their flows. The fields stored in each entry of the location table are: node_ID of node n, geographical position of the node (LA), timestamp, type of entry (cached or updated by location update protocol), flags to indicate if this entry corresponds either to the HGC or the VGC of the node_ID.

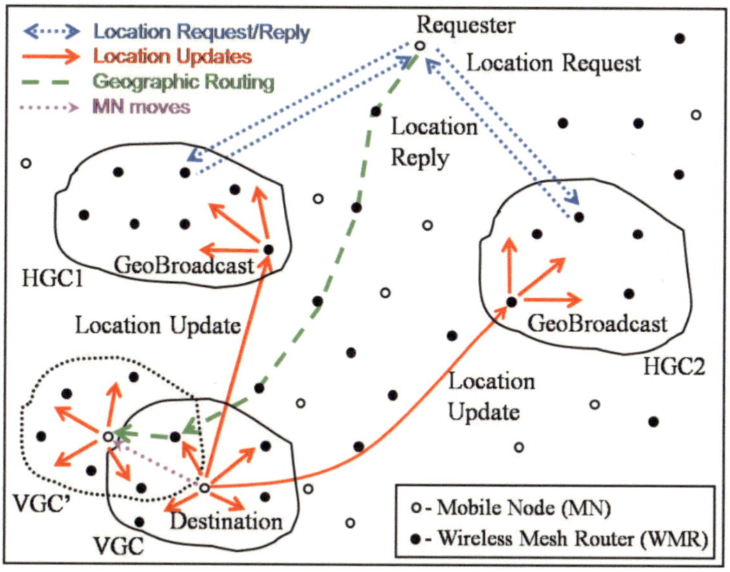

Fig. 1. Example of the operation of VIMLOC when using two hash functions

3.1.4 VIMLOC procedures

3.1.4.1 Location server selection

Location server selection is the procedure by which location servers for node n are selected. It is supposed that all WMRs inside a GeoCluster of node n are servers that store its location information. In particular, all WMRs inside the VGC and HGC_i (i=1…k) are considered as database servers. Note that it does not just correspond to the region around node n, but also to those around each of the positions obtained by applying the hash functions to the node_ID of node n. The size of GeoClusters is defined in accordance with reliability requirements of the network so that each GeoCluster maintains an approximately constant number of location servers (WMRs). It is assumed that routing will allow reaching the location servers inside a cluster.

3.1.4.2 Location server update

Node *n* initiates updates of its location servers in the following cases: 1) the network is turned on, 2) a node joins the mesh, 3) a node moves, 4) a node does not move, but soft-state refreshing is needed. Location updates are initiated by a moving node depending on the chosen scheme. In fact, there is a number of schemes that can be used for initiation of location updates, for instance, distance-based, state-based, and timer-based, among others (Wong & Leung, 2000).

The procedure of updates (or refreshing) of location servers inside the VGC is different from that of location servers inside HGCs. Inside the VGC, node *n* broadcasts updates to location servers inside the neighborhood. For HGCs, node *n* sends geobroadcast updates (Seada & Helmy, 2006) throughout the network to the positions obtained by means of the hash functions of node *n* (as explained above). That is, first, geographic routing is used so that the location update message reaches any server inside an HGC, then, the message is geobroadcasted inside the HGC to update the location information of all location servers of node *n* inside this cluster. Note that for HGCs, a location update message is sent by node *n* to all HGCs in parallel to increase system robustness, although it leads to additional overhead. Furthermore, to control the overall location update overhead of VIMLOC, a "lazy" location update procedure is applied. That is, when a MN moves, WMRs inside the VGC of the MN are updated more frequently than WMRs in HGCs. In other words, different thresholds are used in the selected scheme for triggering updates in the VGC and HGCs, e.g., different distance values in case the distance-based scheme is chosen. When a node does not move, the refreshing procedure of all HGCs and VGC is periodically carried out. In summary, the VGC of a MN is refreshed more often than its HGCs or the VGC of a WMR (as it is static).

3.1.4.3 Location request

The location request procedure is as follows. Firstly, a source node looks up the destination node_ID in the local table by checking its cache and by checking if the current node acts as location server of the destination. If an entry is not found, the source node calculates all hash functions and selects the closest one. This approach is applied to decrease the location request overhead. Location requests are sent to the best (e.g., the closest one) HGC using geoanycast (Seada & Helmy, 2006). Other options, like sending the request to all HGCs at the same time, would have other overhead-robustness trade-offs. Geographic routing is used until a location request message reaches any server (WMR) inside the HGC. The first server inside the HGC, receiving the geoanycasted request replies.

Note that the above-described VIMLOC protocol can run in parallel to any position-based routing algorithm, such as greedy forwarding (Camp, 2006) or restricted directional flooding (Ko & Vaidya, 2000). The location scheme may as well be used in conjunction with hierarchical routing approaches, i.e., those that combine both position routing for wide area routing and non-position-based algorithms for local area routing (e.g., Terminodes (Blazevic et al., 2001), Ballistic geographical routing (Rousseau et al., 2008)). However, the location scheme should be slightly modified in this case, as illustrated in (Rousseau et al., 2008).

The next subsection provides a detailed description of VIMLOC when combined with a geographic routing protocol.

3.1.5 Operation of VIMLOC in combination with geographic routing

In this subsection, the operation of VIMLOC is explained with the help of Fig. 1. When a MN first joins the network, it is loosely attached (i.e., ad-hoc mode is used) to all WMRs from which it receives beacons and selects the best one at any time instant, e.g., by choosing the least loaded. From then on, the MN periodically sends its location to its HGCs by geobroadcasting location update packets (solid lines in Fig. 1), as explained in section 3.1.4.2. When a MN moves, its VGC moves together with it (VGC' in Fig. 1), i.e., WMRs inside the neighborhood of the node change. In this way, the VGC enables packet diverting to compensate the potentially outdated information received from distant servers (or servers not updated recently). In this way, fine-grained mobility is supported. Besides, the timer of the entries in the location table of those WMRs that are not anymore in the VGC of the MN allows removing stale entries.

When a source node (Requester in Fig. 1) gets a packet from an application that must be sent to the destination node_ID, it uses VIMLOC to obtain the LA of the destination. The packet is in the buffer of the source/requester node while the node is obtaining the corresponding LA of the destination. By applying all the hash functions to the destination ID, the source node (requester) obtains the central location of all the HGCs of the destination node. Among them, it may choose the closest one or it may select a subset (or all) of them, mainly depending on the overhead-response time trade-off. In particular, in Fig. 1, the request is simultaneously sent to both HGCs (dotted lines).

The location request is geoanycasted, that is, once any of the WMRs inside the HGC receives the request, it sends the reply back to the source node, and it is not further forwarded inside the HGC. After receiving the position reply, the source node puts the location information of the destination node into the packet header and sends the data packet through intermediate WMRs to the destination node using the underlying geographic routing protocol (the dashed line in Fig. 1).

When an intermediate WMR receives a packet, it first checks whether the destination LA is its own or the address of a MN attached to the WMR. If this is the case, the packet is delivered. Otherwise, the WMR checks whether the destination node_ID is among its location table entries (only entries with flags corresponding to VGC are checked) to appropriately divert the packet, if needed. In other words, it is checked whether the packet has reached a location server inside the VGC of the destination node. In this case, the destination LA field in the packet header is overwritten with the value obtained from the entry in the location table corresponding to the destination node_ID. Then, geographic forwarding eventually delivers the packet to the correct destination inside the VGC, even if the information initially used by the source node was a bit outdated. On the other hand, if there is no entry for the destination node_ID in the location table (e.g., because the packet did not reach the VGC), the packet is forwarded based on its current LA. Note that the same procedure is applied to location replies in case the source/requester node is also moving, i.e., the LA of the source node in a header of a reply packet may be updated by intermediate WMRs while the packet approaches the node.

After both communicating nodes establish a communication, location tracking (Blazevic et al., 2004) is used. Therefore, data packets periodically piggyback the current locations of communicating nodes. If there are no data to send, nodes send location control packets with their location information.

3.2 Proactive and reactive location management schemes

For the comparison with VIMLOC, one proactive and one reactive location management are also considered, as they represent the two main philosophies of operation in location management (Camp, 2006). The proactive scheme under consideration may be classified as a proactive location dissemination scheme (Camp, 2006), e.g., DREAM (Basagni at al., 1998). As all nodes have location databases that store information about all other nodes in the network, it is an *all*-for-*all* approach (Mauve at al., 2001). Moreover, all nodes periodically flood the network so that all WMRs update the LA of that node. Therefore, there are no location requests sent through the network, as they are answered by looking up the local location table.
As for the reactive scheme, e.g., RLS (Kaseman at al., 2002), before a node sends a packet towards a certain destination node_ID, a location request asking for the LA of the destination node is sent to all nodes by flooding the network. The WMR owning this location information (i.e., the WMR to which the destination node is attached) sends a reply back to the requester with its node_ID-to-LA mapping. This approach may be classified as an *all*-for-*some* approach, that is, every node in the network maintains location information on some other nodes in the network (Mauve at al., 2001). In our case, it is assumed that each node only maintains its own location information. And thus, there is no location update procedure in the scheme.
The Click environment is used (Kohler et al., 1999) for the implementation of these three protocols (VIMLOC, proactive, and reactive schemes). The wireless mesh networking framework, including the testbed and some implementation issues of the protocols, is presented in the next section.

4. Wireless mesh networking framework

This section describes the main implementation choices and the testbed over which all the results presented in this chapter have been obtained, as well as the automated measurement framework that was developed to gather them. It also presents the main parameters that characterize the scenarios under evaluation.

4.1 Wireless mesh networking testbed

An indoors wireless mesh networking testbed was built to evaluate the VIMLOC distributed location management scheme in conjunction with greedy forwarding and to compare it with simple proactive and reactive schemes. The experimental setup includes a 12-node multi-radio backbone WMN, as shown in Fig. 2(a), over an approximate area of 1200 square meters. All nodes run Click 1.6.0 over a Linux kernel 2.6.24. Backbone nodes (WMRs) are built based on a mini-ITX board (Pentium M 1.6 GHz) and mount up to four CM9 wireless cards (802.11abg) with Madwifi driver v0.9.4. One of these cards may be used for offering access to MNs. Notice that antennas are omnidirectional and a link is established between two nodes if they have cards assigned to the same channel. In this way, the topology of the testbed can be easily modified by modifying channel assignment. For simplicity, channels are assigned in the network so that all the links are in different channels in order to minimize contention and interferences. External interference with other wireless networks usually configured in 2.4 GHz band is avoided by configuring the wireless cards to 5 GHz band (i.e., 802.11a mode).
Experiment automation benefits from the capabilities of the EXTREME Testbed® (EXTREME, 2010). The autoconfiguration software provides automation to scenario configuration tasks. It is composed of custom made code and as well as code from various open source software projects.

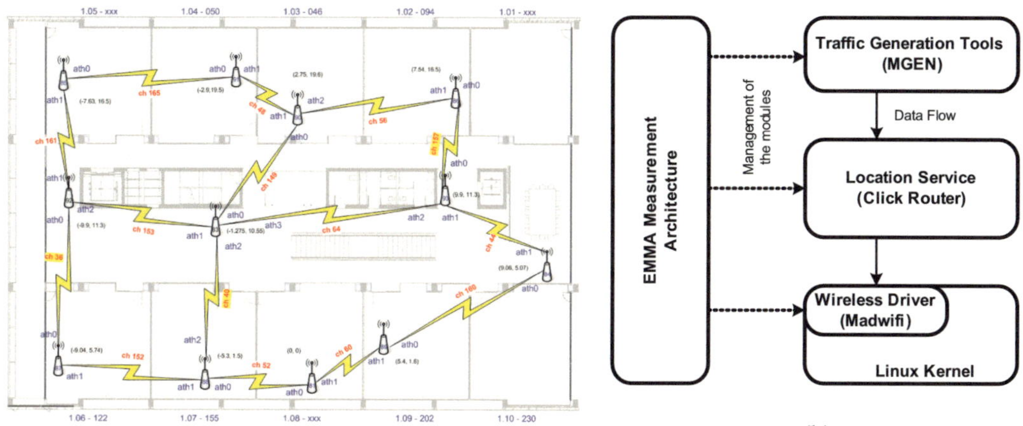

(a) (b)

Fig. 2. (a) Wireless Mesh Networking testbed scheme. Plan of the building showing the positions of backbone nodes, links, and channels configured in each link between nodes, and (b) Software framework in all nodes of the testbed

4.2 Software framework

Fig. 2b presents the software framework in all nodes of the testbed for the evaluation of the location management schemes. The software architecture for location management is based on the Click modular router (Kohler et al., 1999), which is modular and easy to extend. Click was designed to implement networking protocols for flexible and configurable routers. A Click router is composed by generic and simple packet processing elements and a configuration file that defines the interconnection of the processing elements and how the packets flow through the router. We exploit its capabilities by using its elements, but we have also developed new elements for location management. Moreover, developing simple elements allows reusing these elements to implement different location management protocols only by changing the Click configuration file of the Click router.

The main building blocks of the Click stack for the VIMLOC implementation are highlighted in Fig. 3.

In the upper-left part, there is the new User interface. The User interface is in charge of the communication between the user applications (including traffic generation tools) and the Click router. That is, Click generates a new virtual interface that can be used by any legacy user application instead of the real wireless interfaces. All such virtual interfaces in all the nodes of the network are assigned IP addresses that belong to the same subnetwork, and the operating system (OS) sends regular IP packets with the user data to this *fake local* subnetwork. In this way, the OS routing tables and forwarding mechanisms can be bypassed and those implemented in Click can be used instead. However, what the legacy application treats as an IP address is in fact treated as a node_ID by the Click stack. So, the IP packet is encapsulated into what we may call a geographic packet. Its header contains the source and destination IDs (i.e., the IP addresses assigned to the virtual interfaces), source and destination LAs (provided by the location scheme), packet type, and additional information of the location protocol.

The lower part of Fig. 3 shows wireless interfaces that are in charge of the communication between the wireless devices and the Click router. Packets might be sent using unicast MAC

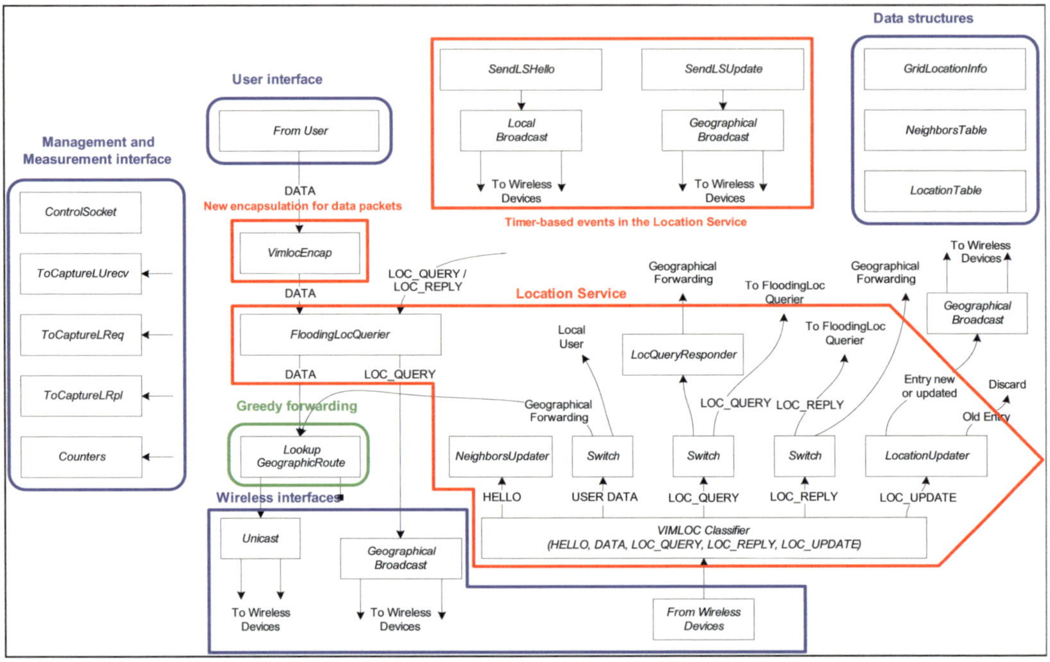

Fig. 3. Software building blocks, including a simplified diagram of the Click stack for the VIMLOC protocol

addresses obtained by means of a geographic forwarding strategy, or in broadcast mode for some packets (e.g., location server updates of HGCs for VIMLOC). Eventually, the geographic packet is encapsulated into an Ethernet packet before delivering it to the wireless card driver.

The core of the VIMLOC (location service engine) and geographic routing implementation is presented in the central part of Fig. 3. In the current implementation, greedy forwarding is used as geographic routing protocol, i.e., a node forwards the packet to its neighboring node that makes the most forward progress (Camp, 2006) in terms of distance towards the destination. It comes mainly from the one implemented by the Grid project (Grid project, 2003) with some adaptations for the location management service and the testbed.

Firstly, there are some important data objects in the central part: the *LocationInfo* element keeps the LA of the WMR, and the *NeighborsTable* and the *LocationTable* elements gather the information of the location database.

Secondly, there are the processing elements. (1) The *FloodingLocQuerier* element is in charge of receiving geographic data packets, and it also looks for the LA of the destination WMR. If there is no information in the location database for this WMR, it starts the location query request procedure by sending location requests to the appropriate location servers in the HGC or the VGC according to the location server selection procedures. It also receives the location replies for the packets waiting in the queue to be transmitted. Once the LA is resolved, it sends the packet to the following processing element. (2) The *LookupGeographicRoute* implements greedy forwarding, it receives a packet for a destination LA, and it looks up in the location database of the neighboring WMR that is closer to the destination LA. (3) The *VIMLOC Classifier* dispatches the received packets to the suitable element that processes it. Some of them will be used locally to populate the location database (e.g., location update packets), and others will be forwarded geographically or/and processed locally depending on whether they already reached or not the destination region towards which the packet is sent. (4) The *NeighborsUpdater* element populates the *NeighborsTable* with information in the incoming packets of the neighboring WMRs in the VGC. (5) The *LocQueryResponder* element answers the location queries if the node processing the packet belongs to an HGC or the VGC of the destination node. (6) The *LocationUpdater* element populates the *LocationTable* with the mapping between node_IDs and their LAs.

In parallel, VIMLOC has some timer-based events. Mainly, it sends location updates to the VGC in local broadcast mode and to the HGC in geobroadcast mode through the *SendLSHello* and *SendLSUpdate* elements, respectively.

This is only a high-level picture of the implementation details. There are other important elements to check if the packet reached the HGC or VGC; to send geobroadcast packets; to send geoanycast packets; and to calculate the regions of the HGC and VGC (i.e., the hash functions). There is also the Management and Measurement Interface that is in charge of the communication between the Click router and the management and measurement tools. It generates the counter files and packet capture files (see Section 4.6) according to the management messages received from the measurement tools.

Simple proactive and reactive location protocols have also been implemented in the testbed to compare their performance with that of the VIMLOC protocol.

The same Click environment for the implementation of these two protocols was used to obtain comparable results. Notice also that Click allows modifying the nodes so that they implement one or the other scheme by simply changing the configuration. In this way, one avoids having to develop the whole processing path from scratch. Furthermore, and due to the modularity and flexibility of the Click Modular Router, the Click stack for each of these two protocols is a simplified version of that of VIMLOC. For instance, in our implementation of the simple proactive dissemination scheme, the concept of VGC and HGC is missing. It just floods location updates periodically to every node in the network. As for location replies, they are answered by just querying the local database, i.e., there is *no location request procedure* for the services. This translates into a simplified Click protocol stack where all the processing elements related with the location queries and the location replies are missing.

Similarly, in the simple reactive scheme, there is *no periodical location update procedure* and WMRs flood a location query to the entire network in case a node initiates a communication and needs the location for another node. As a result, nodes do not have the location database to maintain location update entries. Only the node with that destination node_ID answers with a location reply. Thus, the generation and processing of the location updates is missing from the Click protocol stack.

Click configuration files are different for each node in the testbed. To ease the deployment and management of location schemes, a mechanism to generate the Click configuration file for each node in the testbed has been developed. Fig. 4a illustrates how it works. In particular, a generic template configuration file and different variable files are used for each node. A generation tool uses the generic template as a model and generates all the Click configuration files according to the values in the variable files. In Fig. 4b, an example of this variable file is shown. It contains the specific values for each node.

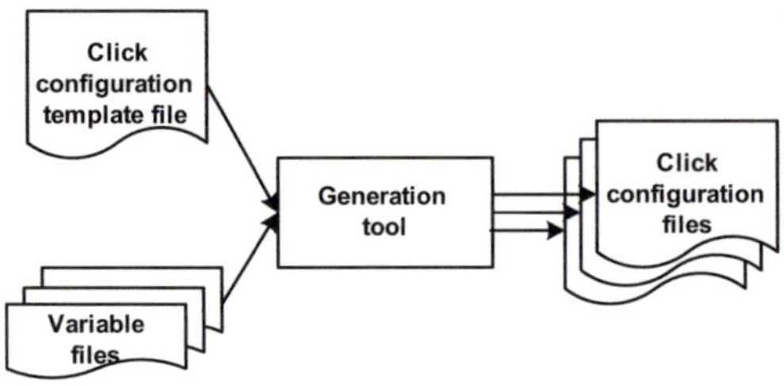

(a)

```
eid_dev     =  click181            locaddr_lat =  0
eid_mac     =  00:01:01:01:01:81   locaddr_lon =  0
eid_addr    =  10.1.1.181
eid_net     =  10.1.1.0/24         ctrl_port   =  10181

athx_num    =  2                   rtes_num    =  0
ath0_dev    =  click-01-0          locaddr_num =  0
ath0_mac    =  00:09:09:81:81:81   rtes_mode   =  dynamic
ath1_dev    =  click-01-1
ath1_mac    =  00:10:10:81:81:81   loc_update  =  yes-locupdate
```

(b)

Fig. 4. (a) Operation of dynamic Click configuration files, and (b) Example of variable file

4.3 Parameters defining the test environment

All average values presented in this paper are based on 30 replications of the experiment. The duration of each replication is 120s. Background traffic does not generate any control traffic for location management. For each link, one bi-directional constant bit rate UDP flow is generated. That is, the destination is the neighboring node through that link. Packet rate is varied to study the operation under different traffic loads in the network. Thus, low load corresponds to no background traffic at all. High load corresponds to a packet rate of 850 pkts/s. This value was chosen to reach network saturation conditions. Finally, medium load corresponds to 550 pkts/s. A packet size of 1000 bytes was chosen. The duration of background flows is the same as that of the replication.

Besides, given the interest on evaluating distributed location management, reference flows were also generated, i.e., those causing location requests. A new data flow is started every 10s by each node towards a random destination. However, as the focus of this chapter is on assessing the control overhead, data traffic is just used to trigger the location requests and is not actually sent through the network. Considering the duration of 120s and the size of the network (12 nodes), 144 requests are generated in the network in each replication. The number of retransmissions in each link is configured to three and the link rate is fixed to 54 Mbps. The transmission power is fixed to 50mW.

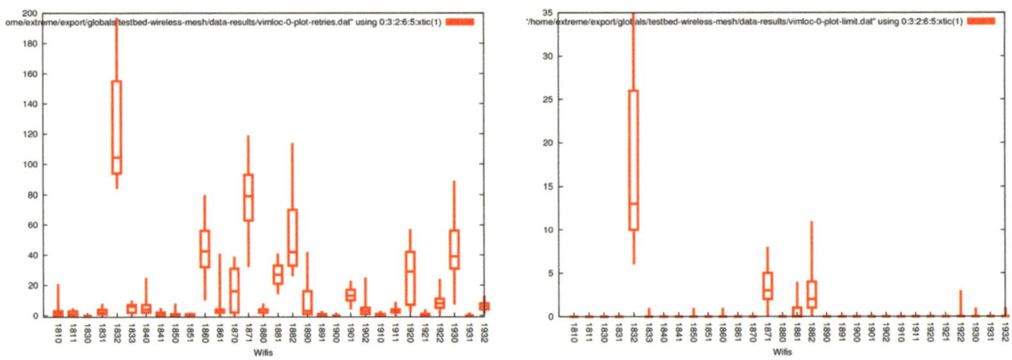

(a) (b)

Fig. 5. Box plot of (a) number of retransmissions and (b) number of times retransmissions were exhausted for all the wireless cards of the testbed under low load conditions

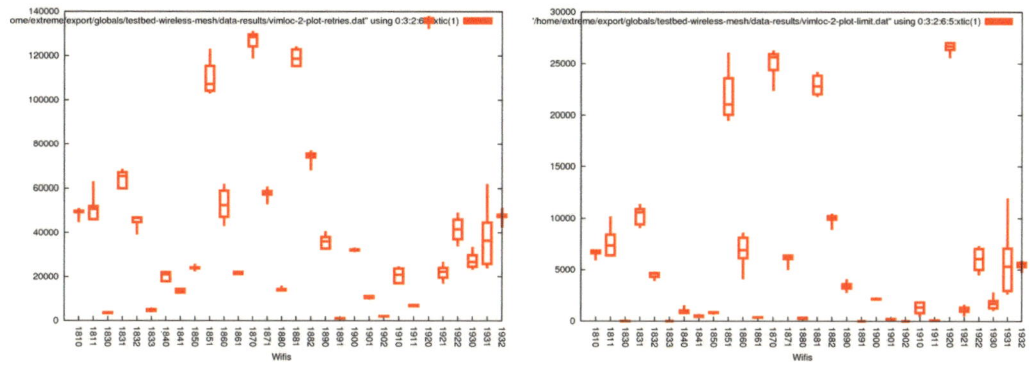

(a) (b)

Fig. 6. Box plot of (a) number of retransmissions and (b) number of times retransmissions were exhausted for all wireless cards of the testbed under high load conditions

Note that some additional work to evaluate the link quality of the testbed has been carried out. This work is used for representing the link quality status when performing an experiment, and is used to understand the outcome of the experiment results. Fig. 5a shows the box plot of the number of retransmissions and Fig. 5b the box plot of the number of times retransmissions were exhausted in each wireless card of the testbed under low load conditions. Hence, in the X-axis, 1831 means node 183 and wireless card number 1. In the Y-axis, the number of total layer 2 retransmissions is plotted. The main goal is to check the status of the network during the time while any developed network protocol is evaluated. For instance, it helps detecting anomalous operation of a certain link, meaning that the antenna fell or new furniture was added, thus modifying the path propagation conditions. These figures also help to differentiate losses in channels and other causes (e.g. buffer overflow at the nodes).

The same graphs were obtained under various conditions showing the same trend, excepting for the substantial increase in the Y-axis values when load is increased (see Fig. 6). In fact, when load is increased the total number of losses in the network substantially grows because more packets traverse each wireless link, hence generating more contention, buffer overflows, and collisions. As observed in Fig. 6, the scenario with high load represents really adverse conditions (saturated network) for the operation of the network, which should never be reached if it is appropriately managed. In any case, we also evaluate our location service in this scenario to test its robustness when many losses occur in the network.

4.4 Measurement framework

The Extreme MeasureMent Architecture (EMMA) (Portoles et al., 2006) is used to define and control the traffic characteristics (e.g., source-destination pairs, packet rates, packet sizes) as well as important events during experiment runs (e.g., new flow, end of flow). EMMA provides the basis for handling the required number of experiment replications to obtain statistically significant results. All traffic is generated using the MGEN v4.2b6 tool.

Additionally, some enhancements were developed to gather, parse, and present the results. More specifically, the following functionalities were added:

- Random Flow Generation: EMMA was expanded to randomly generate source-destination pairs for reference traffic. To send reference traffic in a random manner from each node in the network, the implementation was expanded with the choice of random destination node, hence generating random source-destination pairs. The set of potential destination nodes are all the nodes in the network except for the source node.

- Interaction between the Click framework and EMMA: Counters were introduced at key points in the Click stack to count and/or dump packets containing relevant information to calculate the parameters of interest (see Section 4.6). The Click and the EMMA frameworks were synchronized through the use of Click handlers. Two Click handlers are launched from the EMMA code indicating the beginning and the end of each experiment replication. After the reception of these events, the Click framework updates the information in its counter files so that all the packets taken into account are within the duration window of the experiment replication. This was done for all nodes and for the three location management schemes under evaluation.
- Post-processing parsers: A series of post-processing scripts download all the experiment data from the local storage of each node to a central server through an out-of-band control network. Data stored can be managed by post-processing parsers either to obtain results related to the whole set of repetitions of the experiment or to obtain results separately from each repetition in the experiment. Their purpose is to generate statistical values of the assessed parameters.
- Plot automation functionalities: Due to the increase in the amount of tests and parameters, plot generation was automated in EMMA. Graph results are stored in a fixed location in the central server. Furthermore, another observed requirement was to generate the same graphs in different formats. This would imply repeating the execution of all the post-processing parsers. With high amounts of data, this task can be highly time-consuming. To avoid this, an input parameter was added to EMMA so that it can be used only as a plot generation facility.

4.5 Practical implementation issues of the location schemes

The following options are initially chosen to ease the implementation of VIMLOC. Each node has two HGCs and one VGC to provide the diverting capability. Thus, two different hash functions provide two geographical positions that correspond to the centers of the HGCs. Each HGC is composed of two WMRs and the VGC is composed of the neighbors of a certain node. For the location update procedure, the timer-based scheme is applied. The update interval of WMRs inside HGCs is twice that of the VGC (i.e., 10s and 5s, respectively). There is no retransmission in the location update procedure. In the location request procedure, a requester sends location queries to two HGCs in parallel. And, in the current setup, each node sent one request every 10s. Entries in location tables expire after 22s. Periodic Hello packets for discovering the coordinates of neighbors are sent every 10s by each node.

4.6 Gathering of measurement results

For the gathering of measurement results, counters for sent and forwarded packets, as well as packet capture tools to obtain information on received packets, have been implemented for the three above protocols. They were implemented by means of Click and EMMA and deployed in backbone nodes (WMRs).

As an example, the diagram presented in Fig. 7 illustrates the approach to gather results for the procedure of the location update of the VIMLOC scheme (VIMLOC-LU) in a WMR.

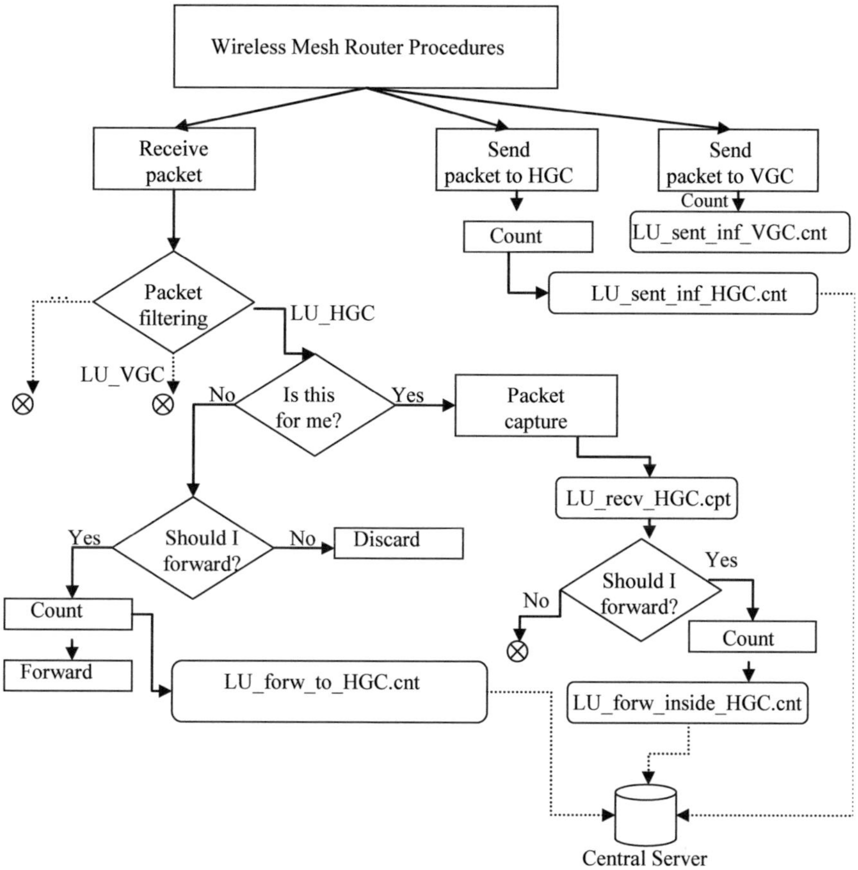

Fig. 7. Approach to gather measurement results (the VIMLOC-LU procedure)

In particular, for the VIMLOC-LU procedure, a counter for packets sent towards the HGCs is installed in each interface of a WMR (LU_sent_inf_HGC.cnt), i.e., a packet sent to two different HGCs in parallel is counted as two packets since two interfaces are used, although packets have the same sequence number. A similar counter is implemented for packets sent towards the VGC (LU_sent_inf_VGC.cnt). There are also two more counters to make the distinction between packets forwarded to HGCs and packets geobroadcasted inside HGCs (LU_forw_to_HGC.cnt and LU_forw_inside_HGC.cnt).

Note that all information about sent/forwarded packets gathered by these counters in each WMR is eventually transferred to the central server. An example of the structure of a counter file stored in the central server that merges counter files coming from various nodes is shown in Table 1.

Node_ID (1..12)	Number of sent/forwarded packets
5	15
8	32
...	....

Table 1. Structure of a merged counter file stored in the central server

Besides, packet capture tools are used to collect information on the behavior of WMRs when acting as location server for one or more nodes. Location updates received when the WMR belongs to an HGC are logged in the LU_recv_HGC.cpt file and in the LU_recv_VGC.cpt file when it belongs to a VGC. An example of the structure of a capture file saved in the central server (after merging files coming from different WMRs) is shown in Table 2.

Node_ID	Source_ID	Seq #	TTL
8	7	1	7
8	2	1	4
8...	7	2	7
9	3	1	5
9...	7	1	6
...	...	...	...

Table 2. Structure of a capture file

Thus, a capture file contains information about Node_ID (the ID of the WMR that sent its capture file to the server), Source_ID (the node for which the WMR acts as location database), sequence number (of the received location update message), and TTL.

The approach to gather results for the location request-reply procedures of VIMLOC (VIMLOC-LR), the LU procedure of the proactive scheme, and the LR procedure for the reactive scheme is similar to the above-described one for VIMLOC-LU procedure.

In particular, for the VIMLOC-LR procedure, in the implementation scenario, a source node (requester) sends a location request with the same sequence number in parallel to both HGCs consisting of two WMRs. As a result, a counter for sent location request packets (LR_send_inf.cnt) is installed in each interface of a WMR. Besides, for the calculation of performance parameters, it is needed to obtain a counter just for those sent location request packets that have different sequence number (LR_sent.cnt). There is also a counter for forwarded request packets towards an HGC (LR_forw.cnt).

When a location request packet reaches the first server inside HGC (the replier) and a location reply packet is sent to the requester, the replier changes packet type and source ID in the packet fields though TTL does not take its initial value. To count sent and forwarded location replies towards the requester, two more counters are implemented (LRpl_forw.cnt and LRpl_sent.cnt). Besides, a capture file is also generated (LRpl_recv.cpt) for received location replies. The LRpl_recv.cpt file just keeps information about the first reply packet that reached the requester, i.e., the second reply packet (with the same sequence number) obtained during the same LR procedure (from the second HGC) is discarded and is not dumped to the file.

For the proactive scheme, having just the LU procedure, three counters are needed, namely, LU_sent.cnt, LU_sent_inf.cnt, LU_forw_int.cnt, and a capture file for received location update packets (LU_recv.cpt).

Correspondingly, for the reactive scheme, having just the LR procedure, three counters for sent/forwarded location requests (LR_sent.cnt, LR_sent_inf.cnt, LR_forw_inf.cnt) and two counters for sent/forwarded location replies (LRpl_sent.cnt and LRpl_forw.cnt) were deployed, and one capture file for received location reply packets was generated (LRpl_recv.cpt).

Note that the structure of counter and capture files is similar to the one presented in Tables 1 and 2, respectively.

5. Parameters assessed

Based on the gathered measurement results, the post-processing scripts calculate performance parameters to compare the three location management schemes, namely VIMLOC, the proactive scheme and the reactive scheme. In particular, the following performance parameters have been defined for this purpose.

5.1 Success Rate (SR)

For the *LU procedure,* the SR is the fraction of LU packets (out of the total number of updates) successfully delivered to the nodes acting as location servers for the originator of the LU, that is, all WMRs for the proactive scheme, and WMRs inside HGCs and VGCs for VIMLOC.
For VIMLOC, the SR for LUs is calculated based on the captured files and counters for the chosen implementation options (Section 4.6) as follows:

$$SR_{LU\_VIMLOC} = \frac{S_{LU\_recv\_HGC} + S_{LU\_recv\_VGC}}{S_{LU\_sent\_\inf\_HGC} + S_{LU\_sent\_\inf\_VGC} + S_{LU\_forw\_inside\_HGC}} \cdot 100\% , \quad (1)$$

where $S_{LU\_recv\_HGC}$ is the number of LU packets received by WMRs inside HGCs (the number of entries in the table of the LU_recv_HGC.cpt file), $S_{LU\_recv\_VGC}$ is the number of LU packets received by WMRs inside VGC (the number of entries in the table of the LU_recv_VGC.cpt file), $S_{LU\_sent\_inf\_HGC}$ is the total number of LU packets (including packets with the same sequence number) sent through all node interfaces to HGCs (i.e., the sum of the number of sent packets in all entries of the table in LU_sent_inf_HGC.cnt), $S_{LU\_sent\_inf\_VGC}$ is the total number of LU packets sent through node interfaces to VGCs (the sum of the number of sent packets in all entries of the table in LU_sent_inf_VGC.cnt), $S_{LU\_forw\_inside\_HGC}$ is the total packets of forwarded LU packets inside HGCs (the sum of the number of forwarded LU packets in all entries of the table in LU_forw_inside_HGC.cnt).
For the proactive scheme, the SR for LUs is defined as:

$$SR_{LU\_PRO} = \frac{S_{LU\_recv}}{(N-1)S_{LU\_sent}} \cdot 100\% , \quad (2)$$

where S_{LU_recv} is the number of LU packets received by all WMRs (the number of entries in the LU_recv.cpt table), S_{LU_sent} is the total number of LU packets with different sequence number sent by WMRs (the sum of the number of sent packets in all entries of the table in the LU_sent.cnt), and N is the number of nodes in the network (12 in the current testbed implementation).
For the *LR procedure,* the SR is defined as the fraction of requests whose reply is successfully delivered to the requesting node.
For VIMLOC, SR for the LR procedure is calculated as

$$SR_{LR\_VIMLOC} = \frac{S_{LRpl\_recv}}{S_{LR\_sent}} \cdot 100\% , \quad (3)$$

where S_{LRpl_recv} is the total number of received location reply packets (the number of entries in the file LRpl_recv.cpt), S_{LR_sent} is the total number of sent location request packets that

have different sequence number (the sum of the number of sent packets in all entries of the table in LR_send.cnt).
For the reactive scheme, the SR for LR procedure is calculated by the same formula.

5.2 Communication Complexity (CC)

The CC is the average number of one-hop transmissions required 1) to update the position of a node (it applies to VIMLOC protocol and the proactive scheme), or 2) to look up the position of a node (the LR procedure), which applies to VIMLOC and the reactive scheme. Correspondingly, for VIMLOC, the CC for the LU procedure can be calculated as:

$$CC_{LU\_VIMLOC} = S_{LU\_sent\_inf\_HGC} + S_{LU\_sent\_inf\_VGC} + S_{LU\_forw\_to\_HGC} + S_{LU\_forw\_inside\_HGC}, \quad (4)$$

where $S_{LU\_forw\_to\_HGC}$ is the total number of forwarded packets towards HGCs (the sum of the number of forwarded packets in all entries of the table in LU_forw_to_HGC.cnt) and the rest of parameters was already defined above.
For the proactive scheme, the CC for the LU procedure is

$$CC_{LU\_PRO} = S_{LU\_sent\_inf} + S_{LU\_forw\_inf}, \quad (5)$$

$S_{LU\_sent\_inf}$ is the total number of sent LU packets (the sum of the number of sent packets throughout in all entries of the table in LU_sent_inf.cnt), $S_{LU\_forw\_inf}$ is the total number of forwarded LU packets (the sum of the number of sent packets in all entries of the table in LU_forw_inf.cnt).
For VIMLOC, the CC for the LR procedure is defined as:

$$CC_{LR\_VIMLOC} = S_{LR\_sent\_inf} + S_{LR\_forw} + S_{LRpl\_sent} + S_{LRpl\_forw}, \quad (6)$$

where $S_{LR\_sent\_inf}$ is the total number of sent location requests (the sum of the number of sent packets in all entries of the table in LR_sent_inf.cnt), S_{LR_forw} is the total number of forwarded LR packets (the sum of the number of sent packets in all entries of the table in LR_forw.cnt), S_{LRpl_sent} is the total number of sent location replies (the sum of the number of sent packets in all entries of the table in LRpl_sent.cnt), S_{LRpl_forw} is the total number of forwarded LRpl packets (the sum of the number of sent packets in all entries of the table in LRpl_forw.cnt).
For the reactive scheme, the CC for the LR procedure is calculated by means of the same formula.

5.3 Overall Overhead (OO)

The OO is the total amount of bytes sent to the network for a certain procedure, and this is calculated as the addition of the number of packets of each type (i.e., update, request, or reply), multiplied by their respective size (P_{size}).
Thus, for VIMLOC and the proactive scheme, the OO for the LU procedure is defined as

$$OO_{LU} = CC_{LU} \cdot P_{size}, \quad (7)$$

and for VIMLOC and the reactive scheme, the OO for the LR procedure is defined as

$$OO_{LR} = (S_{LRpl\_sent\_inf} + S_{LRpl\_forw\_inf})P_{LR\_size} + (S_{LRpl\_sent} + S_{LRpl\_forw})P_{LRpl\_size}. \quad (8)$$

5.4 Efficiency Factor (EF)

The EF is defined for both the LU and the LR procedures as the ratio between the number of "useful" one-hop transmissions and the number of overall one-hop transmissions. The number of *useful one-hop transmissions* is equal to the number of hops of the most efficient path followed to deliver a packet to the appropriate location server node. The number of overall one-hop transmissions means the total number of hops used in the LU/LR procedure, i.e., this is the CC. With this parameter, we try to capture the how inefficient in terms of wasted transmissions (e.g., by flooding) each of the schemes.

The expression for calculation of the EF is the same for both the LU and LR procedures and can be defined as follow:

$$EF_{LU/LR} = \frac{UHT_{LU/LR}}{CC_{LU/LR}}, \tag{9}$$

where $UHT_{LU/LR}$ is the number of useful one-hop transmissions for LU or LR procedures. It is determined from the corresponding tables of capture files, where the number of "useful" hops traversed by successfully received location update/reply packets may be counted by means of the initial and final values of TTL carried in packets. In the case of VIMLOC, this parameter is the sum of the number of useful one-hop transmissions for packets received by nodes belonging to VGCs and HGCs.

5.5 State Volume (SV)

The SV is measured as the average number of entries in the location database of a node. And it is defined just for the proactive scheme (from the LU_recv.cpt file) and VIMLOC (from the LU_recv_HGC.cpt and the LU_recv_VGC.cpt files), since the reactive scheme does not contain a location database.

5.6 Successful Communication Complexity (SCC)

The SCC is the average number of one-hop transmissions required to have a successful reception of either 1) one location update packet for the proactive scheme and the VIMLOC-LU procedure, or 2) one location reply packet for the reactive scheme and VIMLOC-LR procedure.

6. Results and discussions

This section presents the results of the experimental evaluation of the parameters defined above. Boxplots are used to present the minimum, 25 percentile, median, 75 percentile, and maximum. Additionally, the curve for each scheme represents the average values for the various background loads under test. Note that VIMLOC was designed for medium/large scale environments. However, it is being tested in a small scale testbed and the results presented are expected to improve with respect to flooding-based approaches as the size of the network increases.

6.1 Robustness of VIMLOC

The SR is used in this section to compare the robustness of the three mechanisms. Let us first recall that robustness refers to the ability of the mechanism to carry out an LU or LR procedure even in the presence of impairments in the network. Such impairments may come from the variability of the wireless medium or the potential losses due to the background load introduced in the experiments.

Fig. 8a presents the comparison of the SR for the LU procedure under different background loads in the network. The proactive scheme works perfectly (i.e., SR=100%) for low loads. This is due to the flooding procedure, as there are multiple paths that a flooded update packet may follow to reach each WMR of the network. Thus, even in case one update is lost in one of these paths, there are others that allow reaching each of the WMRs.

The SR for VIMLOC is a bit less than 100% for low loads, because the losses due to fading in the wireless channel make more likely that the update is lost in the unicast part of the geobroadcast procedure. There is just a slight decrease of the SR as load increases for both schemes since background load introduced in the network makes more likely that the packet is lost in all the possible paths that could be used to reach a certain WMR. Anyway, they both show a similar behavior despite their very different operating principles.

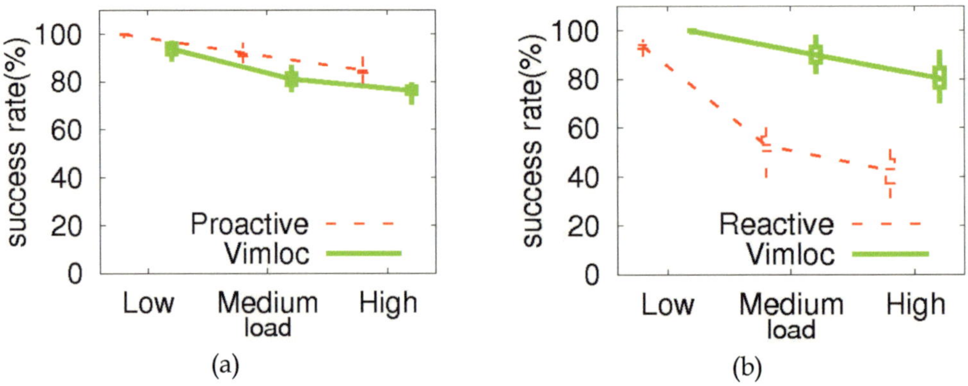

Fig. 8. Success Rate vs. load. (a) Location update (Proactive and VIMLOC) and (b) Location request (Reactive and VIMLOC)

For the request-reply (LR) procedure (Fig. 8b), the SR for VIMLOC is equal to 100%. It decreases for medium and high loads. The reactive mechanism presents SR values smaller than VIMLOC for low loads, but the SR dramatically decreases when load increases. This is because the reply packet for the reactive scheme is sent back in unicast mode to the requester node. And, as explained above for VIMLOC, the loss of a unicast packet is quite likely, especially for medium and high loads. On the other hand, in the current setup, there are two HGCs from which a reply could be received, and thus, even if unicast, the loss probability of at least one of them is smaller. Therefore, the SR is much better for VIMLOC.

6.2 Trade-off between robustness and overhead

The SR allows quantifying the robustness of the mechanisms. However, the way VIMLOC and the flooding-based protocols achieve their respective SR is different in the sense that the latter do it by introducing a huge amount of overhead, which is inherently inefficient. For this reason, we present some results to quantify the trade-off between robustness and the overhead introduced by each of the mechanisms.

The Y-axis of Fig. 9 presents the overall overhead (OO) for each mechanism for the LU procedure (Fig. 9a) or the LR procedure (Fig. 9b). And the X-axis measures how successful each of these procedures was. In Fig. 9a, it presents the probability of having outdated location information at nodes (POI), and it is calculated as 1-*SR* for location updates. In Fig. 9b, it presents the probability of not getting an answer to a request (PNA) and it is measured as 1-*SR* for the location request-reply procedure. Each point in the figure represents the OO for one replication of the experiment. Recall that 30 replications were ran for each of the three loads considered.

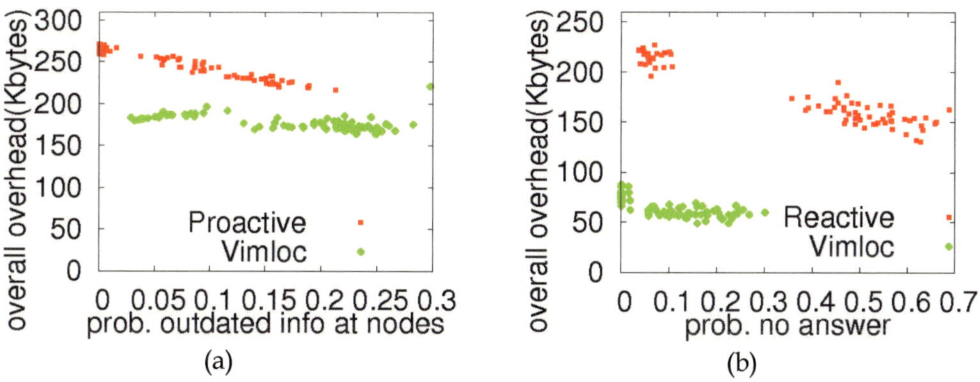

Fig. 9. (a) OO vs. probability of outdated info at nodes (update procedure) and (b) OO vs. probability of no answer (request-reply procedure)

Focusing on Fig. 9a, one observes that VIMLOC generates less overhead than the proactive scheme, but with a bit worse POI values. However, note that the accuracy is twice worse for the proactive scheme, because, with VIMLOC, forwarding can benefit from updates sent to certain key nodes (VGC) that are updated twice as often as in the proactive scheme. As a consequence, they have information which is twice fresher, which is fundamental in highly mobile environments. Achieving the same level of accuracy for the proactive scheme would imply flooding the network twice as often, thus doubling the control overhead and generating more congestion in the network.

With respect to Fig. 9b, one observes that the overhead of the request-reply procedure for the reactive mechanism is much higher than that for VIMLOC, since it follows a flooding-based behavior. Besides, VIMLOC has less probability of not getting an answer to a request than the reactive scheme. This is because the potential loss of unicast packets is compensated by the fact that VIMLOC sends two requests in parallel, thus doubling the chance to receive at least one reply.

6.3 Efficiency of VIMLOC

The state volume (SV) parameter illustrates the efficiency of VIMLOC and the proactive schemes in terms of number node_ID-to-LA mapping entries stored in the location database. The location database of an arbitrary WMR for VIMLOC contains an entry for a node if the WMR belongs to one of the two HGCs of the node or the WMR belongs to the VGC of the node. As it is seen from Fig. 10, the average number of entries in the location table of a WMR for the proactive scheme is equal to 11. That is, each node contains an entry for all other nodes in the network. And it is 7 entries for VIMLOC. Therefore, the state location information stored in a WMR for VIMLOC is about 40% less than for the proactive scheme. But one should notice that, as explained above, VIMLOC has twice better accuracy. If we assign a similar accuracy for VIMLOC and the proactive scheme, the state information stored in a WMR for VIMLOC is almost three times less in accordance with our measurements. Moreover, taking into account the flooding-based nature of the proactive scheme, it is expected that the difference between the state volume values of the two schemes will significantly increase with the size of the network.

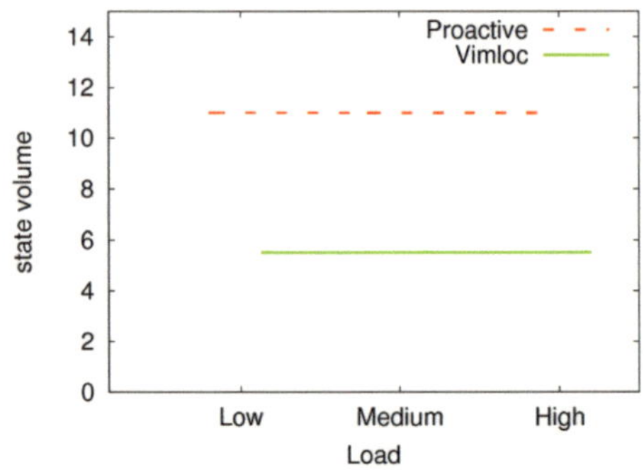

Fig. 10. State Volume vs. load. Location update (Proactive and VIMLOC)

On the other hand, the efficiency factor (EF) illustrates the efficiency of the LU/LR procedure of VIMLOC, the proactive, and the reactive scheme. And it is calculated as the ratio between 1) the number of one hop transmissions really involved in generating a successful location update/request-reply delivery and 2) the total number of one hop transmissions. Out of the total number of one-hop transmissions caused by flooding in the proactive scheme, just around 50 % are used to deliver successful location updates, as one may observe in Fig. 11a. Thus, half of the total one-hop transmissions are not useful in the LU procedure of the proactive scheme. On the other hand, the useful amount of one-hop transmissions for VIMLOC for the LU procedure is close to 100% for low loads and it decreases to 80% for high loads. But even for high loads, VIMLOC is 30% more efficient than the proactive scheme.

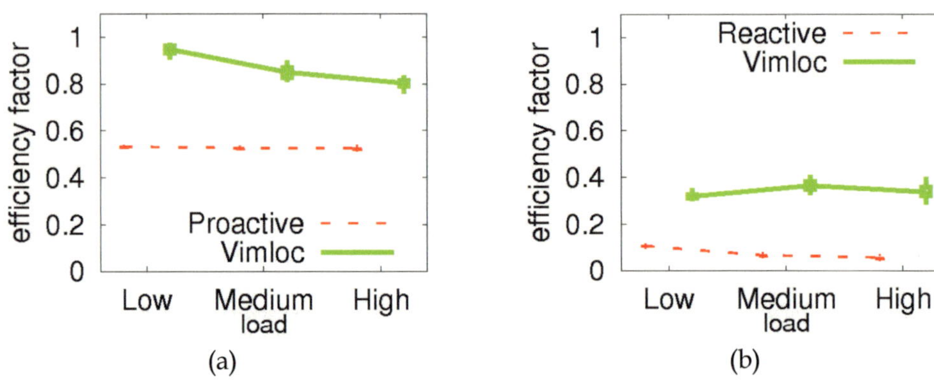

Fig. 11. Efficiency Factor (EF) vs. load. (a) Location update (Proactive and VIMLOC) and (b) Location request (Reactive and VIMLOC)

Fig. 11b shows that more than 90% of one-hop transmissions caused by flooding in the LR procedure of the reactive scheme are useless. In this sense, the VIMLOC EF (around 35%) is more than three times more efficient than the reactive scheme. And these values seem to be constant for all loads tested. The remaining 65%, corresponding to useless transmissions, is the price paid for the high success rate of the VIMLOC-LR procedure (Fig. 8b), since the location request is sent in parallel to two HGCs. In fact, if two successful replies arrive at the requesting node, just the first one received is taken into account for EF calculations. Thus, VIMLOC is much more efficient in terms of the number of useful one-hop transmissions than the proactive and the reactive schemes.

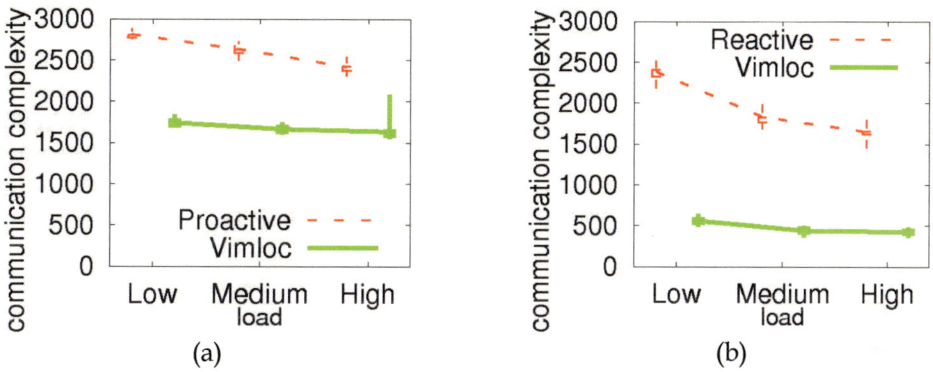

Fig. 12. Communication Complexity (CC) vs. load. (a) Location Update (Proactive and VIMLOC) (b) Location request-reply (Reactive and VIMLOC)

The communication complexity (CC) shows how many one-hop transmissions (on average) it takes to deliver one location update (Fig. 12a) or request-reply (Fig. 12b). As it is shown in Fig. 12a, VIMLOC generates 45% less one-hop transmissions than the proactive scheme for low loads in the LU procedure, whilst having twice higher accuracy. At the same time, the CC for VIMLOC is almost four times smaller than that of reactive schemes for the LR procedure, as illustrated in Fig. 12b. The difference between values slightly decreases with background load due to packet losses in the LU/LR procedure. But in any case, the advantage of VIMLOC is substantial.

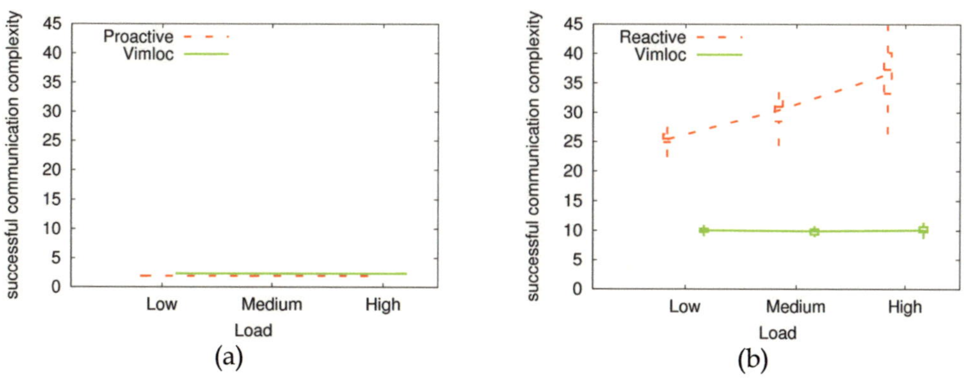

Fig. 13. Successful communication complexity (SCC) vs. load. (a) Location Update (Proactive and VIMLOC) (b) Location request-reply (Reactive and VIMLOC)

Successful communication complexity (SCC) shows how many one-hop transmissions (on average) it takes to deliver one successful location update (Fig. 13a) or request-reply (Fig.

13b). As it is shown in Fig. 13a, curves for VIMLOC and the proactive schemes almost coincide and do not change with load. However, as emphasized above, VIMLOC has twice higher accuracy. If levels of accuracy are the same, VIMLOC shows a reduction of around 30% in SCC, according to our measurements.

SCC for VIMLOC is approximately between one third and one fourth that of the reactive schemes for the loads tested (Fig. 13b). Furthermore, this difference is expected to substantially increase with the size of the network due to the better scaling properties of VIMLOC with respect to the reactive scheme.

7. Conclusion

This chapter presents VIMLOC, which stands for VIrtual home region Multi-hash LOCation management service. VIMLOC is a novel distributed location management scheme that works in conjunction with any position-based routing scheme. Up to our knowledge, this is the first location service specially designed for large-scale WMNs. In this sense, it exploits: 1) the stable and non-power-constrained nature of the backbone of a wireless mesh network, 2) the use of geographic positions, and 3) multiple hash functions. This work also seems to be the first implementation, experimental evaluation, and comparison over self-organized wireless networks (including ad hoc and mesh networks) of various location management schemes. More specifically, a comparison of the performance of VIMLOC with canonical proactive and reactive location mechanisms was carried out in a 12-node testbed. The experimental results show that VIMLOC outperforms these two latter approaches. First, the accuracy of location information is not compromised even though VIMLOC does not flood the network. Second, the state volume stored at each WMR does not grow linearly with the size of the network. Finally, VIMLOC provides robustness across a range of different workloads environments. Besides, it also has much better scaling properties, which mainly comes as a consequence of exploiting geographic information. Overall, this renders VIMLOC a promising solution for location management when wireless mesh networks are used to provide broadband wireless access.

Future work will mainly consist in assessing other parameters, such as reaction time, and to provide mechanisms to automatically tune the parameters already evaluated to the particular network scenario in which VIMLOC is deployed.

8. Acknowledgement

This work was supported in part by the Spanish Ministry of Science and Innovation under grant number TEC2008-06826 (ARTICO), by the Catalan Regional Government under grant 2009SGR-940, and by the European Commission project WIP under contract 27402.

9. References

Akyildiz, I.F. & Wang, X. (2005). A Survey on Wireless Mesh Networks, *IEEE Communications Magazine*, September 2005.

Basagni, S.; Chlamtac, I.; Syrotiuk, V.R. & Woodward, B.A. (1998). A Distance Routing Effect Algorithm for Mobility (DREAM). *Proceedings of the MobiCom´98*, pp. 76-84, Dallas, October, 1998.

Blazevic, L.; Buttyan, L.; Capkun, S.; Giordano, S.; Hubaux, J.-P. & Le Boudec, J.-Y. (2001). Self-organisation in Mobile Ad Hoc Networks: the approach of Terminodes, *IEEE Communications Magazine*, June 2001.

Blazevic, L.; Le Boudec, J-Y. & Giordano S. (2004). A location-based routing method for Mobile Ad-hoc Networks. *IEEE Transactions on mobile computing*, Vol. 3, No. 4, October-December, 2004.

Camp, T. (2006). Location services in Mobile Ad-hoc Networks (2006), *Handbook of algorithms for wireless networking and mobile computing*. 14: 319-341, 2006.

Cheng, H.; Cao, J. & Cheng, H.-H., (2007). GrLS: Group-based location services in Mobile Ad Hoc Network. *Proceedings of IEEE ICC-07*, 24-28 June, 2007.

Grid project (2003). Grid Ad hoc Networking Project. Information available at: http://www.pdos.lcs.mit.edu/grid

Derhab, A. & Badache, N. (2008). Balancing the tradeoffs between scalability and availability in mobile ad hoc networks with a flat hashing-based location service. *Ad Hoc Netw.* 6, 7, September, 2008.

EXTREME, (2010). EXTREME Testbed® of the CTTC. More information available at: http://www.cttc.cat/en/project/EXTREME.jsp

Haas, Z.J. & Liang, B. (1999). Ad-Hoc Mobility management with Uniform Quorum Systems, *IEEE/ACM Trans.* Net., Vol. 7, No.2, April 1999.

Hu, W.; Zou, S. & Cheng, S. (2007). Performance analysis of location management schemes in WiMAX Mesh Network, *Proceedings of Communications and Networking conference (CHINACOM'07)*, China, 22-24 Aug, 2007.

Hu, W.; Zou, S. & Cheng, S. (2009). A Novel Location Management Scheme for Wireless Mesh Networks, *Int. Journal of Distributed Sensor Networks*, Vol. 5, Is. 1, January, 2009.

Kasemann, M.; Fusler, H; Hartenstein, H & Mauve, M. (2002). A reactive location service for mobile ad hoc network. *Technical report TR-02-014*, Department of Science, University of Mannheim, November, 2002.

Karp, B. N. & Kung, H.T (2000). GPRS: Greedy Perimeter Stateless Routing for Wireless Networks, *Proceeding of the MobiCom-2000*, pp. 243-254, Boston, USA, August 2000.

Kies, W. Hierarchical location services for Mobile Ad-hoc Networks (2003). *Master's thesis*. Department of Computer Science, University of Mannheim, Germany, 2003.

Kies W.; Fusler, H.; Widmer, J. & Mauve M. (2004). Hierarchical location services for Mobile Ad-hoc Networks. *Mobile Computing and Communications Review*, Vol. 1, No. 2, 2004.

Ko, Y.B. & Vaidya, N.H. (2000). Location-Aided Routing (LAR) in Mobile Ad Hoc Network, *ACM/Baltzer WINET J.*, vol. 6, no. 2, pp. 307-21, 2000.

Kohler, E. et al. (1999). The Click Modular Router. *Operating Systems Review*, 34 (5), *Proceedings of the 17th Symposium on Operating Systems Principles*, pp. 217-231, December 1999.

Li et al (2005). A Scalable Location Service for Geographic Ad Hoc Routing. *Mechatronics and Automation, IEEE Int. Conf.*, Vol. 2 , pp. 831-836, August, 2005.

Mauve, M.; Widmer, J. & Hartenstein, H. (2001). A survey on position-based routing in mobile ad hoc networks, *IEEE Network*, 15: 30-39, December, 2001.

Portolés-Comeras, M.; Requena-Esteso, M.; Mangues-Bafalluy, J. & Cardenete, M., (2006). EXTREME: Combining the ease of management of multi-user experimental facilities and the flexibility of proof of concept testbeds, *Proc. TRIDENTCOM 2006*, Barcelona (Spain), March 1-3, 2006.

Rousseau, F.; Theoleyre, F.; Duda, A.; Krendzel, A.; Mangues-Bafalluy, J. & Requena-Esteso, M., (2008). Geo-mobility and Location Service in Spontaneous Wireless Mesh Networks, *Proceedings of ICT-Mobile Summit 2008*, Stockholm, Sweden, June 2008.

Saltzer, J. (1993). On the naming and binding of network destinations, *RFC 1498*, M.I.T. Laboratory for Computer Science, August, 1993.

Seada, K. & Helmy, A., (2006). Geographic services for wireless networks. *Handbook of algorithms for wireless networking and mobile computing*. 15: 343-364, 2006.

Stojmenovic, I. (1999). A scalable quorum based location update scheme for routing in ad hoc wireless networks. *Technical Report, TR-99-09,* University of Ottawa, September 1999.

Wong, V. W.-S. & V Leung, C. M., (2000). Location Management for Next-Generation Personal Communications Networks. *IEEE Network*, pp. 18-24, September/October 2000.

Wu, X. (2005). VPDS: Virtual Home Region based Distributed Position Service in Mobile Ad Hoc Networks. *Proceedings of 25-th IEEE Conf. on Distributed Computing Systems* (ICSCS-2005), 2005.

11

Secure Routing in Wireless Mesh Networks

Jaydip Sen
Innovation Lab, Tata Consultancy Services Ltd.
India

1. Introduction

Wireless mesh networks (WMNs) have emerged as a promising concept to meet the challenges in next-generation networks such as providing flexible, adaptive, and reconfigurable architecture while offering cost-effective solutions to the service providers (Akyildiz et al., 2005). Unlike traditional Wi-Fi networks, with each *access point* (AP) connected to the wired network, in WMNs only a subset of the APs are required to be connected to the wired network. The APs that are connected to the wired network are called the *Internet gateways* (IGWs), while the APs that do not have wired connections are called the *mesh routers* (MRs). The MRs are connected to the IGWs using multi-hop communication. The IGWs provide access to conventional clients and interconnect ad hoc, sensor, cellular, and other networks to the Internet as shown in Fig. 1.

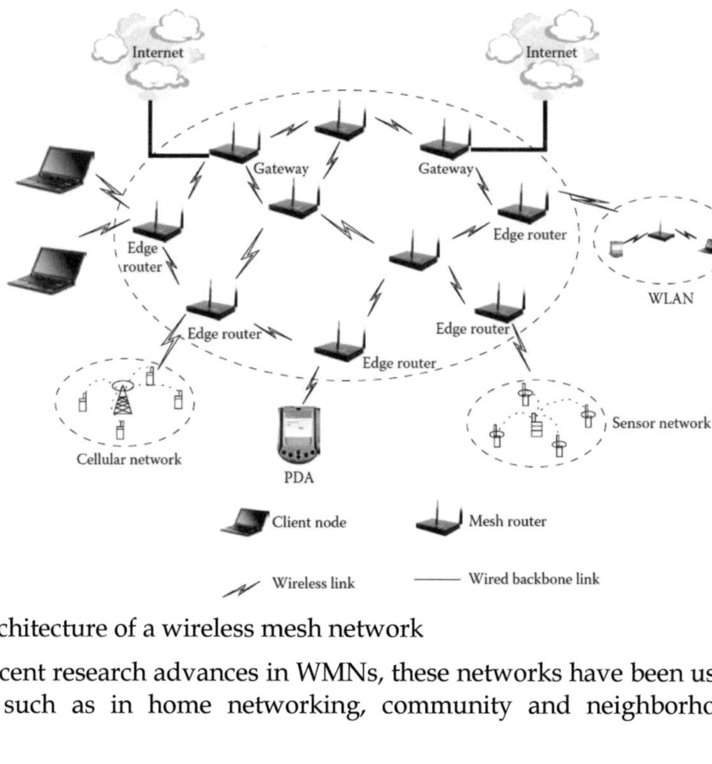

Fig. 1. The architecture of a wireless mesh network

Due to the recent research advances in WMNs, these networks have been used in numerous applications such as in home networking, community and neighborhood monitoring, security surveillance systems, disaster management and rescue operations etc (Franklin et al., 2007). As there is no wired infrastructure to deploy in the case of WMNs, they are considered cost-effective alternative to *wireless local area networks* (WLANs) and backbone networks to mobile clients. The existing wireless networking technologies such as IEEE 802.11, IEEE 802.15, IEEE 802.16, and IEEE 802.20 are used in the implementation of WMNs.

As WMNs become an increasingly popular replacement technology for last-mile connectivity to the home networking, community and neighborhood networking, it is imperative to design an efficient resource management system for these networks. Routing is one of the most challenging issues in resource management for supporting real-time applications with stringent *quality of service* (QoS) requirements. However, most of the existing routing protocols for WMNs are extensions of protocols originally designed for *mobile ad hoc networks* (MANETs) and thus they perform sub-optimally. Moreover, most routing protocols for WMNs are designed without security issues in mind, where the nodes are all assumed to be honest. In practical deployment scenarios, this assumption does not hold. In a community-based WMN, a group of MRs managed by different operators form an access network to provide last-mile connectivity to the Internet. As with any end-user supported infrastructure, ubiquitous cooperative behavior in these networks cannot be assumed *a priori*. Preserving scarce access bandwidth and power, as well as security concerns may induce some selfish users to avoid forwarding data for other nodes, even as they send their own traffic through the network. The selfish behavior of an MR degrades the performance of a WMN since it increases the latency in packet delivery and packet drops and decreases the network throughput. In addition, some nodes may also launch malicious packet dropping attacks. Therefore, enforcing cooperation among the nodes in WMNs becomes a critical issue and a routing protocol should make use of such a cooperation enforcement scheme in order to ensure efficiency in packet forwarding and minimizing packet drops (Dong, 2009). To enforce cooperation among nodes and detect malicious and selfish nodes in self-organizing networks such as MANETs, various collaboration schemes have been proposed in the literature (Santhanam et al., 2008). Most of these proposals are based on trust and reputation frameworks which attempt to identify misbehaving nodes by an appropriate detection and decision making system, and then isolate or punish them. Unfortunately, most of these schemes are not directly applicable for WMNs due to inherent differences in characteristics between MANETs and WMNs. Efficient, reliable and secure routing protocols for WMNs are clearly in demand.

Keeping this in mind, this chapter provides a comprehensive overview of security issues in WMNs and then particularly focuses on secure routing in these networks. First, it identifies security vulnerabilities in the *medium access control* (MAC) and the network layers. Various possibilities of compromising data confidentiality, data integrity, replay attacks and offline cryptanalysis are also discussed. Then various types of attacks in the MAC and the network layers are discussed. In the MAC layer, attacks such as passive eavesdropping, link layer jamming (Law et al., 2005; Brown et al., 2006), MAC spoofing, replay attacks (Mishra et al., 2002) are discussed in detail. In the network layer, two broad categories of attacks are identified: (i) attacks on the control plane and (ii) attacks on the data plane. Among the attacks on the control plane, rushing attack (Hu et al., 2003a), wormhole attack (Hu et al., 2003b), blackhole attack (Al-Shurman et al., 2004), grayhole attack (Sen et al., 2007), Sybil attack (Newsome et al., 2004) are discussed. The data plane attacks are launched by the selfish and malicious nodes which lead to degradation in the network performance (Zhong et al., 2005; Salem et al., 2003). After enumerating the various types of attacks on the MAC and the network layer, the chapter briefly discusses on some of the preventive mechanisms for those attacks. After the preliminary discussion on various attacks and their countermeasures, the chapter focuses on its major issue- security in routing. It first identifies the major security requirements for design of a routing protocol in WMNs. Then various existing secure routing protocols for self-organizing networks such as ARAN (Sanzgiri et al., 2002), SAODV (Zapata et al., 2002), SRP (Papadimitratos et al., 2002), SEAD (Hu et al., 2002b), ARIADNE (Hu et al., 2002a), SEAODV (Li et al., 2011) etc. are discussed. All these protocols are compared in terms of their relative performance and their areas of application. After discussing these existing mechanisms, the chapter presents two novel secure routing protocols that detect selfish nodes in WMNs and isolate those nodes from the network activities so as to maximize the network throughput while providing desired QoS of the user application (Sen, 2010a; Sen, 2010b).

The organization of the chapter is as follows. In Section 2, we discuss various security vulnerabilities in different layers of the protocol stack of a WMN. Attacks at the physical, MAC, network, and transport layers are discussed in detail, and the countermeasures to defend against such attacks are briefly presented. In Section 3, several routing challenges in WMNs are highlighted. Section 4 presents some of the well-known existing security mechanisms for routing in WMNs. These protocols are also compared with respect to their capabilities in defending against different attacks in the network layer of WMNs. In Section 5, two novel routing protocols for WMNs are presented. These protocols can guarantee application QoS in addition to identifying malicious and selfish nodes in the network. Section 6 concludes the chapter while identifying some open issues and future research directions in designing secure routing protocols for WMNs.

In summary, the chapter makes the following contributions:

- It proposes threat models and security goals for secure routing in WMNs.
- It identifies various possible attacks on different layers of a WMN.
- It demonstrates how attacks against MANETs and peer-to-peer networks can be adapted into powerful attacks against WMNs.
- It makes security analysis of some of the major existing routing protocols fro WMNs.
- It presents various defense mechanisms to counter the well-known attacks on the routing protocols of WMNs.
- It presents two novel routing protocols for WMNs. These protocols enhance the routing efficiency and the application QoS while providing security in routing.
- It identifies some open research problems in the area of secure routing in WMNs.

2. Security Vulnerabilities in WMNs

Several vulnerabilities exist in the protocols foe WMNs. These vulnerabilities can be exploited by the attackers to degrade the performance of the network. The nodes in a WMN depend on the cooperation of the other nodes in the network. Consequently, the MAC layer and the network layer protocols for these networks usually assume that the participating nodes are honest and well-behaving with no malicious or dishonest intentions. In practice, however, some nodes in a WMN may behave in a selfish manner or may be compromised by malicious users. The assumed trust and the lack of accountability due to the absence of a central administrator make the MAC and the network layer protocols vulnerable to various types of attacks. In this section, a comprehensive discussion on various types of attacks in different layers of the protocol stack of a WMN is provided.

2.1 Physical layer attacks

The physical layer is responsible for frequency selection, carrier frequency generation, signal detection, modulation, and data encryption. As with any radio-based medium, the possibility of jamming attacks in this layer of WMNs is always there. Jamming is a type of attack which interferes with the radio frequencies that the nodes use in a WMN for communication (Shi et al., 2004). A jamming source may be powerful enough to disrupt communication in the entire network. Even with less powerful jamming sources, an adversary can potentially disrupt communication in the entire network by strategically distributing the jamming sources. An intermittent jamming source may also prove detrimental as some communications in WMNs may be time-sensitive. More complex forms of radio jamming attacks have been studied in (Xu et al., 2005), where the attacking devices do not obey the MAC layer protocols.

2.2 MAC layer attacks

Different types of attacks are possible in the MAC layer of a WMN. Some of the major attacks at this layer are: passive eavesdropping, jamming, MAC address spoofing, replay, unfairness in allocation, pre-computation and partial matching etc. These attacks are briefly described in this subsection.

i. **Passive eavesdropping:** the broadcast nature of transmission of the wireless networks makes these networks prone to passive eavesdropping by the external attackers within the transmission range of the communicating nodes. Multi-hop wireless networks like WMNs are also prone to internal eavesdropping by the intermediate hops, whereby a malicious intermediate node may keep the copy of all the data that it forwards without the knowledge of any other nodes in the network. Although passive eavesdropping does not affect the network functionality directly, it leads to the compromise in data confidentiality and data integrity. Data encryption is generally employed using strong encryption keys to protect the confidentiality and integrity of data.

ii. **Link layer jamming attack:** link layer attacks are more complex compared to blind physical layer jamming attacks. Rather than transmitting random bits constantly, the attacker may transmit regular MAC frame headers (no payload) on the transmission channel which conforms to the MAC protocol being used in the victim network (Law et al., 2005). Consequently, the legitimate nodes always find the channel busy and back off for a random period of time before sensing the channel again. This leads to the denial-of-service for the legitimate nodes and also enables the jamming node to conserve its energy. In addition to the MAC layer, jamming can also be used to exploit the network and transport layer protocols (Brown et al., 2006). Intelligent jamming is not a purely transmit activity. Sophisticated sensors are deployed, which detect and identify victim network activity, with a particular focus on the semantics of higher-layer protocols (e.g., AODV and TCP). Based on the observations of the sensors, the attackers can exploit the predictable timing behavior exhibited by higher-layer protocols and use offline analysis of packet sequences to maximize the potential gain for the jammer. These attacks can be effective even if encryption techniques such as *wired equivalent privacy* (WEP) and *WiFi protocol access* (WPA) have been employed. This is because the sensor that assists the jammer can still monitor the packet size, timing, and sequence to guide the jammer. Because these attacks are based on carefully exploiting protocol patterns and consistencies across size, timing and sequence, preventing them will require modifications to the protocol semantics so that these consistencies are removed wherever possible.

iii. **Intentional collision of frames:** a collision occurs when two nodes attempt to transmit on the same frequency simultaneously (Wood et al., 2002). When frames collide, they are discarded and need to be retransmitted. An adversary may strategically cause collisions in specific packets such as acknowledgment (ACK) control messages. A possible result of such collision is the costly exponential back-off. The adversary may simply violate the communication protocol and continuously transmit messages in an attempt to generate collisions. Repeated collisions can also be used by an attacker to cause resource exhaustion. For example a naïve MAC layer implementation may continuously attempt to retransmit the corrupted packets. Unless these retransmissions are detected early, the energy levels of the nodes would be exhausted quickly. An attacker may cause unfairness by intermittently using the MAC layer attacks. In this case, the adversary causes degradation of real-time applications running on other nodes by intermittently disrupting their frame transmissions.

iv. **MAC spoofing attack:** MAC addresses have long been used as the singularly unique layer-2 network identifiers in both wired and wireless LANs. MAC addresses which are globally unique have often been used as an authentication factor or as a unique identifier for granting varying levels of network privileges to a user. This is particularly common in 802.11 WiFi networks. However, today's MAC protocols (802.11) and network interface cards do not provide any safeguards that would prevent a potential attacker from modifying the source MAC address in its transmitted frames. On the contrary, there is often full support in the form of drivers from manufacturers, which makes this particularly easy. Modifying MAC addresses in transmitted frames is referred to as MAC spoofing, and can be used by attackers in a variety of ways. MAC spoofing enables the attacker to evade *intrusion detection systems* (IDSs) that are in place. Further, today's network administrators often use MAC addresses in access control lists. For example, only registered MAC addresses are allowed to connect to the access points. An attacker can easily eavesdrop on the network to determine the MAC addresses of legitimate devices. This enables the attacker to masquerade as a legitimate user and gain access to the network. An attacker can even inject a large number of bogus frames into the network to deplete the resources (in particular, bandwidth and energy), which may lead to denial of services for the legitimate nodes.

v. **Replay attack:** the replay attack, often known as the *man-in-the-middle* attack (Mishra et al., 2002), can be launched by external as well as internal nodes. An external malicious node (not a member of WMN) can eavesdrop on the broadcast communication between two nodes (*A* and *B*) in the network as shown in Fig. 2. It can then transmit legitimate messages at a later stage of time to gain access to the network resources. Generally, the authentication information is replayed where the attacker deceives a node (node *B* in Fig. 2) to believe that the attacker is a legitimate node (node *A* in Fig. 2). On a similar note, an internal malicious node, which is an intermediate hop between two communicating node, can keep a copy of all relayed data. It can then retransmit this data at a later point in time to gain the unauthorized access to the network resources.

vi. **Pre-computation and partial matching attack:** unlike the above-mentioned attacks, where MAC protocol vulnerabilities are exploited, these attacks exploit the vulnerabilities in the security mechanisms that are employed to secure the MAC layer of the network. Pre-computation and partial matching attacks exploit the cryptographic primitives that are used at MAC layer to secure the communication. In a pre-computation attack or *time memory trade-off attack* (TMTO), the attacker computes a large amount of information (key, plaintext, and respective ciphertext) and stores that information before launching the attack. When the actual transmission starts, the attacker uses the pre-computed information to speed up the cryptanalysis process. TMTO attacks are highly effective against a large number of cryptographic solutions. On the other hand, in a partial matching attack, the attacker has access to some (cipher text, plaintext) pairs, which in turn decreases the encryption key strength, and improves the chances of success of the brute force mechanisms. Partial matching attacks exploit the weak implementations of encryption algorithms. For example, the IEEE80.11i standard for MAC layer security in wireless networks is prone to the sensor hijacking attack and the man-in-the-middle attack that exploit the vulnerabilities in IEEE802.1X. DoS attacks on the four-way handshake procedure in IEEE 80.211i.

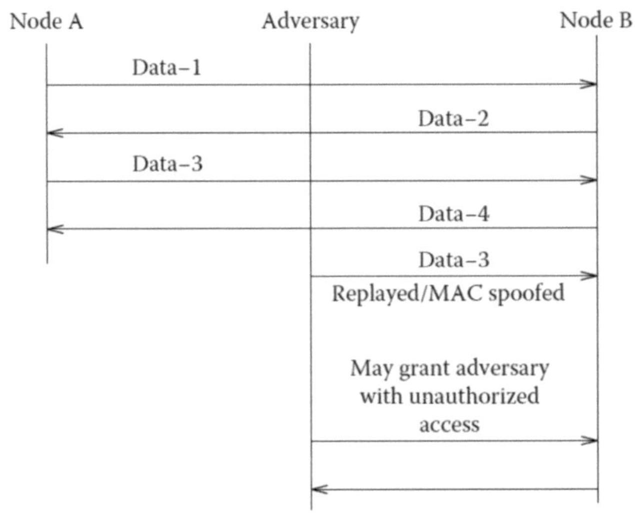

Fig. 2. Illustration of MAC spoofing and replay attacks

DoS attacks may also be launched by exploiting the security mechanisms. For example, the IEEE 802.11i standard for MAC layer security in wireless networks is prone to the sensor hijacking attack and the man-in-the-middle attack, exploiting the vulnerabilities in IEEE 802.1X, and DoS attack, exploiting vulnerabilities in the four-way handshake procedure in IEEEE 802.11i.

2.3 Network layer attacks

The attacks on the network layer can be divided into *control plane attacks* and *data plane attacks,* and can be active or passive in nature. Control plane attacks generally target the routing functionality of the network layer. The objective of the attacker is to make routes unavailable or force the network to choose sub-optimal routes. On the other hand, the data plane attacks affect the packet forwarding functionality of the network. The objective of the attacker is to cause the denial of service for the legitimate user by making user data undeliverable or injecting malicious data into the network. We first consider the network layer control plane attacks, and then the network layer data plane attacks.

i. **Control plane attacks:** *Rushing* attacks (Hu et al., 2003a) targeting the on-demand routing protocols (e.g., AODV) were among the first exposed attacks on the network layer of multi-hop wireless networks. Rushing attacks exploit the route discovery mechanism of on-demand routing protocols. In these protocols, the node requiring the route to the destination floods the *route_request* (RREQ) message, which is identified by a sequence number. To limit the flooding, each node only forwards the first message that it receives and drops remaining messages with the same sequence number. To avoid collisions of the messages, the protocol specifies a specific amount of delay between the receiving of a route request message by a particular node, and its forwarding by the same node. The malicious node launching the rushing attack forwards the RREQ message to the target node before any other intermediate node from the source to destination. This can easily be achieved by ignoring the specified delay. Consequently, the route from the source to the destination includes the malicious node as an intermediate hop, which can then drop the packets of the flow thereby launching a data plane DoS attack.

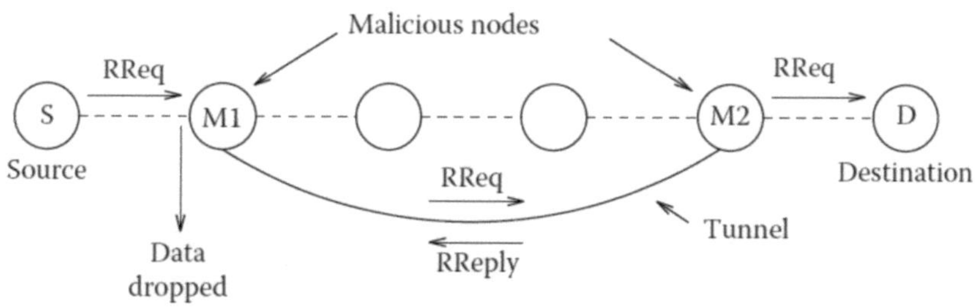

Fig. 3. Illustration of wormhole attack launched by nodes M1 and M2

A *wormhole* attack has a similar objective albeit it uses a different technique (Hu et al., 2003b). During a wormhole attack, two or more malicious nodes collude together by establishing a tunnel using an efficient communication medium (i.e., wired connection or high-speed wireless connection etc.), as shown in Fig. 3. During the route discovery phase of the on-demand routing protocols, the RREQ messages are forwarded between the malicious nodes using the established tunnel. Therefore, the first RREQ message that reaches the destination node is the one forwarded by the malicious nodes. Consequently, the malicious nodes are added in the path from the source to the destination. Once the malicious nodes are included in the routing path, these nodes either drop all the packets resulting in a complete DoS attack, or drop the packets selectively to avoid detection.

A *blackhole* attack (or *sinkhole* attack) (Al-Shurman et al., 2004) is another attack that leads to denial of service in WMNs. It also exploits the route discovery mechanism of on-demand routing protocols. In a blackhole attack, the malicious node always replies positively to a RREQ, although it may not have a valid route to the destination. Because the malicious node does not check its routing entries, it will always be the first to reply to the RREQ message. Therefore, almost all the traffic within the neighborhood of the malicious node will be directed towards the malicious node, which may drop all the packets, resulting in denial of service. Fig. 4 shows the effect of a blackhole attack in the neighborhood of the malicious node where the traffic is directed towards the malicious node. A more complex form of the attack is the cooperative blackhole attack where multiple nodes collude together, resulting in complete disruption of routing and packet forwarding functionality of the network. The cooperative blackhole attack and the prevention mechanism have been studied in (Ramaswamy et al., 2003).

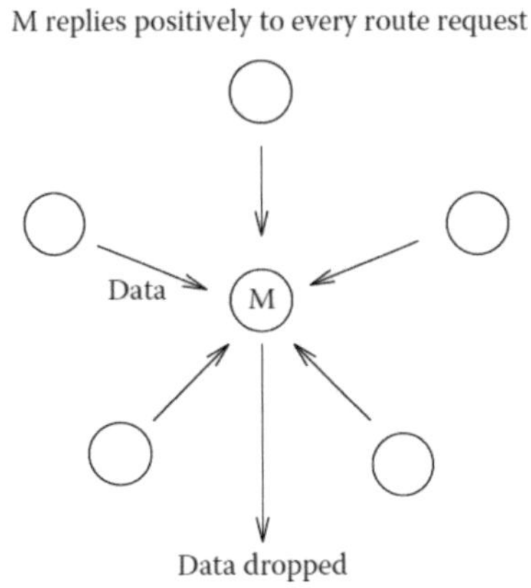

Fig. 4. Illustration of blackhole attack launched by node *M*

A *grayhole* attack is a variant of the blackhole attack (Sen et al., 2007). In a blackhole attack, the malicious node drops all the traffic that it is supposed to forward. This makes detection of the malicious node a relatively easier task. In a grayhole attack, the adversary avoids the detection by dropping the packets selectively. A grayhole does not lead to complete denial of service, but it may go undetected for a longer duration of time. This is because the malicious packet dropping may be considered congestion in the network, which also leads to selective packet loss.

A *Sybil* attack is the form of attack where a malicious node creates multiple identities in the network, each appearing as a legitimate node (Newsome et al., 2004). A Sybil attack was first exposed in distributed computing applications where the redundancy in the system was exploited by creating multiple identities and controlling considerable system resources. In the networking scenario, a number of services like packet forwarding, routing, and collaborative security mechanisms can be disrupted by the adversary using a Sybil attack. Following form of the attack affects the network layer of WMNs, which are supposed to take advantage of the path diversity in the network to increase the available bandwidth and reliability. If the malicious node creates multiple identities in the network, the legitimate nodes will assume these identities to be distinct nodes and will add these identities in the list of distinct paths available to a particular destination. When the packets are forwarded to these fake nodes, the malicious node that created the identities processes these packets. Consequently, all the distinct routing paths will pass through the malicious node. The malicious node may then launch any of the above-mentioned attacks. Even if no other attack is launched, the advantage of path diversity is diminished, resulting in degraded performance.

In addition to the above-mentioned attacks, the network layer of WMNs are also prone to various types of attack such as: *route request (RREQ) flooding attack, route reply (RREP) loop attack, route re-direction attack, fabrication attack, network partitioning* attack etc. RREQ flooding is one of the simplest attacks in which a malicious node tries to flood the entire network with RREQ message. As a consequence, this causes a large number of unnecessary broadcast communications resulting in energy drains and bandwidth wastage in the network. A *routing loop* is a path that goes through the same nodes over and over again. As a result, this kind of attack will deplete the resources of every node in the loop and will lead to isolation of the destination node.

Fig. 5 describes two instances where *route re-direction attack* has been launched by a malicious node *M*. In case *A*, the malicious node *M* tries to initiate the attack by modifying the mutable fields in the routing messages. These mutable fields include hop count, sequence numbers and other metric-related fields. The malicious node *M* could divert the traffic through itself by advertising a route to the destination with a larger *destination sequence number* (DSN) than the one it received from the destination. In case *B*, route re-direction attack may be launched by modifying the metric field in the AODV routing message, which is the hop-count field in this case. The malicious node *M* simply modifies the hop count field to zero in order to claim that it has a shorter path to the destination.

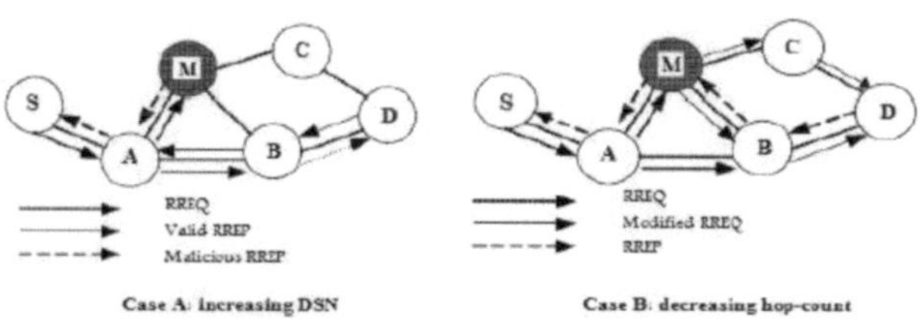

Fig. 5. Illustration of route re-direction attack

An adversary may fabricate false routing messages in order to disrupt routing in the network. For example, a malicious node may fabricate a *route error* (RERR) message in the AODV protocol. This may result in the upstream nodes re-initiating the route request to the unreachable destination so as to discover and establish alternative routes to them leading to energy and bandwidth wastage in the network. In a network partitioning attack, the malicious nodes collude together to disrupt the routing tables in such a way that the network is divided into disconnected partitions, resulting in denial of service for a certain network portion. Routing loop attacks affect the packet-forwarding capability of the network where the packets keep circulating in loop until they reach the maximum hop count, at which stage the packets are simply dropped.

ii. **Data plane attacks:** data plane attacks are primarily launched by selfish and malicious (compromised) nodes in the network and lead to performance degradation or denial of service of the legitimate user data traffic. The simplest of the data plane attacks is *passive eavesdropping*. Eavesdropping is a MAC layer attack. Selfish behavior of the participating WMN nodes is a major security issue because the WMN nodes are dependent on each other for data forwarding. The intermediate-hop selfish nodes may not perform the packet-forwarding functionality as per the protocol. The selfish node may drop all the data packets, resulting in complete denial of service, or it may drop the data packets selectively or randomly. It is hard to distinguish between such a selfish behavior and the link failure or network congestion. On the other hand, malicious intermediate-hop nodes may inject junk packets into the network. Considerable network resources (bandwidth and packet processing time) may be consumed to forward the junk packets, which may lead to denial of service for legitimate user traffic. The malicious nodes may also inject the maliciously crafted control packets, which may lead to the disruption of routing functionality. The control plane attacks are dependent on such maliciously crafted control packets. The malicious and selfish behaviors of nodes in WMNs have been studied in (Zhong et al., 2005; Salem et al., 2003).

2.4 Transport layer attacks

The attacks that can be launched on the transport layer of a WMN are flooding attack and de-synchronization attack. Whenever a protocol is required to maintain state at either end of a connection, it becomes vulnerable to memory exhaustion through flooding. An attacker may repeatedly make new connection request until the resources required by each connection are exhausted or reach a maximum limit. In either case, further legitimate requests will be ignored. De-synchronization refers to the disruption of an existing connection (Wood et al., 2002). An attacker may, for example, repeatedly spoof messages to an end host causing the host to request the retransmission of missed frames. If timed correctly, an attacker may degrade or even prevent the ability of the end hosts to successfully exchange data causing them instead to waste energy attempting to recover from errors which never really exist.
Table 1 presents various types of vulnerabilities in different layers of a WMN and their respective defense mechanisms.

Layer	Attacks	Defense Mechanism
Physical	Jamming Device tampering	Spread-spectrum, priority messages, lower duty cycle, region mapping, mode change
MAC	Collision	Error-correction code
	Exhaustion	Rate limitation
	Unfairness	Small frames
Network	Spoofed routing information & selective forwarding	Egress filtering, authentication, monitoring
	Sinkhole	Redundancy checking
	Sybil	Authentication, monitoring, redundancy
	Wormhole	Authentication, probing
	Hello Flood	Authentication, packet leashes by using geographic and temporal info
	Ack. flooding	Authentication, bi-directional link authentication verification
Transport	Flooding De-synchronization	Client puzzles Authentication
Application	Logic errors Buffer overflow	Application authentication Trusted computing

Table 1. Attacks on different layers of a WMN and their countermeasures

3. Routing Challenges in WMNs

In this section, some of the important challenges in designing routing protocols for WMNs are discussed. A typical architecture of a hierarchical WMN is presented in Fig. 1. At the top layer, are the *Internet gateways* (IGWs) which are connected to the wired Internet. They form the backbone infrastructure for providing Internet connectivity to the elements in the second level. The entities at the second level are called wireless *mesh routers* (MRs) that eliminate the need for wired infrastructure at every MR and forward their traffic in a multi-hop fashion towards the IGW. At the lowest level are the *mesh clients* (MCs) which are the wireless devices of the users. Internet connectivity and peer-to-peer communications inside the mesh are two important applications for a WMN. Therefore, design of an efficient and low-overhead routing protocol that avoids unreliable routes, and accurately estimate the end-to-end delay of a flow along the path from the source to the destination is a major challenge. Some of the major challenges in designing routing protocol for WMNs are discussed below:

i. **Measuring link reliability**: it has been observed that in wireless ad hoc networks like WMNs, nodes receiving broadcast messages introduce communication gray zones (Lundgren et al., 2002). In such zones, data messages cannot be exchanged although the hello messages reach the neighbors. This leads to disruption in communication among the nodes. Since the routing protocols such as AODV and WMR (Xue et al., 2003) relay on control packets like RREQ, these protocols are highly unreliable for estimating the quality of wireless links. Due to communication gray zone problem, nodes that are able to send and receive bi-directional RREQ packets sometimes cannot send/receive data packets at high rate. These fragile links trigger link repairs resulting in high control overhead.
ii. **End-to-end delay estimation:** an important issue in a routing protocol is end-to-end delay estimation. Current protocols estimate end-to-end delay by measuring the time taken to route route request (RREQ) and route reply (RREP) packets along the given path. However, RREQ and RREP packets are different from normal data packets and hence they are unlikely to experience the same levels of delay and loss as data packets. It has been observed through simulation that a RREP-based estimator overestimates while a hop-count-based estimator underestimates the actual delay experienced by the data packets (Kone et al., 2007). The reason for the significant deviation of a RREP-based estimator from the actual end-to-end delay is interference of signals. The RREQ packets are flooded in the network resulting in a heavy burst of traffic. This heavy traffic causes inter-flow interference in the paths. The unicast data packets do not cause such events. Moreover, as a stream of packets traverse along a route, due to the broadcast nature of wireless links, different packets in the same flow interfere with each other resulting in per-packet delays. Since the control packets do not experience per-packet delay, the estimates based on control packet delay deviate widely from the actual delay experience by the data packets.
iii. **Reduction of control overhead:** since the effective bandwidth of wireless channels vary continuously, reduction of control overhead is important in order to maximize throughput in the network. Reactive protocols such as AODV and DSR use flooding of RREQ packets for route discovery. This consumes a high proportion of the network bandwidth and reduces the effective throughput. An important challenge in designing a routing protocol for WMNs is to optimize the communication and computation overhead of the control messages so that the bandwidth of the wireless channels may be used for applications as efficiently as possible. Security and privacy issues bring another dimension of complexity. The goal of the protocol designer would be to design the security framework in such as way that it involves minimum computational and communication overhead.

4. Secure Routing Protocols for WMNs

Extensive work has been done in the area of secure unicast routing in multi-hop wireless networks (Hu et al., 2002a; Hu et al., 2002b; Sanzgiri et al., 2002; Marti et al., 2000; Papadimitratos et al., 2003a; Awerbuch et al., 2002; Awerbuch et al., 2005). As mentioned in Section 2.3, attacks on routing protocols can target either the route establishment process or the data delivery process, or both. Ariadne (Hu et al., 2002a) and SRP (Papadimitratos et al., 2003a) propose to secure on-demand source routing protocols by using hop-by-hop authentication techniques to prevent malicious packet manipulations on the route discovery process. SAODV (Zapata et al., 2002), SEAD (Hu et al., 2002b), and ARAN (Sanzgiri et al., 2002) propose to secure on-demand distance vector routing protocols by using one-way hash chains to secure the propagation of hop counts. The authors in (Papadimitratos et al., 2003b) propose a secure link state routing protocol that ensures the correctness of link state updates with digital signatures and one-way hash chains. To ensure correct data delivery, (Marti et al., 2000) proposes the *watchdog* and *pathrater* techniques to detect adversarial nodes by having each node monitor if its neighbors forward packets correctly. SMT (Papadimitratos et al., 2003a) and Ariadne (Hu et al., 2002a) use multi-hop routing to prevent malicious nodes from selectively dropping data. ODSBR (Awerbuch et al., 2002; Awerbuch et al., 2005) provides resilience to colluding Byzantine attacks by detecting malicious links based on end-to-end acknowledgment-based feedback technique. In HWMP (Bahr, 2006; Bahr, 2007), the on-demand node allows two mesh points (MPs) to communicate using peer-to-peer paths. This model is primarily used if nodes experience a changing environment and no root MP is configured. While the proactive tree building mode is an efficient choice for nodes in a fixed network topology, HWMP does not address security issues and is vulnerable to a numerous attacks such as RREQ flooding attack, RREP routing loop attack, route re-direction attack, fabrication attack, tunnelling attack etc (Li et al., 2011). LHAP (Zhu et al., 2003) is a lightweight transparent authentication protocol for wireless ad hoc networks. It uses TESLA (Perrig et al., 2000) to maintain the trust relationship among nodes, which is not realistic due to TESLA's delayed key disclosure period. In LHAP, simply attaching the TRAFFIC key right after the raw message is not secure since the traffic key has no relationship with the message being transmitted.

In contrast to secure unicast routing, work studying security problems specific to multicast routing in wireless networks is particularly scarce, with the notable exception of the work by (Roy et al., 2005) and BSMR (Curtmola et al., 2007). The work in (Roy et al., 2005) proposes an authentication framework that prevents outsider attacks in tree-based multicast protocol, MAODV (Royer et al., 2000), while BSMR (Curtmola et al., 2007) complements the work in (Roy et al., 2005) and presents a measurement-based technique that addresses insider attacks in tree-based multicast protocols.

A key point to note is that all of the above existing work in either secure unicast or multicast routing considers routing protocols that use only basic routing metrics, such as hop-count and latency. None of them consider routing protocols that incorporate high-throughput metrics, which have been shown to be critical for achieving high performance in wireless networks. On the contrary, many of them even have to remove important performance optimizations in existing protocols in order to prevent security attacks. There are also a few studies (Papadimitratos et al., 2006; Zhu et al., 2006) on secure QoS routing in wireless networks. However, they require strong assumptions, such as symmetric links, correct trust evaluation on nodes, ability to correctly determine link metrics despite attacks etc. In addition, none of them consider attacks on the data delivery phase. The work presented in (Dong, 2009) is the first of its kind that encompasses both high performance and security as goals in multicast routing and considers attacks on both path establishment and data delivery phases.

As mentioned in Section 2.3, wireless networks are also subject to attacks such as rushing attacks and wormhole attacks. Defenses against these attacks have been extensively studied in (Hu et al., 2003b; Hu et al., 2003a; Eriksson et al., 2006; Hu et al., 2004). RAP (Hu et al., 2003a) prevents the rushing attack by waiting for several flood requests and then randomly selecting one to forward, rather than always forwarding only the first one. Techniques to defend against wormhole attacks include *packet leashes* (Hu et al., 2003b) which restricts the maximum transmission distance by using time or location information. Truelink (Eriksson et al., 2006) which uses MAC level acknowledgments to infer whether a link exists between two nodes, and the work in (Hu et al., 2004) that relies on directional antennas are two mechanisms for defense against the wormhole attack.

In the following sub-sections, some of the well-known security protocols for routing in WMNs are presented. These protocols are extensions of base routing protocols like AODV, DSR etc. and use cryptographic mechanisms for ensuring node authentication, message integrity and message confidentiality.

4.1 Authenticated Routing for Ad Hoc Networks (ARAN)

Authenticated routing for ad hoc networks (ARAN) protocol (Sanzgiri et al., 2002), is an on-demand routing protocol that makes use of cryptographic certificates to offer routing security. It takes care of authentication, message integrity, and non-repudiation, but expects a small amount of prior security coordination among the nodes. In (Sanzgiri et al., 2002), vulnerabilities and attacks specific to AODV and DSR protocols are discussed and the two protocols are comapred with the ARAN protocol.

During the route discovery process of ARAN, the source node brodcasts *route_request* (RREQ) packets. The destination node, on receiving the RREQ packets, responds by unicasting back a reply packt, called the *route_reply* (RREP) packet. The ARAN protocol uses a preliminary cryptographic certification process, followed by an end-to-end route authentication process, which ensures secure route establishment. The protocol requires the use of a trusted certificate server *T*, whose public key is known to all the nodes in the network. End-to-end authentication is achieved by the source by having it verify that the intended destination was indeed reached. The source trusts the destination to choose the return path. The protocol is briefly discussed below.

Issue of certificates: ARAN utilizes an authenticated trusted server whose public key is known to all legitimate nodes in the network. The protocol assumes that keys are generated *a priori* by the server and distributed to all nodes in the network. It does not specify any specific key distribution algorithm. On joining the network, each node receives a certificate from the trusted server. The certificate received by a node *A* from the trusted server *T* looks like the following:

$$T \rightarrow A : cert_{A} = [IP_A, K_{A+}, t, e]K_{T-} \quad (1)$$

In (1), IP_A, K_{A+}, *t*, *e* and K_{T-} represent the IP address of node *A*, the public key of node *A*, the time of creation of the certificate, the time of expiry of the certificate, and the private key of the server, respectively.

End-to-end route authentication: the main goal of the end-to-end route authentication process is to ensure that the packets reach the current intended destination from the source node. The source node S broadcasts a RREQ (i.e. route discovery) packet destined to the destination node D. The RREQ packet contains the packet identifier (*route discovery process* (RDP)), the IP address of the destination (IP_D), the certificate of the source node S ($Cert_S$), the current time (t) and a nonce N_S. The process can be denoted as in (2), where, K_{S-} is the private key of the source node S.

$$S \rightarrow broadcasts := [RDP, IP_D, Cert_S, N_S, t]K_{S-} \quad (2)$$

Whenever the source sends a route discovery message, it increments the value of the nonce. Nonce is a counter used in conjunction with the time-stamp in order to make the nonce recycling easier. When a node receives an RDP packet from the source with a higher value of the source's nonce than that in the previously received RDP packets from the same source node, it makes a record of the neighbor from which it received the packet, encrypts the packet with its own certificate, and broadcasts it further. The process is represented in (3) below:

$$A \rightarrow broadcasts := [[RDP, IP_D, Cert_S, N_s, t]K_{s-}]K_{A-}, Cert_A \quad (3)$$

An intermediate node B on receiving an RDP packet from node A removes its neighbor's certificate, inserts its own certificate, and broadcast the packet further. The destination node, on receiving an RDP packet, verifies node S's certificate and the tuple (N_S, t) and then replies with the *route reply* (REP). The destination unicasts the REP packet to the source node along the reverse path as in (4):

$$D \rightarrow X := [REP, IP_S, Cert_D, N_S, t]K_{D-} \quad (4)$$

In (4), node X is the neighbor of the destination node D, which had originally forwarded the RDP packet to node D. The REP packet follows the same procedure on the reverse path as that followed by the route-discovery packet. An error message is generated if the time-stamp or nonce does not match the requirements or if the certificate fails. The error message looks similar to the other packets except that the packet identifier is replaced by the ERR message.
In summary, ARAN is a robust protocol in the presence of attacks such as unauthorized participation, spoofed route signaling, fabricated routing messages, alteration of routing messages, securing shortest paths, and replay attacks. However, since ARAN uses public-key cryptography for authentication, it is particularly vulnerable to DoS attacks based on flooding the network with bogus control packets for which signature verifications are required. As long as a node can't verify signature at required speed, an attacker can force that node to discard some fraction of the control packets it receives.

4.2 Secure Efficient Ad Hoc Distance Vector (SEAD) routing protocol

Secure efficient ad hoc distance vector (SEAD) (Hu et al., 2002b) is a secure and proactive ad hoc routing protocol based on the *destination-sequenced distance vector* (DSDV) routing protocol (Perkins et al., 1994). This protocol is mainly designed to overcome security attacks such as DoS and resource consumption attacks. The operation of the routing protocol does not get affected even in the presence of multiple uncoordinated attackers corrupting the routing tables. The protocol uses a one-way hash function and does not involve any asymmetric cryptographic operation. The basic idea of SEAD is to authenticate the sequence number and metrics of a routing table update message using hash chain elements. The receiver also authenticates the sender ensuring that the routing information originates from the correct node. The source of each routing update message is also authenticated so as to prevent creation of a routing loop by an attacker launching an impersonation attack.

In the following, first a brief description of the base DSDV protocol is given followed by a discussion on the enhancements proposed in the SEAD protocol.

Distance vector routing: distance vector routing protocols belong to the category of table-driven routing protocols. Each node maintains a routing table containing the list of all known routes to various destination nodes in the network. The metric used for routing is the distance measured in terms of hop-count. The routing table is updated periodically by exchanging routing information. An alternative to this approach is *triggered updates,* in which each node broadcasts routing updates only if its routing table gets altered. The DSDV protocol for ad hoc wireless networks and WMNs uses *sequence number* tags to prevent the formation of loops, to counter the count-to-infinity problem, and for faster convergence. When a new route update packet is received for a destination, the node updates the corresponding entry in its routing table only if the sequence number on the received update is greater than that recorded with the corresponding entry in the routing table. If the received sequence number and the previously recorded sequence number are both equal, but if the routing update has a new value for the routing metric (distance in number of hops), then in this case also the update is effected. Otherwise, the received update packet is discarded. DSDV uses triggered updates (for important routing changes) in addition to the regular periodic updates. A slight variation of DSDV protocol known as *DSDV sequence number* (DSDV-SQ), initiates triggered updates on receiving a new sequence number update.

One-way hash function: SEAD uses authentication to differentiate between updates that are received from non-malicious nodes and malicious nodes. This minimizes the chances of resource consumption attacks caused by malicious nodes. SEAD uses a one-way hash function for authenticating the updates. A one-way hash function (H) generates a one-way hash chain (h_1, h_2, ...). The function H maps an input bit-string of any length to a fixed length bit-string, that is, $H : (0, 1)^* \rightarrow (0, 1)^{\rho}$, where ρ is the length in bits of the output bit-string. To create a one-way hash chain, a node generates a random number with initial value $x \in (0, 1)^{\rho}$. h_0, the first number in the hash chain is initialized to x. The remaining values in the chain are computed using the general formula $h_i = H(h_{i-1})$ for $0 \le i \le n$, for some n. The way one-way hash function incorporates security into the existing DSDV-DQ routing protocol will now be explained. The SEAD protocol assumes an upper bound on the metric used. For example, if the metric used is distance, then the upper bound value $m - 1$ defines the maximum diameter (maximum of lengths of all the routes between a pair of nodes) of the ad hoc wireless network or the WMN. Hence, the routing protocol assumes that no route of length greater than m hops exists between any two nodes.

If the sequence of values calculated by a node using the hash function H is given by (h_1, h_2,... h_n), where n is divisible by m, then for a routing table entry with sequence number i, let $k = \frac{k}{m} - i$. If the metric j (distance) used for that routing table entry is, $0 \le j \le m - 1$, then the value of h_{km+j} is used to authenticate the routing update entry for that sequence number i and that metric j. Whenever a route update message is sent, the node appends the value used for authentication along with it. If the authentication value used is h_{km+j}, then the attacker who tries to modify this value can do so only if he/she knows h_{km+j-1}. Since it is a one-way hash chain, calculating h_{km+j-1} becomes impossible. An intermediate node, on receiving this authenticated update, calculates the new hash value based on the earlier updates (h_{km+j-1}), the value of the metric, and the sequence number. If the calculated value matches with the one present in the route update message, then the update is done. Otherwise, the received update is just discarded.

SEAD avoids routing loops unless the loop contains more than one attacker. This protocol could be implemented easily with slight modifications to the DSDV protocol. The use of one-way hash chain to verify the authentication largely reduces the computational complexity. Moreover, the protocol is robust against multiple uncoordinated attacks. The main disadvantage is that a trusted entity is needed in the network to distribute and maintain the verification element of every node since the verification element of a hash chain is detached by a trusted entity. This leads to a single-point of failure in the protocol. If the trusted entity is compromised, the entire network becomes vulnerable. In addition, the protocol is vulnerable in situations where an attacker uses the same metric and sequence number which has been used in a recent update message and sends a new routing update.

4.3 Security-Aware Ad Hoc Routing (SAR) protocol

The *security-aware ad hoc routing* (SAR) protocol (Yi et al., 2001) uses security as one of the key metrics in path finding and provides a framework for enforcing and measuring the attributes of the security metric. This framework also enables the use of different levels of security for different applications that use SAR for routing. In WMNs, communication between two end nodes through possibly multiple nodes is based on the fact that the end nodes trust the intermediate nodes. SAR defines *level of trust* as a metric for routing and as one of the attributes for security to be taken into consideration. In SAR, security metric is embedded into the RREQ packet and the forwarding behavior of the protocol is implemented with respect to the RREQs. The intermediate nodes receive an RREQ packet with a particular security metric or trust level. The protocol ensures that a node can only process the packet or forward it if the node itself can provide the required security or has the required authorization or trust level. If the node cannot provide the required security, the RREQ is dropped. If an end-to-end path with the required security attributes can be found, a suitably modified RREP is sent from an intermediate node or the destination node. The routing protocol based on the level of trust is explained using Fig. 6.

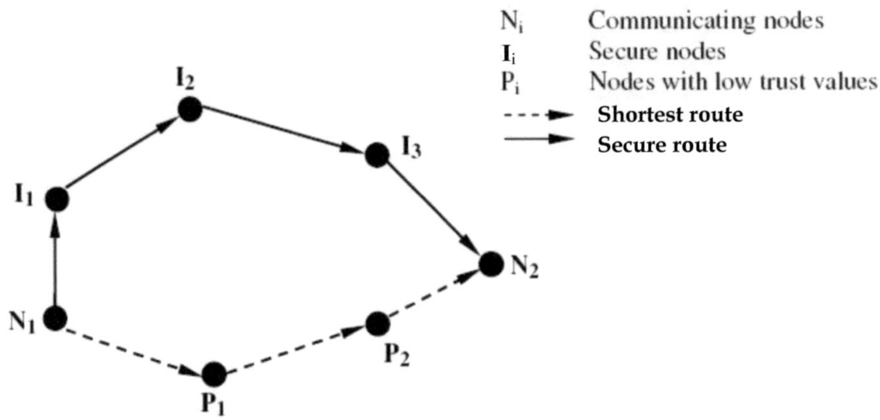

Fig. 6. Illustration of use of trust metric of nodes in routing

As shown in Fig. 6, two paths exist between the nodes N_1 and N_2 who want to communicate with each other. One of these paths is shorter which passes through private nodes (P_1 and P_2) whose trust levels are low. Hence, the protocol chooses a longer but secure path which passes through secure nodes I_1, I_2, and I_3.
The SAR protocol can be explained using any one of the traditional routing protocols. In this Section, SAR protocol has been explained using AODV protocol (Perkins et al., 1999). In the AODV protocol, the source node broadcasts a *route_request* (RREQ) packet to its neighbors. An intermediate node, on receiving a RREQ packet, forwards it further if it does not have a route to the destination. Otherwise, it initiates *route_reply* (RREP) packet back to the source node using the reverse path traversed by the RREQ packet. In SAR, a certain level of security is incorporated into the packet-forwarding mechanism. Here, each packet is associated with a security level which is determined by a number calculation method (explained later in this section). Each intermediate node is also associated with a certain level of security. On receiving a packet, the intermediate node is also associated with a certain level of security. On receiving a packet, the intermediate node compares its level of security with that defined for the packet. If node's security level is less than that of the packet, the RREQ is simply discarded. If it is greater, the node is considered to be a secure node and is permitted to forward the packet in addition to being able to view the packet. If the security level of the intermediate node and the received packet are found to be equal, then the intermediate node will not be able to view the packet (which can be ensured using a proper authentication mechanism); it just forwards the packet further.
Nodes of equal level of trust distribute a common key among themselves and with those nodes having higher levels of trust. Hence, a hierarchical level of security could be maintained. This ensures that an encrypted packet can be decrypted (using the common key) only by nodes of the same or higher levels of security compared to the level of security of the packet. Different levels of trust can be defined using a number calculated based on the level of security required. It can be calculated using a number of methods. Since timeliness, in-order delivery of packets, authenticity, authorization, integrity, confidentiality, and non-repudiation are some of the desired characteristics of a routing protocol, a suitable number can be defined for the trust level for nodes and packets based on the number of such characteristics taken into account.
The SAR protocol can be easily incorporated into the traditional routing protocols for ad hoc wireless networks and WMNs. It could be incorporated into both on-demand and table-driven routing protocols. The SAR protocol allows the application to choose the level of security it requires. But the protocol requires different keys for different levels of security. This tends to increase the number of keys required when the number of security levels used increases.

4.4 Secure Ad Hoc On-Demand Distance Vector (SAODV) routing protocol

In this section, a secure version of the AODV protocol will be described that plugs some well-known vulnerabilities of the routing protocol. Before presenting the secure version, a brief discussion of the base AODV protocol is presented.
Ad hoc on-demand distance vector (AODV) routing protocol: it is a reactive routing protocol (Perkins et al., 1999; Perkins et al., 2003) for MANETs and WMNs that maintains routes only between nodes which need to communicate. The routing messages do not contain information about the whole routing path, but only about the source and the

destination. Therefore, routing messages do not have an increasing size. It uses destination sequence numbers to specify how fresh a route is (in comparison to the others), which is used to grant loop freedom.
Whenever a node needs to send a packet to a destination for which it has no 'fresh enough' route (i.e., a valid route entry for the destination whose associated sequence number is at least as great as the one contained in any RREQ that the node has received for that destination), it broadcasts an RREQ message to its neighbors. Each node that receives the broadcast message sets up a reverse route towards the originator of the RREQ, unless it has a 'fresher' one (Fig. 7). When the intended destination (or an intermediate node that has a 'fresh enough' route to the destination) receives the RREQ, it replies by sending an RREP. It is important that the only mutable information in an RREQ and in an RREP is the hop-count (which is being monotonically increased at each hop). The RREP is unicast back to the originator of the RREQ (Fig. 8).

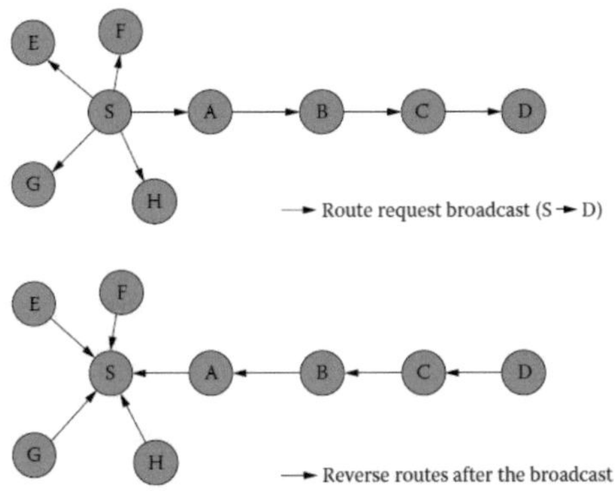

Fig. 7. Route request in AODV. *S* and *D* are the source and destination nodes respectively

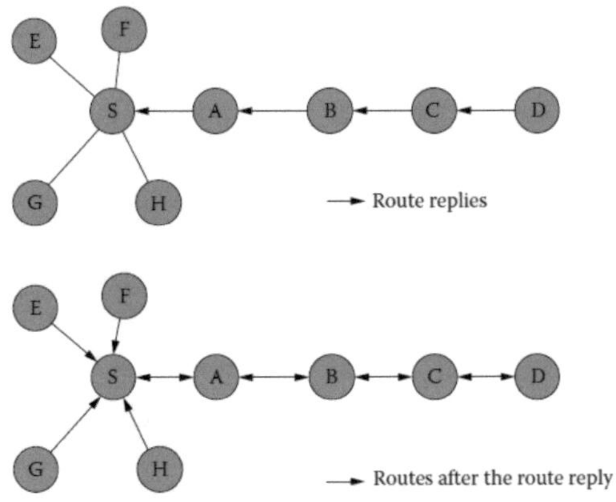

Fig. 8. Route reply in AODV. *S* and *D* are the source and destination nodes respectively

At each intermediate node, a route to the destination is set unless the node has a 'fresher' route than the one specified in the RREP). In the case that the RREQ is replied to by an intermediate node (and if the RREQ had set this option), the intermediate node also sends an RREP to the destination. In this way, it can be granted that the node path is being set up bi-directionally. In the case that a node receives a new route (by an RREQ or by an RREP) and the node already has a route 'as fresh' as the received one, the shortest one will be updated. Optionally, *route_reply acknowledgment* (RREP-ACK) message may be sent by the originator of the RREQ to acknowledge the receipt of the RREP. An RREP-ACK message has no mutable information. In addition to these routing messages, a *route_error* (RERR) message is used to notify the other nodes that certain nodes are not reachable anymore due to link breakage. When a node re-broadcasts an RERR, it only adds the unreachable destinations to which the node might forward messages. Therefore, the mutable information in an RERR is the list of unreachable destinations and the counter of unreachable destinations included in the message. It is predictable that, in each hop, the unreachable destination list may not change or become a subset of the original one.

Because AODV has no security mechanisms, malicious nodes can perform many attacks just by not following the protocol. A malicious node *M* can carry out the following attacks (among many others) against AODV:

- Impersonate a node *S* by forging an RREQ with its address as the originator address.
- When forwarding an RREQ generated by node *S* to discover a route to node *D*, reduce the hop count field to increase the chances of being in the route path between *S* and *D* so that it can analyze the traffic between them.
- Impersonate a node *D* by forging an RREP with its address as a destination address.
- Impersonate a node by forging an RREP that claims that the node is the destination.
- Selectively drop certain RREQs and RREPs and data packets. This kind of attack is especially hard even to detect because transmission errors have similar effect.
- Forge an RERR message pretending it is the node *S* and send it to its neighbor *D*. The RERR message has a very high destination sequence number (*dsn*) for one of the unreachable destination, say, *U*. This might cause *D* to update the destination sequence number corresponding to *U* with the value *dsn* and, therefore, future route discoveries performed by *D* to obtain a route to *U* will fail (because *U*'s destination sequence number will be much smaller than the one stored in *D*'s routing table).
- According to the AODV specification (Perkins et al., 1999), the originator of an RREQ can put a much bigger destination sequence number than the real one. In addition, sequence numbers wrap around when they reach the maximum value allowed by the field size. This allows a very easy attack, where an attacker is able to set the sequence number of a node to any desired value by just sending two RREQ messages.

To plug these vulnerabilities the secure version of the AODV protocol is now presented.

Secure ad hoc on-demand distance vector (SAODV) routing protocol: this protocol has been proposed to secure the AODV protocol (Zapata et al. 2002). The idea behind SAODV is to use a signature to authenticate most of the fields of RREQs and RREPs and to use hash chains to authenticate the hop-count. SAODV designs signature extensions to AODV. Network nodes authenticate AODV routing packets with an SAODV signature extension, which prevents certain certain impersonation attacks. In SAODV, an RREQ packet includes a *route request single signature extension* (RREQ-SSE). The initiator chooses a maximum hop count, based on the expected network diameter, and generates a one-way hash chain of length equal to the maximum hop count plus one. This one-way hash chain is used as a metric authenticator, much like the hash chain within SEAD protocol (Hu et al., 2002b). The initiator signs the RREQ and the anchor of this hash chain; both this signature and the anchor are included in the RREQ-SSE. In addition, the RREQ-SSE includes an element of the hash chain based on the actual hop count in the RREQ header. For sake of explanation, we call this value the *hop-count authenticator* (HCA). For example, if the hash chain values h_0, h_1, ….., h_N were generated such that $h_i = H[h_{i+1}]$, then the hop-count authenticator h_i corresponds to a hop count of $N - i$.

With the exception of the hop-count field and HCA, the fields of the RREQ and RREQ-SSE headers are immutable and therefore can be authenticated by verifying the signature in the RREQ-SSE extension. To verify the hop-count field in the RREQ header, a node can follow the hash chain to the anchor. For example, if the hop-count field is i, then HCA should be $H^i[h_N]$. Because the length (N) and the anchor (h_N) of this hash chain are included in the RREQ-SSE and authenticated by the signature, a node can follow the hash chain and ensure that $h_N = H^{N-i}[HCA]$.

When forwarding an RREQ in SAODV, a node first authenticates the RREQ to ensure that each field is valid. It then performs duplicate suppression to ensure that it forwards only a single RREQ for each route discovery. The node then increments the hop-count field in the RREQ header, hashes the HCA, and re-broadcasts the RREQ, together with its RREQ-SSE extension. When the RREQ reaches the target, the target checks the authentication in the RREQ-SSE. If the RREQ is valid, the target returns an RREP as in AODV. A *route reply single signature extension* (RREP-SSE) provides authentication for the RREP. As in the RREQ, the only mutable field is the hop-count; as a result, the RREP is secured in the same way as the RREQ. In particular, an RRE-SSE has a signature covering the hash chain anchor together with all RREP fields except the hop count. The hop-count is authenticated by an HCA, which is also a hash chain element; an HCA h_i corresponds to a hop-count of $N - i$.

A node forwarding an RREP checks the signature extension. If the signature is valid, then the forwarding node sets its routing table entry for the RREP's original source, specifying that packets to that destination should be forwarded to the node from which the forwarding node heard the RREP. For example, in Fig. 9, when node B forwards the RREP from node C, it sets its next hop for destination node D to C.

$$S \rightarrow * : \left\langle (RREQ, id, S, seq_S, D, oldseq_D, h_0, N)_{K_S^-}, o, h_N \right\rangle$$

$$A \rightarrow * : \left\langle (RREQ, id, S, seq_S, D, oldseq_D, h_0, N)_{K_S^-}, 1, h_{N-1} \right\rangle$$

$$B \rightarrow * : \left\langle (RREQ, id, S, seq_S, D, oldseq_D, h_0, N)_{K_S^-}, 2, h_{N-2} \right\rangle$$

$$C \rightarrow * : \left\langle (RREQ, id, S, seq_S, D, oldseq_D, h_0, N)_{K_S^-}, 3, h_{N-3} \right\rangle$$

$$D \rightarrow C : \left\langle (RREP, D, S, seq_D, S, lifetime, h_0', N)_{K_D^-}, o, h_N' \right\rangle$$

$$C \rightarrow B : \left\langle (RREP, D, S, seq_D, S, lifetime, h_0', N)_{K_D^-}, 1, h_{N-1}' \right\rangle$$

$$B \rightarrow A : \left\langle (RREP, D, S, seq_D, S, lifetime, h_0', N)_{K_D^-}, 2, h_{N-2}' \right\rangle$$

$$A \rightarrow S : \left\langle (RREP, D, S, seq_D, S, lifetime, h_0', N)_{K_D^-}, 3, h_{N-1}' \right\rangle$$

Fig. 9. Route discovery in SAODV protocol. Node S is discovering a route to node D

SAODV allows replies from intermediate nodes through the use of a *route reply double signature extension* (RREP-DSE). An intermediate node replying to an RREQ includes an RREP-DSE. The idea here is that to establish a route to the destination, an intermediate node must have previously forwarded an RREP from the destination. If the intermediate node has stored the RREP and the signature, it can then return the same RREP if the sequence number in that RREP is greater than the sequence number specified in the RREQ. However, some of the fields of that RREP, in particular the life-time field, are no longer valid. As a result, a second signature, computed by the intermediate node, is used to authenticate this field.
To allow replies based on routing information from an RREQ packet, the initiator includes a signature suitable for an RREP packet through the use of an RREQ-DSE. Conceptually, the RREQ-DSE is an RREQ and RREP rolled into one packet. To reduce overhead, SAODV uses the observation that the RREQ and RREP fields substantially overlap. In particular, the RREQ-DSE needs to include some flags, a prefix size, and some reserved fields, together with a signature valid for an RREP using those values. When a node forwards an RREQ-DSE, it caches the route and the signature in the same way as if it had forwarded an RREP.
SAODV also uses signatures to protect the *route error* (RERR) message used in route maintenance. In SAODV, each node signs the RERR it transmits, whether it's originating the RERR or forwarding it. Nodes implementing SADOV don't change their destination sequence number information when receiving an RERR because the destination doesn't authenticate the destination sequence number. Fig. 10 shows an example of SAODV route maintenance.

$$B \rightarrow A: (RERR, D, seq_D)_{K_B^-}$$
$$A \rightarrow S: (RERR, D, seq_D)_{K_A^-}$$

Fig. 10. Route maintenance in SAODV protocol.

4.5 Secure Routing Protocol (SRP)

Papadimitratos et al. (Papadimitratos et al., 2002) have proposed a *secure routing protocol* (SRP) that can be applied to several existing routing protocols (in particular to DSR (Johnson et al., 2007)). It is an on-demand source routing protocol that captures the basic features of reactive routing. The packets in SRP have extension headers that are attached to RREQ and RREP messages. The protocol doesn't attempt to secure RERR packets; instead it delegates the route-maintenance function of the secure route maintenance portion of the *secure message transmission protocol*. SRP uses a sequence number in the RREQs and RREPs to ensure freshness, but this sequence number can only be checked at the target. SRP requires a security association only between communicating nodes and uses this security association to authenticate RREQs and RREPs through the use of *message authentication codes* (MACs). At the target, SRP can detect any modifications of the RREQs, and at the source node, it can detect modifications of the RREPs. In the following, the protocol is discussed briefly.
In SRP, *route requests* (RREQs) generated by a source node S are protected by *message authentication codes* (MACs) computed using a key shared with the target T. Requests are broadcast to all the neighbors of S. Each neighbor that receives a request for the first time appends its identifier to the request and re-broadcasts it. The intermediate nodes also perform the same actions. The MAC in the request is not checked because only S and T know the key being used to compute it. When the request reaches the target T, its MAC is checked by T. If it is valid, then it is assumed by the target that all adjacent pairs of nodes on

the path of the RREQ are neighbors. Such paths are called valid or plausible routes. The target T replaces the MAC of a valid RREQ by a MAC computed with the same key that authenticates the route. This is then sent back (upstream) to S using the reverse route. For example, an RREQ that reaches an intermediate node X_j is of the following form:

$$msg_{S,T,rreq} = (rreq, S, T, id, sn, X_1, X_2 ....X_j, mac_S) \quad (5)$$

In (5), *id* is a randomly generated route identifier, *sn* is a session number and mac_S is a MAC on *(rreq, S, T, id, sn)* computed by S using a key shared with T, X_1, ..., X_p, T is a discovered route, then the *route reply* (RREP) of the target T has the following form for all intermediate nodes X_j, $1 \le j \le p$:

$$msg_{S,T,rrep} = (rrep, S, T, id, sn, X_1, X_2, ...X_p, mac_T) \quad (6)$$

In (6), mac_T is a MAC computed by T with the key shared with S on the message field preceding it. Intermediate nodes should check the RREP header (including its *id* and *sn*) and that they are adjacent with two of their neighbors on the route before sending the RREP upstream.

SRP doesn't attempt to prevent unauthorized modification of fields that are ordinarily modified in the course of forwarding these packets. For example, a node can freely remove or corrupt the node list of an RREQ packet that it forwards. Since SRP requires a security association between communicating nodes, it uses extremely lightweight mechanisms to prevent other attacks. For example, to limit flooding, nodes record the rate at which each neighbor forwards the RREQ packets and gives priority to REQUEST packets sent through neighbor that less frequently forward REQUEST packets. Such mechanisms can secure a protocol when few attackers are present. However, such techniques provide secondary attacks, such as sending forged RREQ packets to reduce the effectiveness of a node's authentic RREQs. In addition, such techniques exacerbate the problem of greedy nodes. For example, a node that doesn't forward RREQ packets ordinarily achieves better performance because it is generally less congested, and it doesn't need to use its battery power to forward packets originated by other nodes. In SRP, a greedy node retains these advantages, and in addition, gets a higher priority when it initiates route discovery.

4.6 ARIADNE: A secure on-demand routing protocol for ad hoc networks

Ariadne (Hu et al., 2002a) is a secure on-demand routing protocol based on the *dynamic source routing* (DSR) protocol (Johnson et al., 2007). The protocol can withstand node compromise and relies only on highly efficient symmetric key cryptography. Ariadne can authenticate routing message using one of the three schemes: (i) shared secret between each pair of nodes, (ii) shared secrets between communicating nodes combined with broadcast authentication using TESLA (Perrig et al., 2001), and (iii) digital signatures. In this section, we discuss Ariadne with TESLA, an efficient broadcast authentication scheme that requires loose time synchronization. Using pair-wise shared keys the protocol avoids the need for time synchronization but at the cost of higher key-setup overhead. Ariadne discovers routes in a reactive (on-demand) manner through route discovery and uses them to source route data packets to their destinations. Each forwarding node helps by performing route maintenance to discover problems with each selected route.

Route discovery: The protocol design is explained in two stages: (i) a mechanism is presented that lets the target node verify the authenticity of the RREQ, and (ii) an efficient per-hop hashing technique is described that verifies whether any node is missed from the node list in the RREQ. In the following, we assume that the initiator node *S* performs a route discovery for target node *D* and that they share the secret keys K_{SD} and K_{DS}, respectively for message authentication in each direction.

i. *Target authenticates route request*: To convince the target of the legitimacy of each field in an RREQ, the initiator simply includes a *message authentication code* (MAC) computed with the key K_{SD} over unique data - for example, a timestamp. The target can easily verify the route requestor's authenticity and freshness using the shared key K_{SD}. In a route discovery, the initiator wants to authenticate each individual node in the node list of the RREP. A secondary requirement is that the target can authenticate each node in the node list of the RREQ so that it will return an RREP only along paths that contain legitimate nodes. Each hop authenticates the new information in the RREQ using its current TESLA key. The target node buffers the RREP until intermediate nodes can release the corresponding TESLA keys. The TESLA security condition is verified at the target node, and the target includes a MAC in the RREP to certify that security condition was met.

ii. *Per-hop hashing*: Authenticating data in routing messages isn't sufficient because an attacker could remove a node from the node list in an RREQ. One-way hash functions are used to verify that no hop was omitted - an approach that is called *per-hop hashing*. To change or remove a previous hop, an attacker must either hear an RREQ without that node listed or must be able to invert the one-way hash function. For efficiency, the authenticator may be included in the hash value passed in the RREQ. Fig. 11 shows an example of Ariadne route discovery.

$$
\begin{aligned}
&S: h_0 = MAC_{K_{SD}}(REQUEST, S, D, id, ti)\\
&S \to *: \langle REQUEST, S, D, id, ti, h_0, (), () \rangle\\
&A: h_1 = H[A, h_0], \langle M_A = MAC_{K_{Ati}}(REQUEST, S, D, id, ti, h_1, (A), ()) \rangle\\
&A \to *: \langle REQUEST, S, D, id, ti, \boldsymbol{h_1}, \mathbf{(A)}, \boldsymbol{M_A} \rangle\\
&B: h_2 = H[B, h_1], \langle M_B = MAC_{K_{Bti}}(REQUEST, S, D, id, ti, h_2, (A, B), (M_A)) \rangle\\
&B \to *: \langle REQUEST, S, D, id, ti, \boldsymbol{h_2}, (\mathrm{A}, \mathbf{B}), (M_A, \boldsymbol{M_B}) \rangle\\
&C: h_3 = H[C, h_2], \langle M_C = MAC_{K_{Cti}}(REQUEST, S, D, id, ti, h_3, (A, B, C), (M_A, M_B)) \rangle\\
&C \to *: \langle REQUEST, S, D, id, ti, \boldsymbol{h_3}, (A, B, \mathbf{C}), (M_A, M_B, \boldsymbol{M_C}) \rangle\\
&D: M_D = \langle MAC_{K_{DS}}(REPLY, D, S, ti, (A, B, C), (M_A, M_B, M_C)) \rangle\\
&D \to C: \langle REPLY, D, S, ti, (A, B, C), (M_A, M_B, M_C), \boldsymbol{M_D}, () \rangle\\
&C \to B: \langle REPLY, D, S, ti, (A, B, C), (M_A, M_B, M_C), M_D, (\boldsymbol{K_{Cti}}) \rangle\\
&B \to A: \langle REPLY, D, S, ti, (A, B, C), (M_A, M_B, M_C), M_D, (K_{Cti}, \boldsymbol{K_{Bti}}) \rangle\\
&A \to S: \langle REPLY, D, S, ti, (A, B, C), (M_A, M_B, M_C), M_D, (K_{Cti}, K_{Bti}, \boldsymbol{K_{Ati}}) \rangle
\end{aligned}
$$

Fig. 11. Route discovery in Ariadne. Initiator *S* attempts to discover a route to target *D*. The bold font indicates changed message fields relative to the previous similar message.

Route maintenance: Route maintenance in Ariadne is based on the DSR protocol. A node forwarding a packet to the next hop along the source route returns an RERR to the packet's original sender if it is unable to deliver the packet to the next-hop after a limited number of retransmission attempts. The mechanisms for securing RERRs are discussed in the following. However, the case in which attackers to not send the RERRs is not considered. To prevent unauthorized nodes from sending RERRs, a mechanism should be in place in which the sender needs to authenticate the RERR messages. Each node on the return path to the source node forwards the RERR message. If the authentication is delayed - for example, when TESLA is used - each node that will be able to authenticate the RERR message buffers it until it can be authenticated.

Avoiding routing misbehavior: Ariadne protocol described above is vulnerable to an attacker that happens to be along the discovered route. In particular, a mechanism should be there that is able to determine whether the intermediate nodes forward the packets that they are requested to forward. To avoid the continued use of malicious routes, the routes are chosen based on their prior performance in packet forwarding. The scheme relies on feedback about which packets were successfully delivered. The feedback can be received either through an extra end-to-end network layer message or by exploiting properties of the transport layers, such as TCP with *selective acknowledgments* (Mathis et al., 1996). This feedback approach is somewhat similar to the one used in IPv6 for neighbor unreachability detection (Narten et al., 2007). A node with multiple routes to a single destination can assign a fraction of packets that it originates to be sent along each route. When a substantially smaller fraction of packets sent along any particular route is successfully delivered, the node can begin sending a smaller fraction of its overall packets to that destination along that route.

4.7 Security Enhanced AODV protocol

A *security enhanced AODV* (SEAODV) routing protocol has been proposed in (Li et al., 2011) that employs Blom's key pre-distribution scheme (Blom, 1985) to compute the *pair-wise transient key* (PTK) through the flooding of enhanced *hello* message and subsequently uses the established PTK to distribute the *group transient key* (GTK). PTK and GTK are used for authenticating unicast and broadcast routing messages respectively. In WMNs, a unique PTK is shared by each pair of nodes, while GTK is shared secretly between the node and all its one-hop neighbors. A *message authentication code* (MAC) is attached as the extension to the original AODV routing message to guarantee the message's authenticity and integrity in a hop-by-hop fashion. Since SEAODV uses Blom's key pre-distribution scheme, for the benefit of the readers, a brief discussion on the key pre-distribution scheme is presented in the following before the secure routing protocol is discussed.

Blom's key pre-distribution scheme: Blom's key pre-distribution is applied for implementing key exchange process (Blom, 1985; Du et al., 2003). Blom's t secure key pre-distribution scheme is as follows. Blom's pre-distribution scheme is based on $(N, t + 1)$ *maximum distance separable* (MDS) linear codes (MacWilliams et al., 1977). In this scheme, before a network is deployed, a central authority first constructs a $(t + 1) \times N$ public matrix P over a finite field $GF(q)$, where N is the network size. Then, the central authority selects a random $(t + 1) \times (t + 1)$ symmetric matrix S over $GF(q)$, where S is secret and only known to the central authority. An $N \times (t + 1)$ matrix $A = (S \, . \, P)^T$ is computed, where $(.)^T$ denotes the transpose operator. The central authority pre-loads the i-th row and i-th column of P to node i, for $i = 1, 2, \ldots n$. When node i and j need to establish a shared key, they first exchange their columns of P, and then node i computes a key K_{ij} as the product of its own row of A and j-th column of P, and node j computes K_{ji} as the product of its own row of A and the i-th column of P. Since S is symmetric, it is easy to see that:

$$K = A \cdot P = (S \cdot P)^T \cdot P = P^T \cdot S^T \cdot P = P^T \cdot S \cdot P = (A \cdot P)^T = K^T \tag{7}$$

The node pair (i, j) uses $K_{ij} = K_{ji}$ as the shared key. The Blom scheme has a *t-secure property*. It implies that in a network of N nodes, the collusion of less than t +1 nodes cannot reveal any key shared by other pairs of nodes. This is because as least t rows of A and t columns of P are required to solve the secret symmetric matrix S. The memory cost per node in the Blom scheme is t + 1. To guarantee perfect security in a WMN with N nodes, the (N - 2)-secure Blom scheme should be used, which means the memory cost per node is N - 1. Hence Blom scheme can provide strong security in networks of small size.

SEAODV protocol: SEAODV is built on AODV protocol. It requires each node in the network to maintain two key hierarchies. One is the broadcast key hierarchy, which includes all the broadcast keys from its active one hop neighbors. The other hierarchy is called unicast hierarchy, which stores all secret pair-wise keys that this node shares with its one hop neighbors. Every node uses keys in its broadcast routing messages (e.g., RREQ messages) from its one hop neighbors and applies secret pair-wise keys in the unicast hierarchy to verify the incoming messages, such as the RREP messages. Various features of the protocol are now described.

i. **Enhanced hello messages:** in AODV, hello message is broadcast by each node in its one-hop neighborhood. In SEAODV, two *enhanced hello* messages are defined following the idea presented in (Jing et al., 2004). Each node embeds its column of the public matrix P into its enhanced *hello RREQ* message. Since each column of P can be regenerated by applying the seed (a primitive element of $GF(q)$) from each node, every node only needs to store the seed in order to exchange the public information of matrix P. To guarantee bi-directional links, the neighboring nodes who receive *hello RREQ* reply with an enhanced *hello RREP*.

ii. **Exchange public *Seed_P* and GTK using enhanced hello message:** during the key pre-distribution phase, every legitimate node in the WMN knows and stores the public *Seed_P* (seed of the column of public matrix P) and the corresponding private row of the generated matrix A. The entire exchange process is depicted in three steps: (a) exchange of Seed_P of public matrix P, (b) derivation of PTK, and (c) exchange of GTK. In the *exchange of Seed_P phase*, each node looks for its public *Seed_P* from its key pool, and broadcasts the enhanced *hello RREQ* message. On completion of this step, each node in the network possesses the public *Seed_P* of all of its one-hop neighbors. In the *derivation of PTK phase*, each node uses the *Seed_P* it received from its neighbors and the node's corresponding private row of matrix A to compute PTK. On completion of this step, every node has stored the public *Seed_P* of its neighbors and has derived the PTK it shares with each of its one-hop neighbors. In the *exchange of GTK phase*, upon receiving hello RREQ from node X, node Y (node X's neighbor) encrypts *GTK_Y* with its private *PTK_Y* and unicasts the corresponding *hello RREP* message back to X. The encrypted *GTK_Y* is also attached in the unicast *hello RREP* message. Once X receives *hello RREP* from Y, X applies its private *PTK_X* to decrypt the *GTK_Y* and stores it in the database. The same process applies to node Y as well. Eventually, every node possesses the GTK keys from all its one-hop neighbors and the group of secret pair-wise PTK keys that it shares with each of its one-hop neighbor.

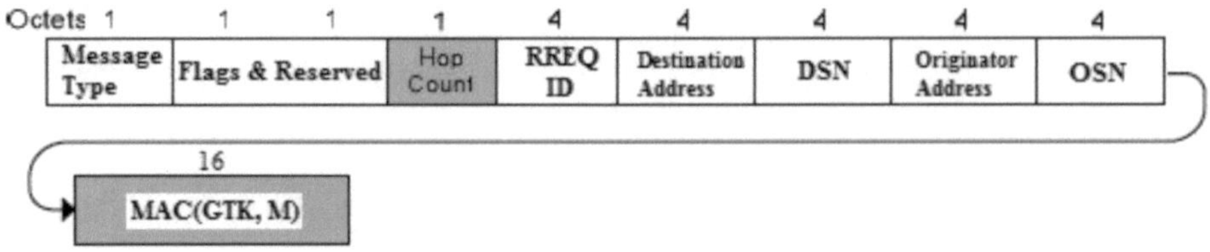

Fig. 12. The structure of RREQ message in SEAODV protocol

iii. **Securing route discovery:** in order to ensure hop-by-hop authentication, each node must verify the incoming message from its one-hop neighbors before re-broadcasting or unicasting the messages. The trust relationship between each pair of nodes relies on the shared GTK and PTK of the nodes. Route discovery process of SEAODV is similar to that of AODV, except for a MAC extension appended to the AODV message. The structure of the RREQ in SEAODV is presented in Fig. 12. The MAC is computed for message *M* using GTK of the node which needs to broadcast a RREQ to its one-hop neighbors. When a node wants to discover a route to a designated destination, it broadcasts the modified RREQ message to its neighbors. The receiving node computes the corresponding MAC value of the received message if the node possesses the GTK of the sender. The receiving node then compares the computed MAC with the one it received. If there is a match, the received RREQ is considered to be authentic and unaltered. The receiving node then updates the mutable field (hop-count in RREQ) and its routing table, and subsequently sets up the reverse path back to the source by recording the neighbor from which it received the RREQ. Finally, the node computes a MAC of the updated RREQ with its GTK and attaches the MAC value to the end of the RREQ before the message is re-broadcast to its neighbors.

iv. **Securing route setup:** the destination node or an intermediate node generates a modified RREP and unicasts it back to the next hop from which it received the RREQ. Since the RREP message is authenticated at each hop using PTKs, an adversary has no opportunity to re-direct the traffic. Before unicasting the modified RREP back to the originator of the RREQ, the node first needs to check its routing table to identify the next hop from which it received the broadcast RREQ. The node then applies PTK that it shares with the identified next hop to compute the *MAC (PTK, M)* and affixes this MAC to the end of RREP as shown in Fig. 13.

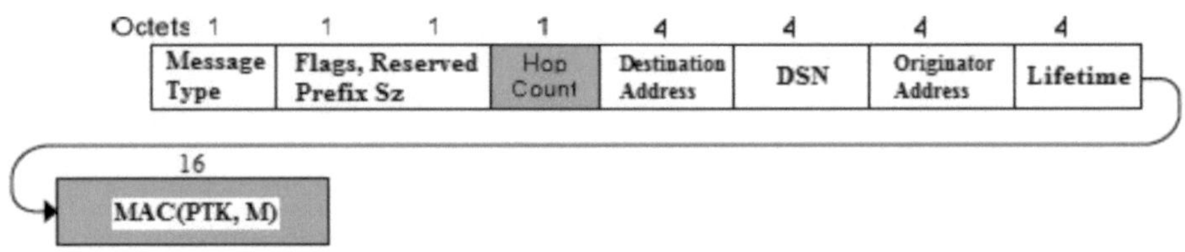

Fig. 13. The structure of RREP message in SEAODV protocol

Upon receiving the RREP from node *Y*, node *X* checks whether *PTK_YX* is in its group PTK. If it is, then node *X* computes *MAC'(PTK_XY, M)* and compares it with the *MAC(PTK_YX, M)* it received from node *Y*. If *MAC'(PTK_XY, M)* matches *MAC(PTK_YX, M)*, the received RREP is considered authentic. Node *X* then updates the hop-count field in the RREP and its own routing table, sets up the forwarding path towards the destination. Node *X* also searches the appropriate PTK that it shares with its next hop to which the new RREP is going to be forwarded to the source. Node *X* then uses the PTK to construct the new MAC and appends it to the new RREP message. Otherwise, the received RREP is deemed to be unauthentic and hence dropped.

v. **Securing route maintenance:** a node generates an RERR message if it receives data packet destined to another node for which it does not have an active route in its routing table or the node detects a broken link for the next hop of an active route or a node receives a RERR message from a neighbor for one or more active routes. The structure of a modified RERR message is presented in Fig. 14. The MAC field in the modified RERR message is computed by applying the node's GTK on the entire RERR packet. On receiving the broadcast RERR message from node *Y*, node *X* first checks whether it has the *GTK_Y*. If it has, node *X* then computes *MAC'(GTK_Y, M')* and compares it with the received MAC. If the two MACs match, node *X* searches its routing table and tries to identify the affected routes (a new group of unreachable destinations) that use node *Y* as its next-hop based on the unreachable destination list received from *Y*. If no routes in node *X*'s routing table is affected, *X* simply drops the RERR message and starts listening to the channel again. Node *X* also discards the RERR message if it fails to find the *GTK_Y* or the *MAC'(GTK_Y, M')* does not match the one received from node *Y*.

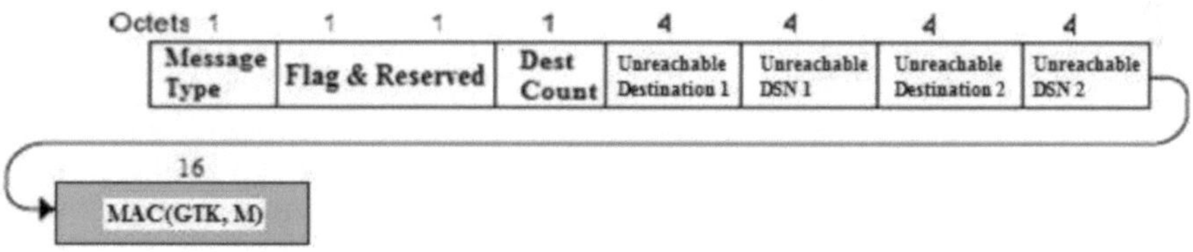

Fig. 14. The structure of RERR message in SEAODV protocol

Security analysis of SEAODV: SEAODV is vulnerable to RREQ flooding attack. However, since it authenticates RREQs from nodes that are in the list of active one-hop neighbors, the detection of the attack will be fast. Since GTKs and PTKs are used to secure the broadcast and unicast messages, and integrity of the messages are protected by MACs, the protocol is robust against RREP routing loop attack and route re-direction attack. RERR fabrication attack has minimal impact on SEAODV protocol, since a receiving node authenticates RERR messages coming from its active one-hop neighbors only. Since a malicious node can only forward the replayed RERR messages coming from the receiving node's one-hop neighbors, launching of RERR fabrication attack becomes particularly difficult.

5. Some novel secure routing protocols for WMNs

In this section, two novel routing protocols for WMNs are presented that can satisfy application QoS requirements in addition to providing security in routing. The first protocol is based on a reliable estimation of the available bandwidth in wireless links and a robust estimation of the end-to-end delay on a routing path. The protocol, while satisfying the application QoS, detects selfish nodes in the network and isolates them from the network activities so that energy of the nodes and the precious bandwidth of the wireless links are optimally utilized. The second protocol is based on an algorithm for detection of selfish nodes in a WMN that uses statistical theory of inference and clustering techniques to make a robust and reliable classification of the nodes based on their packet forwarding activities. It also introduces some additional fields in the packet header for AODV protocol so that detection accuracy is increased. In the following sub-sections the two protocols are discussed in detail.

5.1 A secure and efficient routing protocol for WMNs

A secure and efficient routing protocol for WMNs has been proposed in (Sen, 2010a) that can handle stringent quality of service (QoS) requirements of real-time applications. There are several key contributions of the work: (i) It provides an accurate estimation of the end-to-end delay in a routing path; the estimated value is then used to check whether the routing can guarantee the application QoS. (ii) It computes a link quality estimator and utilizes it in route selection. (iii) It provides a framework for reliable estimation of available bandwidth in a routing path so that flow admission with guaranteed QoS can be made. (iv) It helps in identifying and isolating selfish nodes.

The protocol is a reactive routing protocol, in which during the routing discovery phase, each intermediate node uses an admission control scheme to check whether the flow can be admitted or not. If a flow is admitted, an entry is created for the flow in a table (called the flow table) maintained locally by the node. The important components of the protocol are described below:

i. **Estimating reliability of routing paths:** every node estimates the reliability of each of its wireless links to its one-hop neighbor nodes. For computing the reliability of a link, the number of control packets that a node receives in a given time window is used as a base parameter. An *exponentially weighted moving average* (EWMA) method is used to update the link reliability estimate. If the percentage of control packets received by a node over a link in the last interval of measurement of link reliability is N_t, and if N_{t-1} is the historical value of the link reliability before the last measurement interval, $\alpha = 0.5$ is the weighting parameter, the updated link reliability (R) is computed using (8):

$$R = \alpha * N_t + (1-\alpha) * N_{t-1} \quad (8)$$

Every node maintains estimates of the reliability of each of its links with its neighbors in a *link reliability table*. The reliability for an end-to-end routing path is computed by taking the average of the reliability values of all the links on the path. Computation of the link reliability values is based on the RREQ packets on the reverse path and the RREP packets on the forward path. The use of routing path with the highest reliability reduces the overhead of route repair and makes the routing process more efficient.

ii. **Use of network topological information in route discovery:** the protocol makes use of the knowledge of network topology by utilizing selective flooding of control messages in a portion of the network. In this way, broadcasting of control messages is avoided and thus the chances of network congestion and disruption of the flows in the network are reduced. If both the source and the destination are under the control of the same mesh router (Fig. 15), the flooding of the control messages are confined within the portion of the network served by the mesh router only. However, if the source and the destination are under different mesh routers, the control traffic is limited to the two mesh groups. To reduce the control overhead further and enhance the routing efficiency, the nodes accept broadcast control messages from only those neighbors which have link reliability greater than 0.5 (i.e., on the average 50% of the control packets sent from those nodes have been received by the node). This ensures that paths with less reliability are not discovered, and hence not considered for routing.

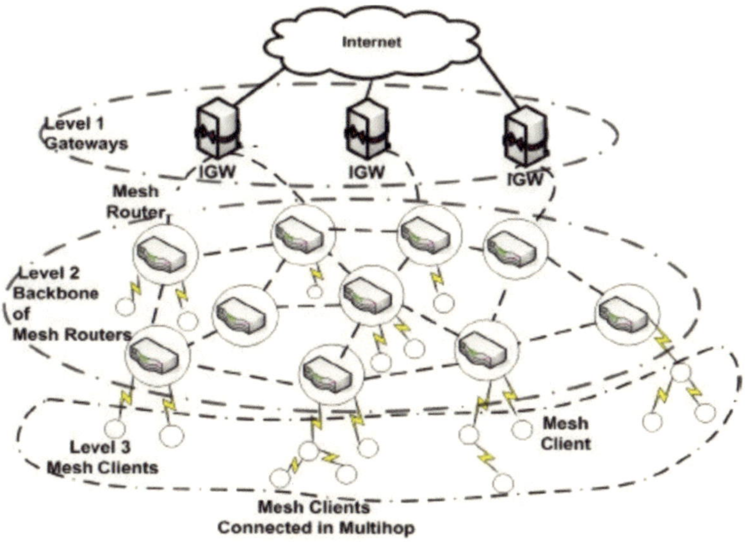

Fig. 15. The hierarchical architecture of a WMN

iii. **Estimating end-to-end delay in a routing path:** for addressing the issue of differential delays experienced by the control and data packets, the protocol makes use of some *probe packets* during the route discovery phase. When a source node receives RREP packets from the destination in response to its RREQ, it stores in a table, the records for all the RREP packets together with the path through which the packets have arrived at it. However, instead of randomly selecting a path to send probe packets to the destination, the packets are sent along the path from which RREP messages have arrived at the source first. This ensures that the probe packets are sent along the path which is likely to induce less end-to-end delay, resulting in a better performance of the protocol. The probe packets are identical to the data packets so far as their size, priority, and flow rates are concerned. The objective of sending probe packets is to simulate the data flow and observe the delay characteristics in the routing path. The number of probe packets is kept limited to $2H$ for a path consisting of H hops to make a trade-off between the control overhead and measurement accuracy (Kone et al., 2007). The destination node sets a timer after it receives the first probe packet from the source node. The timer duration is based on the estimated time for receiving all the probe packets and is computed statistically. The destination computes the average delay experienced by all the probe packets it has received, and send the computed value to the source node piggybacking it on an RREP message. If the computed value is within the limit of tolerance of the application QoS, the source selects the route and sends packets through it. If the delay exceeds the acceptable limit, the source selects the next best path (based on the arrival of RREP packets) from its table and tries once again. Since the routing path is set up based on probe packets rather than the naïve RREP packets, the protocol has higher route establishment time. However, since the selected paths have high end-to-end reliability, the delay and the control overhead are reduced because of minimal subsequent route breaks.

iv. **Estimation of available network bandwidth:** the protocol estimates the available bandwidth in a wireless link using its end-to-end delay and the loss of packets due to congestion. The packet loss due to congestion in the link is estimated as follows. In a wireless link packet loss may happen due to tow reasons: (a) loss due to faulty wireless links and (b) loss due to network congestion. The *radio link control* (RLC) layer segments an IP packet into several RLC frames before transmission and reassembles them into an IP packet at the receiver side. An IP packet loss occurs when any RLC frame belonging to an IP packet fails to be delivered. When this happens, the receiver knows that the RLC frames re-assembly has failed and the IP packet has been lost due to wireless error. Meanwhile, the sender detects *retransmission time out* (RTO) of the frame and discards all the RLC frames belonging to the IP packet. This enables the sender to compute packet drop rate in the wireless links. Moreover, using the sequence numbers of the IP packets received at the receiver, it is possible to differentiate the packet loss due to link error and packet loss due to congestion (Yang et al., 2004). For example, while receiving two incoming packets with sequence number i and i +2, if the receiver finds an IP packet assembly failure in RLC layer, the packet with sequence number i+1 is lost due to wireless channel. Once the packet loss ratio due to congestion ($P_{congestion}$) is estimated, the available bandwidth in the wireless link, *estrat*, is computed as follows (Yang et al., 2004):

$$estrat = \frac{PacketSize}{X+Y} \tag{9}$$

In (9), X and Y are given by:

$$X = RTT\sqrt{\frac{2Pcongestion}{3}} \tag{10}$$

$$Y = RTO * \min(1, 3 * \sqrt{\frac{3Pcongestion}{8}} Pcongestion(1 + 32P^2_{congestion}) \tag{11}$$

In (10), RTT is the average round trip time for a control packet. RTO is the retransmission time out for a packet, and is computed using (12):

$$RTO = \overline{RTT} + k * \overline{RTT}_{Var} \tag{12}$$

In (12), $\overline{RTT}$ and $\overline{RTT}_{Var}$ are the mean and variance respectively of RTTs and k is set to 4. This bandwidth estimator is employed to dynamically compute the available bandwidth in the wireless links on a routing path so that the guaranteed minimum bandwidth for the flow is always maintained throughout the application life-time.

v. **Identifying selfish nodes:** the protocol also enforces cooperation among the nodes by identifying the selfish nodes in the network and isolating them. Selfishness is an inherent problem associated with any capacity-constrained multi-hop wireless networks like WMNs. A mesh router can behave selfishly owing to various reasons such as: (a) to obtain more wireless or Internet throughput, or (b) to avoid path congestion. A selfish mesh router increases the packet delivery latency, and also increases the packet loss rate. A selfish node while utilizing the network resources for routing its own packet, avoids forwarding packets for others to conserve its energy.

Identification of selfish nodes is therefore, a vital issue. Several schemes have been proposed in the literature to mitigate the selfish behavior of nodes in wireless networks, such as credit-based schemes, reputation-based schemes, and game theory-based scheme (Santhanam et al., 2008). However, to keep the overhead of computation and communication at the minimum, the protocol employs a simple mechanism to discourage selfish behavior and encourage cooperation among nodes. To punish the selfish nodes, each node forwards packets to its neighbor node for routing only if the link reliability of the latter is greater than a threshold value (say, 0.5). Since the link reliability of a selfish node is 0, the packets arriving from this node will not be forwarded. Therefore, to keep link reliability higher than the threshold, each node has to participate and cooperate in routing. The link reliability serves dual purpose of enhancing reliability and enforcing node cooperation in the network.

vi. **QoS violation and recovery:** the protocol detects failure to guarantee QoS along a path with the help of reservation timeouts in flow tables records maintained in the nodes, by detection of non-availability of minimum bandwidth as estimated along its outbound wireless link. Failure to guarantee QoS may occur in three different scenarios. In the first case, a node receives a data packet for which it does not find a corresponding record in its flow table. This implies that a reservation time-out has happened for that flow. The node, therefore, sends a *route error* (RERR), to the source which re-initiates route discovery. In the second scenario, a destination node detects from its flow table records that the data packets received have exceeded the maximum allowable delay (T_{max}). To restore the path, the destination broadcasts a new RREP back to the source, and the source starts re-routing the packets via the same path on which RREP has traversed. In the third case, an intermediate node on the routing path may find that the estimated bandwidth (using (9)) in its forwarding link is less than the guaranteed minimum (B_{min}) value. In this case, the intermediate node sends an RERR to the source which re-initiates the route discovery process. The real-time estimation of the bandwidth in the next-hop wireless link at each node on the routing path makes the protocol more robust and reliable compared to most of the existing routing protocols for WMNs. For example, the similar protocol presented in (Kone et al., 2007) does not employ any bandwidth estimation mechanism at intermediate nodes, and therefore, cannot ensure delivery of all packets for every admitted flow in the network.

5.2 A Trust-based protocol for selfish nodes detection in WMNs

To address the issue of selfish nodes in a WMN, a scheme has been proposed that uses local observations in the nodes for detecting node misbehavior (Sen, 2010b). The scheme is applicable for on-demand routing protocol like *ad hoc on-demand distance vector* (AODV) protocol, and uses statistical theory of inference and clustering techniques to make a robust and reliable classification (cooperative or selfish) of the nodes based on their neighbors. In addition, the scheme introduces additional fields in the packet header of AODV packets so that detection accuracy is increased. Since the security protocol works on AODV protocol, a brief description of AODV protocol is given before the protocol is described for the benefit of the readers.

AODV protocol and modeling of the state machine: AODV routing protocol uses an on-demand approach for finding routes to a destination node. It employs destination sequence numbers to identify the most recent path. The source node and the intermediate nodes store the next-hop information corresponding to each flow of data packet transmission. The source node floods the *route request* (RREQ) packet in the network when a route is not available for the desired destination. It may obtain multiple routes to different destinations from a single RREQ. The RREQ carries the source identifier (*src_id*), the destination identifier (*dest_id*), the source sequence number (*src_seq_num*), the destination sequence number (*dest_seq_num*), the broadcast identifier (*bcast_id*), and the *time to live* (TTL). When an intermediate node receives an RREQ, it either forwards the request further or prepares a *route reply* (RREP) if it has a valid route to the destination. Every intermediate node, while forwarding an RREQ, enters the previous node address and its *bcast_id*. A timer is used to delete this entry in case an RREP is not received before the timer expires. This helps in storing an active path at the intermediate node as AODV does not employ source routing of data packets. When a node receives an RREP packet, information of the previous node from which the packet was received is also stored, so that data packets may be routed to that node as the next hop towards the destination. It is clear that AODV depends heavily on cooperation among the nodes for its successful operation. A selfish node can easily manipulate the protocol to minimize its chances of being included on routes for it is neither the source nor the destination. It may drop or tamper with the RREQ messages to ensure that no routes will ever be selected through it. Alternatively, it may drop, delay, or modify the RREP messages so as to prevent the replies from reaching the source node. The security protocol proposed in this work attempts to detect selfish nodes in a WMN so that these nodes may be isolated from the network. In the following, a *finite state machine* (FSM) model of the AODV protocol is presented which is utilized later for describing the security protocol.

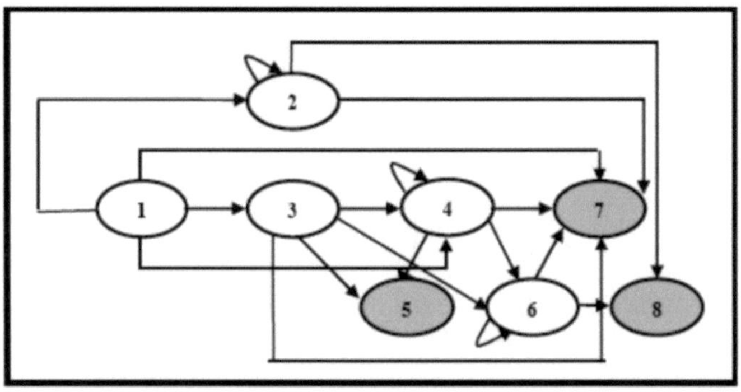

Fig. 16. The finite state machine of a monitored node

Finite state machine model: in the security mechanism, with AODV as the underlying routing protocol, the set of all messages corresponding to a RREQ flooding and the unicast RREP is referred to as a *message unit*. It is clear that no node in the network can observe all the transmission in a message unit. The subset of a message unit that a node can observe is referred to as the *local message unit* (LMU). The LMU for a particular node consists of the messages transmitted by that node, the messages transmitted by all its neighbors, and messages overheard by the node. The detection of selfish nodes is made on the basis of data collected by each node from its observed LMUs. Corresponding to each message transmission in an LMU, a node maintains a record of its sender, and the receiver in its neighborhood. It also keeps record of the neighbor nodes that receive the RREQ broadcast messages sent by the node itself. The messages are assumed to follow the sequence of the

AODV protocol. The finite state machine shown in Fig. 16 depicts various states through which a neighbor node undergoes for each LMU (Wang et al., 2008). The corresponding states for the numbers mentioned in Fig.16 can be found in Table 2.

State	Interpretation
1: init	Initial phase; no RREQ is observed
2: unexp RREP	Receipt of a RREP without RREQ observed
3: rcvd RREQ	Receipt of a RREQ observed
4: fwd RREQ	Broadcast of a RREQ observed
5: timeout RREQ	Timeout after receipt of RREQ
6: rcvd RREP	Receipt of a RREP observed
7: LMU complete	Forwarding of a valid a RREP observed
8: timeout RREP	Timeout after receipt of a RREP

Table 2. The states of the finite state machine for a local message unit (LMU)

To distinguish the finals states, these states are shaded. Every message transmission by a node causes a state transition in each of its neighbor's finite state machine. The finite state machine in one neighbor node gives only a local view of the activities of the node being monitored. It does not, in any way, represent the actual behavior of the monitored node. The collaborative participation of each neighbor node makes it possible to get an accurate global picture regarding the monitored node's behavior. A node whose activity is being monitored by its neighbors is referred to as a *monitored node*, and its neighbors are referred to as a *monitor node*. Each node plays the dual role of a monitor node and a monitored node for each of its neighbors. Each monitor node in the network observes a series of interleaved LMUs for a routing session. Each LMU can be identified by the source-destination pair contained in an RREQ message. Let us denote the kth LMU observed by a monitor node as (s_k, d_k). The pair (s_k, d_k) does not uniquely identify an LMU, because source can issue multiple RREQs for the same destination. However, since the subsequent RREQs have some delays associated with them, we can safely assume that there is only one active LMU (s_k, d_k) in the network at any point of time. At the beginning, a monitored node starts with the state 1 in its finite state machine. As the monitor node(s) observes the behavior of the monitored node by examining the LMUs, it records a sequence of transitions form its initial state 1 to one of its possible final states -- 5, 7 and 8. When a monitor node broadcasts an RREQ, it assumes that the monitored node has received it. The monitor node, therefore, records a state transition 1 → 3 for the monitored node's finite state machine. If a monitor node observes a monitored node to broadcast an RREQ, then a state transition of 3 → 4 is recorded if the RREQ message was previously sent by the monitor node to the monitored node; otherwise a transition of 1 → 4 will be recorded meaning thereby that the RREQ was received by the monitored node from some other neighbor. The transition to a timeout state occurs when a monitor node finds no activity by the monitored node for the concerned LMU before the expiry of a timer. When a monitor node observes a monitored node to forward an RREP, it records a transition to the final state – *LMU complete* (State No 7). At this state, the monitored node becomes a candidate for inclusion on a routing path.
Fig. 17 depicts an example of LMU observed by the node *N* during the discovery of a route from the source node *S* to the destination node *D* indicated by bold lines. Table 3 shows the events observed by node *N* and the corresponding state transitions for each of its three neighbor nodes *X*, *Y* and *Z*. When the final state is reached, the finite state machine

terminates and the corresponding sequences of state transitions are stored by each node for each of its neighbors. When sufficient number of events is collected by a node, a statistical analysis is performed to detect the presence of any selfish nodes in the network.

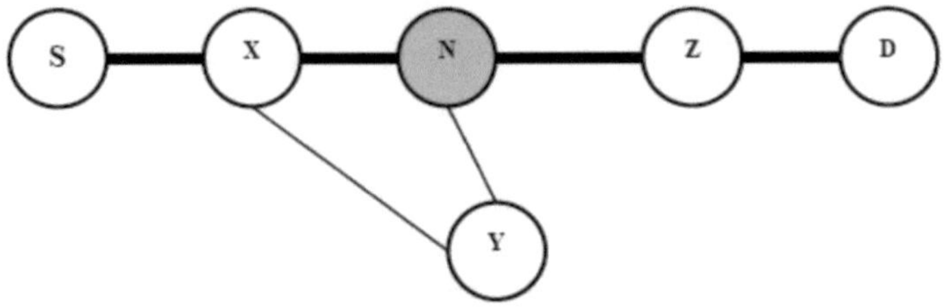

Fig. 17. An example of local message unit (LMU) observed by node *N*

Neighbor	Events	State changes
X	*X* broadcasts RREQ *N* broadcasts RREQ *N* sends RREP to *X* *X* sends RREP to *S* (overheard)	1 → 4 4 → 4 4 → 6 6 → 7
Y	*Y* broadcasts RREQ *N* broadcasts RREQ Timeout	1 → 4 4 → 4 4 → 5
Z	*N* broadcasts RREQ *Z* broadcasts RREQ *Z* sends RREP to *N*	1 → 3 3 → 4 4 → 7

Table 3. The state transitions of the neighbor nodes of node *N*

The security algorithm: As mentioned in the previous section, a monitoring node keeps a record of state transitions in the finite state machine of a monitored node in each LMU. These sequences can be represented as a transition matrix $T = [T_{ij}]$, where T_{ij} is the number of times the transition $i \rightarrow j$ is found. The monitor node invokes a detection algorithm every W seconds using data from the most recent $D = d * W$ seconds of observations, where d is a small integer. The parameter D, called the *detection window*, should be such that it is possible to punish the selfish nodes promptly while maintaining a high level of accuracy.

In the proposed algorithm, a node is assumed to monitor the activities of its R neighbors which are identified by their respective indices 1, 2,….R. Let $T^{(r)} = \left[f_{ij}^{(r)} \right]$ denote the observed transition matrix for the rth neighbor, where $[f_{ij}^{(r)}]$ is the number of transitions from state i to state j observed in the previous detection window. If m is the number of states in the finite state machine in each node, the size of $T^{(r)}$ is m x m. Let $T^{(r)} = [f_{i1}^{(r)}, \ldots f_{im}^{(r)}]$ denote the ith row of the transition matrix $T^{(r)}$, which shows the transitions out of state i at the neighbor node r. If two neighbor nodes r and s have identical distributions corresponding to transitions from state i, then one can write $T_i^{(r)} \equiv T_i^{(s)}$.

To test the hypothesis $T_i^{(r)} \equiv T_i^{(s)}$ the Pearson's $\chi 2$ test is used as follows.

$$\chi^2(i) = \frac{\sum_{l\varepsilon(r,s)} \sum_{j=1}^{m} \left[f_{ij}^{(l)} - \overline{f}_{ij}^{(l)} \right]^2}{\overline{f}_{ij}^{(l)}} \tag{13}$$

$$\overline{f}_{ij}^{(l)} = F_{ij}^{(l)} \frac{f_{ij}^{(r)} + f_{ij}^{(s)}}{F_i^{(r)} + F_i^{(s)}} \tag{14}$$

where $F_i^{(r)}$ and $F_i^{(s)}$ denote total number of transitions for state i in $T^{(r)}$ and $T^{(s)}$ respectively.

If the value of χ^2 exceeds the value of $\chi^2_{m-1,\alpha}$, then the hypothesis $T_i^{(r)} \equiv T_i^{(s)}$ is rejected at confidence interval α. If we write K_i^{rs} for the event that $\chi^2_{(i)} > \chi^2_{m-1,\alpha}$, then the conditional probability $P\left(T_i^{(r)} \equiv T_i^{(s)} \mid B_i^{rs}\right)$ can be taken as a reasonable estimator of the similarity between r and s with respect to the state i. In absence of any prior information, it is reasonable to assume that r and s have no similarity in state i and the probability that the Pearson test rejects its hypothesis to be 0.5 (Wang et al., 2008). In order to evaluate the similarity between r and s for all the m states, (1) is applied to all rows of *T(r)* and *T(s)*. This yields a vector, $\{i = 1,.....,m\}$. From the standard Markovian principle one can write:

$$\begin{aligned} L_{rs} &= P(T^{(r)} \equiv T^{(s)} \mid B^{(rs)}) \\ &= \alpha^{S^{(rs)}} (1-\alpha)^{m-S^{(rs)}} \approx \alpha^{S^{(rs)}} \end{aligned} \tag{14}$$

where

$$S^{(rs)} = \sum_{i=1}^{m} B_i^{(rs)} \tag{15}$$

The lower-order terms in the right hand side of (15) are ignored since $a << 1$. For small value of a , L_{rs} monotonically decreases in $S^{(rs)}$, which, as evident from (15), is the number of rejections of Pearson's hypothesis. Therefore, 1 -- L_{rs} may be taken as the measure of the dissimilarity between the neighbor nodes r and s. In presence of noise in the data, however, it is found that for two nodes r and s which have $L_{rs} \approx 1$, a third node t may cause inconsistency such that $L_{rt} \neq L_{st}$. To avoid this inconsistency in clustering in the proposed algorithm, clustering are not computed on the basis of pair-wise dissimilarity. To compute dissimilarity between r and s, the L values for all neighbors are computed with respect to r and s separately, and the following equation is applied:

$$d_{rs} = 1 - \frac{n_{rs}^2}{n_{r/s} * n_{s/r}} \tag{16}$$

where,

$$n_{rs} = \sum_{t \neq r,s} \min(L_{rt}, L_{st}),$$

$$n_{r/s} = \sum_{t \neq r,s}^{K} L_{rt}$$

$$n_{s/r} = \sum_{t \neq r,s}^{K} L_{st}$$

It may be observed that the computation of d_{rs} does not involve L_{rs} -- the pair-wise similarity index between nodes r and s. In fact, it measures the degree of inconsistency in similarity between r and s with all their neighbors. Since, in the computation, contribution of each neighbor plays its role, d_{rs} presents a robust indicator for dissimilarity between nodes and plays a crucial part in computing the clusters (Wang et al., 2008). For clustering, an *agglomerative hierarchical clustering* technique is used. This is a single-linkage approach in which each cluster is represented by all of the objects in the cluster, and the similarity between two clusters is measured by the similarity of the closest pair of data points belonging to different clusters. The cluster merging process repeats until all the objects are eventually merged to form one cluster (Eddy et al., 1996). After the nodes are clustered into similar sets, the sets are further classified into three groups: (i) a set (G) of cooperative nodes, (ii) a set (B) of selfish nodes, and (iii) a set of nodes whose behavior could not be ascertained. The cooperation score (C_r) of a node is computed as (Wang et al., 2008):

$$C_r = \frac{\sum_{i,j\varepsilon G}^{m} n_{ij}^{(r)}}{|G|} - \frac{\sum_{i,j\varepsilon B}^{m} n_{ij}^{(r)}}{|B|} \tag{17}$$

The set B is most likely to contain the selfish nodes. To reduce false positives (i.e. wrongly identifying a cooperative node as selfish), an ANOVA test is applied. The ANOVA approach computes a probability P_k of the random variation among the mean cooperation scores of k clusters. A lower value of P_k implies that the clusters actually represent distinct differences in their behavior. At each iteration, k clusters are formed and P_k is compared with a pre-defined level of significance β. If $P_k < \beta$, clusters are believed to be reliably reflecting the behavior of the nodes and their classifications are accepted. The cluster with lowest mean cooperation score is assumed to contain the selfish nodes. If $P_k > P_{k-1}$, the neighbor behavior has not been properly reflected in the cluster formation, which has led to the increase in the value of P_k . In this case, all the nodes are classified as cooperative, and the next iteration of the algorithm is executed. The confidence parameter β can be tuned so as to adjust the alacrity of detection of selfish nodes and rate of false positives (Wang et al., 2008). In spite of all the above statistical approaches, there is still a possibility of misclassification. The proposed algorithm further reduces the probability of misclassification by a new cross-checking mechanism. For this purpose, a minor modification is suggested in the packet header for AODV routing. Two additional fields are inserted in the header of an RREQ packet. These fields are: *next_to_source* and *duplicate_flag* to indicate respectively the address of the node that is next hop to the source, and whether the packet is a duplicate packet which has already been broadcasted by some other nodes in the network. In the header of an RREQ packet, in addition to the above two fields, another field called *next_to_destination* is added to indicate the address of the node to which the packet must be forwarded in the reverse path. It has been shown in (Kim et al., 2008), with the above extra fields, it is possible to detect every instance of selfish behavior in a wireless network with 100% detection accuracy, if the following conditions are satisfied: (i) no packet loss lost due to interference, (ii) links are bi-directional, (iii) the nodes are stationary, and (iv)

the queuing delays are bounded. Since all these conditions cannot be guaranteed in a real-world deployment, there will be always some detection inaccuracy.
Table 4 presents a list of vulnerabilities in different layers of the protocol stack of WMNs and the security protocols for defending those attacks. Table 5 compares the secure routing protocols discussed in this chapter with respect to various mechanisms these protocols use.

6. Conclusion

WMNs have become the focus of research in recent years, owing to their great promise in realizing numerous next-generation wireless services. Driven by the demand for rich and high-speed content access, recent research on WMNs has focussed on developing high performance communication protocols, while the security of the proposed protocols have received relatively little attention. However, given the wireless and multi-hop nature of the communication, WMNs are subject to a wide range of security threats. In this chapter, a large number of security issues at various layers of WMNs have been presented with a particular focus on the network layer. In addition, some of the major routing security mechanisms for WMNs currently existing in the literature have been presented and compared with respect to their strengths and weaknesses. A few novel secure routing mechanisms that take into account application QoS while detecting malicious and selfish nodes are also discussed. Although, researchers have done substantial contributions in the area of routing security in WMNs, there are still many challenges that remain to be addressed. First, efficient (i.e., lightweight) and robust authentication protocols for the *mesh routers* (MRs) need to be designed which involves scalable key management techniques. Second, for reliability in routing, energy-aware and secure multi-path routing protocols are in demand. Third issue is on strategic deployment of hop integrity protocols in WMNs. Hop integrity protocols are open to incremental deployment, and the security they provide increases with the number of pairs of hop integrity-equipped mesh routers, because an adversary will have less venues to launch his/her attacks. However, due to hardware/software compatibility and efficiency consideration, it may be worthwhile to consider a strategic deployment scheme. For example, few hotspots in the network may be required to install static hop integrity, in which hop integrity is always turned on; other spots in the network can install dynamic hop integrity, in which hop integrity is randomly turned on and off. Fourth, efficient security mechanism should be designed for defending against *tunnelling* attack, in which two malicious nodes advertise in such a way as if they have a very reliable link between them. This is achieved by tunnelling AODV messages between them. No security scheme exists so far that can detect this attack promptly and efficiently. Fifth, appropriate security protocols should be designed for hybrid networks. In many deployment situations, WMNs are designed to be integrated with other types of networks, such as wired networks and cellular networks. Addressing attacks in hybrid environment also presents an interesting future direction. Such networks are vulnerable to a wider range of attacks than its individual network components. For example, a mesh network for wireless Internet access can be targeted with DDoS attacks launched from the Internet. The scarcity of bandwidth resource on WMNs further exacerbates the severity of such attacks. On the other hand, hybrid networks possess additional resources and opportunities for defending against attacks. For example, WMNs connected to the wired networks, it is possible to leverage the high bandwidth, low latency wired links, and deploy powerful computers on the wired networks to defend against attacks. Sixth, a balanced

Attack	Targeted layer in the protocol stack	Protocols
Jamming	Physical and MAC layers	Frequenscy hopping spread spectrum (FHSS), Direct sequence spread spectrum (DSSS)
Wormhole	Network layer	Packet Leashes (Hu, 2003b)
Blackhole	Network layer	SAR (Yi, 2001)
Grayhole	Network layer	GRAYSEC (Sen, 2007), SAR (Yi, 2001)
Sybil	Network layer	SYIBSEC (Newsome, 2004)
Selective packet dropping	Network layer	SMT (Papadimitratos, 2003a), ARIADNE (Hu, 2002a), Sen (2010a), Sen (2010b)
Rushing	Network layer	ARAN (Sanzgiri, 2002), SAR (Yi, 2001), SEAD (Hu, 2002b), ARIADNE (Hu, 2002a), SAODV (Li, 2001), SRP (Papadimitratos, 2002), SEAODV (Li, 2011)
Byzantine	Network layer	ODSBR (Awerbuch, 2002)
Resource depletion	Network layer	SEAD (Hu, 2002b)
Information disclosure	Network layer	SMT (Papadimitratos, 2003a)
Location disclosure	Network layer	SRP (Papadimitratos, 2002)
Routing table modification	Network layer	ARAN (Sanzgiri, 2002), SAR (Yi, 2001), SRP (Papadimitratos, 2002), SEAD (Hu, 2002b), ARIADNE (Hu, 2002a), SAODV (Li, 2001), SEAODV (Li, 2011)
Repudiation	Application layer	ARAN (Sanzgiri, 2002)
Denial of service	Multi-layer	SRP (Papadimitratos, 2002), SEAD (Hu, 2002b), ARIADNE (Hu, 2002a)
Impersonation	Multi-layer	ARAN (Sanzgiri, 2002), SEAD (Hu, 2002b), SEAODV (Li, 2011)

Table 4. Different attacks on WMN protocol stack and protocols for defending the attacks

Protocol	Secret Keys	MAC	Digital Signature	Hash Chain	Cryptographic mechanism	Assumptions	Verification mechanism
ARAN (Sanzgiri, 2002)	Public and private key pair for each node	-------	-------	-------	Public key cryptography	Trusted certificate server	Public key cryptography verification mechanism
SEAD (Hu, 2002b)	Initial secret key K_N for hash function	-------	-------	Authenticates the sequence number and routing table metric by one-way hash chain	-------	Secure way of delivering initial secret	Hash chain verification
SAR (Yi, 2001)	Public and private key pairs for each node	-------	-------	-------	Public key cryptography	Trust metric computation model is present in the network	Trust verification at each node on routing path
SAODV (Zapata, 2002)	Public and private key pair for each node	-------	Sender uses digital signature to sign the messages	One-way hash chain to authenticate hop-count	-------	Network should have a key distribution system	Digital signature verification system
SRP (Papadimitratos, 2002)	SA between source and destination	MAC computation with K_{ST}	-------	-------	-------	Secure way of delivering the SA	MAC verification mechanism
ARIADNE (Hu, 2002a)	Secret MAC keys between sender and receiver	MAC_{KSD}	-------	TESLA keys authenticate message. It uses hash-chain to generate these keys	-------	Nodes have loosely synchronized clocks	MAC verification mechanism
SEAODV (Li, 2011)	Pairwise transient keys and group transient keys	Broadcast message: MAC_{GTK} Unicast message: MAC_{PTK}	-------	-------	Symmetric key cryptography	Blom's key pre-distribution system is present	MAC verification mechanism
SECROUTE (Sen, 2010a)	Public and private key pair for each node	-------	-------	-------	Public key cryptography	Nodes have watchdog mechanisms	Link reliability verification by packet forwarding statistics analysis
TRUSTROUTE (Sen, 2010b)	Public and private key pair for each node	-------	-------	-------	-------	Nodes have watchdog mechanisms	Agglomerative clustering mechanism to identify selfish nodes

Table 5. Comparative analysis of various secure routing protocols for WMNs

network coding system needs to be designed for high performance secure routing (Ahlswede et al., 2000). Existing network coding systems are vulnerable to a wide range of attacks besides the most well-known *packet pollution attacks* (Yu et al., 2008). Many of the weaknesses of existing system designs lie in their single focus in performance optimizations. A more balanced approach, which can provide improved security guarantees, is crucial for the actual adoption of network coding in real-world applications. A future direction of research is to uncover the security implications of different design and optimization techniques, and explore balanced system designs with network coding that achieve appropriate tradeoffs between security and performance suitable for different application requirements. Finally, multi-layer (i.e. cross-layer) security protocols should be developed that address network vulnerabilities in multiple layers of the protocol stack to provide robust and highest level of protection to mission-critical network deployments.

7. References

Ahlswede, R.; Cai, N.; Li, S.- Y. & Yeung, R. (2000). Network information flow. *IEEE Transactions on Information Theory*, Vol 46, No 4, pp. 1204 - 1216.

Akyildiz, I.F.; Wang, X.; & Wang, W. (2005). Wireless mesh networks: a survey. *Journal of Computer Networks*, Vol 47, No 4, pp. 445 - 487.

Al-Shurman, M.; Yoo, S. & Park, S. (2004). Black hole attack in mobile ad hoc networks. *Proceedings of the 42nd Annual Southeast Regional Conference*, Huntsville, Alabama, USA.

Awerbuch, B.; Holmer, D.; Nita-Rotaru, C. & Rubens, H. (2002). An on-demand secure routing protocol resilient to Byzantine failure. *Proceedings of ACM Workshop on Wireless Security (WiSe)*, ACM Press.

Awerbuch, B.; Curtmola, R.; Holmer, D.; Nita-Rotaru, C. & Rubens, H. (2005). On the survivability of routing protocols in ad hoc wireless networks. *Proceedings of ICST International Conference on Security and Privacy in Communication Networks (SecureComm)*.

Bahr, M. (2006). Proposed routing for IEEE 802.11s WLAN mesh networks. *Proceedings of the 2nd Annual International Wireless Internet Conference (WICON)*, pp. 133 - 144, Boston, MA, USA.

Bahr, M. (2007). Update on the hybrid wireless mesh protocol 80.11s. *Proceedings of the IEEE International Conference on Mobile Ad Hoc and Sensor Systems, (MASS'07)*, pp. 1 - 6.

Blom, R. (1985). An optimal class of symmetric key generation systems. *Proceedings of the EUROCRYPT'84*, pp. 335 - 338.

Brown, T.; James, J. & Sethi, A. (2006). Jamming and sensing of encrypted wireless ad hoc networks. *Proceedings of ACM MOBIHOC'06*.

Curtmola, R. & Nita-Rotaru, C. (2007). BSMR: Byzantine-resilient secure multicast routing in multi-hop wireless networks. *Proceedings of IEEE Communications Society Conference on Sensor, Mesh and Ad Hoc Communications and Networks (SECON)*.

Dong, J. (2009). Secure and robust communication in wireless mesh networks. *Doctoral Thesis*, Purdue University, Indiana, USA.

Du, W.; Deng, J.; Han, Y. S. & Varshney, P. K. (2003). A pair-wise key pre-distribution scheme for wireless sensor networks. *ACM Transactions on Information and System Security*, Vol 8, No 2, pp. 228 - 258.

Eddy, W.F.; Mockus, A. & Oue, S. (1996). Approximate single linkage cluster analysis of large datasets in high dimensional spaces. *Journal of Computational Statistics and Data Analysis*, Vol 23, pp. 29 - 43.

Eriksson, J.; Krishnamurthy, S. V. & Faloutsos, M. (2006). Truelink: a practical countermeasure to the wormhole attack in wireless networks. *Proceedings of IEEE International Conference on Network Protocols (ICNP).*

Franklin, A. A. & C. S. R. Murthy. (2007). An introduction to wireless mesh networks. Book chapter in: *Security in Wireless Mesh Networks*, Zhang, Y.; Zheng, J. & Hu, H. (eds.), CRC Press, pp. 3 - 44.

Hu, L. & Evans, D. (2004). Using directional antennas to prevent wormhole attacks. *Proceedings of ISOC Symposium of Network and Distributed Systems Security (NDSS'04).*

Hu, Y.-C.; Perrig, A. & Johnson, D. (2002a). Ariadne: a secure on-demand routing protocol for ad hoc networks. *Proceedings of ACM Annual International Conference on Mobile Computing (MOBICOM'02)*, pp. 21 - 38, Atlanta, GA, USA.

Hu, Y.-C. ; Johnson, D.B. & Perrig, A. (2002b). SEAD : secure efficient distance vector routing for mobile wireless ad hoc networks. *Proceedings of IEEE Workshop on Mobile Computing Systems and Applications (WMCSA'02)*, pp. 3 - 13.

Hu, Y.-C. ; Perrig, A. & Johnson, D.B. (2003a). Rushing attacks and defense in wireless ad hoc network routing protocols. *Proceedings of the ACM Workshop on Wireless Security (WiSe'03) in conjunction with MOBICOM'03*, pp. 30 - 40.

Hu, Y.-C.; Perrig, A. & Johnson, D.B. (2003b). Packet leashes: a defense against wormhole attacks in wireless ad hoc networks. *Proceedings of IEEE INFOCOM'03.*

Jing, X. & Lee, M. J. (2004). Energy-aware algorithms for AODV in ad hoc networks. *Proceedings of Mobile Computing and Ubiquitous Networking*, pp. 466 - 468, Yokosuka, Japan.

Johnson, D. B. (2007). The dynamic source routing protocol (DSR) for mobile ad hoc networks for IPv4. *IETF Request for Comments, RFC4728.*

Kim, H. J. & Peha, J. M. (2008). Detecting selfish behavior in a cooperative commons. *Proceedings of IEEE DySPAN*, pp. 1 -12.

Kone, V.; Das, S.; Zhao, B. Y. & Zheng, H. (2007). Quorum: quality of service in wireless mesh networks. Journal of Mobile Networks and Applications, Vol 12, No 5, pp. 358 - 369.

Law, Y.; Hoesel, L.; Doumen, J.; Hartel, P. & Havinga, P. (2005). Energy-efficient link-layer jamming attacks against wireless sensor network MAC protocols. *Proceedings of the 3rd ACM Workshop on Security of Ad Hoc and Sensor Networks (SASN'05).*

Li, C.; Wang, Z. & Yang, C. (2011). Secure routing for wireless mesh networks. *International Journal of Network Security*, Vol 13, No 2, pp. 109 - 120.

Lundgren, H.; Nordstrom, E. & Tschudin, C. (2002). The gray zone problem in IEEE 802.11b based ad hoc networks, *M2CR*, Vol 6, No 2, pp. 104 - 105.

MacWillams, F. J. & Sloane, N. J. A. (1977). *The Theory of Error Correction Codes*. North-Holland, New York.

Marti, S.; Guili, T.; Lai, K. & Baker, M. (2000). Mitigating routing misbehavior in mobile ad hoc networks. *Proceedings of ACM Annual International Conference on Mobile Computing (MOBICOM).*

Mathis, M.; Mahdavi, J.; Floyd, S. & Romanow, A. (1996). TCP selective acknowledgment options. *IETF RFC 2018*, October 1996.

Mishra, A. & Arbaugh, W.A. (2002). An initial security analysis of the IEEE 802.1X standard. *Technical Report*, University of Maryland, USA.

Narten, T.; Nordmark, E.; Simpson, W. & Soliman, H. (2007). Neighbor discovery for IP version 6 (IPv6). *IETF RFC 4861*, September 2007.

Newsome, J. ; Shi, E. ; Song, D. & Perrig, A. (2004). The Sybil attack in sensor networks: analysis and defenses. *Proceedings of the 3rd International Symposium on Information Processing in Sensor Networks (IPSN'04)*, pp. 259 - 268.

Papadimitratos, P. & Haas, Z.J. (2002). Secure routing for mobile ad hoc networks. *Proceedings of the SCS Communication Networks and Distributed Systems Modelling and Simulation Conference (CNDS'02).*

Papadimitratos, P. & Hass, Z. J. (2003a). Secure data transmission in mobile ad hoc networks. *Proceedings of ACM Workshop on Wireless Security (WiSe)*, pp. 41 - 50.

Papadimitratos, P. & Hass Z. J. (2003b). Secure link state routing for mobile ad hoc networks. *Proceedings of the Symposium on Applications and the Internet Workshops (SAINT'03 Workshops).*

Papadimitratos, P. & Haas, Z. J. (2006). Secure route discovery of QoS-aware routing in ad hoc networks. *Proceedings of IEEE Sarnoff Symposium.*

Perkins, C.E. & Belding-Royer, E.M. (1999). Ad hoc on-demand distance vector routing. *Proceedings of the IEEE Workshop on Mobile Computing Systems and Applications*, pp. 90 - 100.

Perkins, C. E.; Belding-Royer, E. M. & Das, S. R. (2003). Ad hoc on-demand distance vector (AODV). *Internet Request for Comments*, RFC 3561.

Perkins, C. E. & Bhagwat, P. (1994). Highly dynamic destination-sequenced distance-vector routing (DSDV) for mobile computers. *Proceedings of ACM SIGCOMM*, pp. 234 - 244.

Perrig, A.; Canetti, R.; Tygar, J. D. & Song, D. (2000). Efficient authentication and signing of multicast streams over lossy channels. *Proceedings of the IEEE Symposium on Security and Privacy*, pp. 56 - 73.

Perrig, A.; Canetti, R.; Song, D. & Tygar, D. (2001). Efficient and secure source authentication for multicast. *Proceedings of the Network and Distributed System Security Symposium (NDSS'01).*

Ramaswamy, S.; Fu, Huirong.; Sreekantaradhya, M.; Dixon, J. & Nygard, K.E. (2003). Prevention of cooperative black hole attacks in wireless ad hoc networks. *Proceedings of the International Conference on Wireless networks*, pp. 570 - 575.

Roy, S.; Addada, V. G.; Setia, S. & Jajodia, S. (2005). Securing MAODV: attacks and countermeasures. *Proceedings of IEEE Communications Society Conference on Sensor, Mesh and Ad Hoc Communications and Networks (SECON).*

Royer, E. M. & Perkins, C. E. (2000). Multicast ad-hoc on-demand distance vector (MAODV) routing. *Internet Draft*, July 2000.

Salem, N.B.; Buttyan, L.; Hubaux, J.-P. & Jacobson, M. (2003). A charging and rewarding scheme for packet forwarding in multi-hop cellular networks. *Proceedings of IEEE MOBIHOC'03*, pp. 1324.

Santhanam, L.; Xie, B. & Agrawal, D. (2008). Selfishness in mesh networks: wired multi-hop MANETs. *IEEE Journal of Wireless Communications,* Vol 15, No 4, pp. 16 - 23.

Sanzgiri, K.; Dahill, B.; Levine, B. N.; Shields, C. & Belding-Royer, E. M. (2002). A secure routing protocol for ad hoc networks. *Proceedings of IEEE International Conference on Network Protocols (ICNP'02)*, pp. 78 - 87.

Sen, J. (2010a). An efficient and reliable routing protocol for wireless mesh networks. *Proceedings of the International Conference on Computational Sciences and its Applications (ICCSA'10),* Lecture Notes in Computer Science (LNCS), Springer-Verlag, Heidelberg, Germany, Vol 6018, pp. 246-257, Fukuaka, Japan.

Sen, J. (2010b). A trust-based detection algorithm of selfish packet dropping nodes in a peer-to-peer wireless mesh networks. *Proceedings of the International Conference on Recent Trends in Network Security and Applications,* Communications in Computer and Information Science (CCIS), Springer-Verlag, Heidelberg, Germany, Vol 89, Part 2, pp. 528 - 537.

Sen, J.; Chandra, M. G.; Harihara, S. G.; Reddy, H. & Balamuralidhar, P. (2007). A mechanism for detection of grayhole attack in mobile ad hoc networks. *Proceedings of the 6th IEEE International Conference on Information, Communications, and Signal Processing (ICICS'07),* Singapore.

Shi, E. & Perrig, A. (2004). Designing secure sensor networks. *IEEE Wireless Communication Magazine,* Vol 11, No 6, pp. 38 - 43.

Wang, B. ; Soltani, S. ; Shaprio, J.K. ; Tan, P.-N. & Mutka, M. (2008). Distributed detection of selfish routing in wireless mesh networks. *Technical Report- MSU-CSE-06-19,* Department of Computer Science and Engineering, Michigan State University.

Wood, A. D. & Stankovic, J. A. (2002). Denial of service in sensor networks. *IEEE Computer,* Vol 35, No. 10, pp. 54 - 62.

Xu, W.; Trappe, W.; Zhang, Y. & Wood, T. (2005). The feasibility of launching and detecting jamming attacks in wireless networks. *Proceedings of ACM MobiHoc'05.*

Xue, Q.; & Ganz, A. (2002). QoS routing for mesh-based wireless LANs. *International Journal of Wireless Information Networks,* Vol 9, No 3, pp. 179 - 190.

Yang, F.; Zhang, Q.; Zhu, W. & Zhang, Y.-Q. (2004). End-to-end TCP-friendly streaming protocol and bit allocation for scalable video over wireless Internet. *IEEE Journal on Selected Areas in Communications,* Vol 22, No 22, pp. 777- 790.

Yi, S.; Naldurg, P. & Kravets, R. (2001). Security-aware ad hoc routing for wireless networks. *Proceedings of ACM MobiHoc'01,* pp. 299 - 302.

Yu, Z.; Wei, Y.; Ramkumar, B. & Guan, Y. (2008). An efficient signature-based scheme for securing network coding against pollution attacks. *Proceedings of the IEEE Conference of the IEEE Communications Society (INFOCOMM'08),* Phoenix, AZ, April, 2008.

Zapata, M. G.; & Asokan, N. (2002). Securing ad hoc routing protocols. *Proceedings of ACM Workshop on Wireless Security (WiSe).*

Zhong, S.; Li, L. E.; Liu, Y. G. & Yang, Y. R. (2005). On designing incentive-compatible routing and forwarding protocols in wireless ad-hoc networks: an integrated approach using game theoretical and cryptographic techniques. *Proceedings of IEEE MOBICO'05*, pp. 117 - 131.

Zhu, S.; Xu, S.; Setia, S. & Jajodia, S. (2003). LHAP: a lightweight hop-by-hop authentication protocol for ad-hoc networks. *Proceedings of ICDCS International Workshop on Mobile and Wireless Network*, pp. 749 - 755, Providence, Rhode Island.

Zhu, T. & Yu, M. (2006). A dynamic secure QoS routing protocol for wireless ad hoc networks. *Proceedings of IEEE Sarnoff Symposium*, pp. 1 - 4.

12

Wireless Service Pricing under Multiple Competitive Providers and Congestion-sensitive Users

Andre Nel and Hailing Zhu
University of Johannesburg
South Africa

1. Introduction

With the deregulation of telecommunication industry and the fast development of broadband wireless technologies, i.e., Wireless Mesh Network (WMN), WiFi (802.11g) and WiMAX (802.16), it can be imagined that in the future users can access Internet or other wireless services, e.g., telephony, through diverse wireless service providers (WSPs) and technologies. In this complex networking landscape, moving decision-making from access points to users is a path to achieving system scalability (Zemlianov & de Veciana, 2005). Thus, for users, it is increasingly the case that they have more freedom to choose among several WSPs who provide wireless services instead of being contractually tied to a single WSP. For example, a user wishing to access the Internet via a WiFi hotspot or access point (AP) may find him in a zone covered by several wireless access providers, or he may choose among different transmission platforms: WiFi, WiMAX, 3G, and so on. In such a market, in which multiple WSPs compete for users who are price- and congestion-sensitive, it is important to investigate the economic issues that arise due to the presence of multiple competing service providers.

In such a competitive environment, all players are self-interested in a sense that their actions or reactions in response to others' actions only focus on maximizing their own payoffs. From a WSP's point of view, it has to compete for users with other WSPs while maximizing its profit. From a user's point of view, he aims to maximize his compensated utility by choosing a WSP offering the best trade-off between quality of service (QoS) and price. Our primary goal is to understand how each WSP sets its price in the presence of price-sensitive and congestion-sensitive users and other competing WSPs to maximize its own profit. Note that we focus on the price setting problem among multiple WSPs instead of price discrimination among users. Thus we simply assume that the users are homogeneous in utility functions and willingness to pay.

According to the current design of WMN architectures, a user's requests will be routed to one AP or base station (BS) (in the IEEE802.16 standards APs of the IEEE 802.11 are called base stations) automatically so that the data flows generated by the user's requests can take the most appropriate route in terms minimum hop count or other QoS metrics (i.e., bandwidth, end-to-end delay, and so on). However, from the user's point of view, besides QoS, the price is also an important consideration when the user selects an AP or BS for wireless service delivery. It is generally accepted that the current wireless data network models are flawed in the sense that they fail to capture (Das et al., 2004):

- The utility of the services and network from the user's perspective;
- The impact of user demands on revenue utility from the service providers perspective.

Even though the current design of architectures, algorithms and protocols for WMNs does take users' QoS requirement into account, price competition among WSPs is not taken into consideration. We believe that in the presence of competition among multiple WSPs with different prices resource distributions within the network would be affected significantly. This would in turn affect the engineering design of WMNs and other wireless delivery models. In our pricing model, we assume that users can choose a WSP's AP or BS based on the WSPs' quoted prices and the perceived QoS instead of just being directed automatically to a certain AP or BS by routing protocols.

In order to obtain a return on investment, each service provider needs a pricing strategy to charge its users for the service it offers. Pricing communication network services has been seen as a soft tool to cope with congestion, to control demand, and to induce users to use the network in a desirable way while maximizing service providers' profits. A well-designed dynamic pricing policy allows a service provider to capture the changes of users behavior and network status, and to adjust its prices based on these dynamic changes. In the case of high network utilization, the service provider increases its price, which in turn makes price-sensitive users reduce their demand as a response. Similarly, in the case of low network utilization, the service provider decreases its price to attract more users. With a proper pricing scheme, a service provider and its users are allowed to act individually to express the values that they are willing to charge or pay, and to reach an equilibrium where their individual utilities are maximized simultaneously. Furthermore, in the presence of other competing service providers, each WSP's price is dependent on other service providers' prices and network status, which affect users behavior because utility maximizing users always choose the service provider offering the best combination of price and QoS.

Game theory attempts to model the strategic interactions among self-interested players who must make choices that potentially affect other players' interests. In particular, non-cooperative game theory is primarily used as a typical modeling tool to analyze situations in which players' payoffs depend on the actions of other players. In principle, in a non-cooperative game each player makes his decision independently and attempts to get the most out of the game on the basis that the other player is not cooperating in any way. In this chapter, we discuss an oligopoly, in which multiple WSPs with asymmetric costs providing wireless services with possibly different qualities compete for a group of users through their prices, using a game-theoretical approach. Our objective is to develop a framework to analyze the interaction among multiple competing WSPs and price- and congestion-sensitive users and identify the Nash equilibrium prices.

2. Game theory for oligopoly and its applications to communication network pricing

Game theory aims at modeling situations in which players have to make specific moves[1] that have mutual, possibly conflicting, consequences. In particular, it studies interactions among self-interested players in a way that interaction strategies can be designed to maximize the payoff of a player in a multi-player game. It also enable the development of mechanisms that have certain desirable properties. As its name suggests, the basic concepts of game theory

[1] In the game theory terminology, a move constitutes taking a decision that will have pre-determined consequences.

arose from the study of games such as chess and checkers (Parsons et al., 2002). However, it rapidly became clear that the techniques and results of game theory can be applied to all interactions that occur between self-interested players. The classic game theoretic question asked is: what is the best or the most rational thing a player can do? In most multi-player games, the overall outcome depends critically on the choices made by all players involved. This implies that in order for a player to make a choice that optimizes his payoff, he must reason strategically. That is, the player must take into account the decisions that other players may make, and must assume that they will act rationally so as to optimize their own payoffs. Game theory provides a mathematical framework for formalizing and analyzing these situations and finding the possible results of the games.

Emerging as a tool for modeling and solving economic problems, game theory has also found its way into other domains where conflicting multiple parties have conflicting goals. Naturally, it has been used extensively for studying pricing problems for the Internet, or more generally telecommunication networks, e.g. (Altman & Basar, 1998) (La & Anantharam, 1999) (Altman et al., 2006) (Musacchio & Walrand, 2006). In Internet pricing, the fundamental aspects of multi-party (Internet service providers (ISPs) and users) optimization problems can be captured by game theory. The outcomes of a game are the utilities of every players. The ISPs and the users respectively choose their best strategies (the price for the ISPs and the demand for the users for instance) to get their desired outcomes.

Normally, pricing with a game theoretic approach is related to network resource management problem. Cooperative game theory, which requires signalization or agreements among player, has been used to obtain a Nash bargaining framework to address network issues like resource allocation, network efficiency, fairness and at the same time service provider's revenue maximization and pricing (Yaïche et al., 2000). In (Dziong & Mason, 1996), it is shown that the cooperation between two ISPs benefits both the ISPs and the users. In (La & Anantharam, 2002), La *et al.* propose an algorithm in which the network providers adjust their prices and the users adjust their rates so that an optimal equilibrium is reached, while maintaining proportional fairness.

However, in a wireless service competition market with multiple competing WSPs and a set of users, all players have conflicting interests. On one hand, the WSPs' ultimate goal is to maximize their own revenues. Their attempt to maximize user's satisfaction, system utilization, etc., is merely an approach to achieve this ultimate goal. Hence, in this WSPs and users game, the revenue is modeled as the WSP's payoff. On the other hand, users want to maximize their own satisfaction with minimum expense, given that they have freedom to choose their WSPs and switch from one WSP to another. Then the user's overall satisfaction is modeled as user payoff. Since these two goals are different and even conflict with each other, there is no apparent motivation for WSPs and users to cooperate with each other to achieve a single optimal goal as suggested by cooperative game theory[2].

In contrast to cooperative game theory, non-cooperative game theory is concerned with situations in which players' payoffs (utilities) depend on the actions of other players and in which the players cannot, in principle, sign binding agreements enforceable by third parties. The following sections give a brief introduction to the theory of non-cooperative games and its applications in Internet price competition games.

[2]Given that cooperation by the WSPs is incompatible with most regulatory frameworks, the interaction among the WSPs can also not be modeled by cooperative game theory.

2.1 Non-cooperative games in strategic form and nash equilibrium

Non-cooperative game theory is a powerful tool for solving problems with conflicting goals. In a non-cooperative game, there are a number of players who have potentially conflicting interests, where each player has a set of strategies with associated payoff values, and makes his decision independently and attempts to obtain the best payoff without cooperating any player in any way. The outcome of the game is a set of strategies, each coming from the strategy set of an individual player, that optimizes the payoffs of all players. In the context of wireless data networks, the player are the WSPs and users. In compliance with the practice of game theory, we assume that both WSPs and users, are rational, meaning that their objectives are to maximize their payoffs (or utilities) individually.

Basically, there are two types of representations generally used to represent a game. Strategic form (or normal form) is the basic type used in studying non-cooperative games. Normally, strategic form games deal with the situation where the strategy decision of each player is made at the same time without observing the decision of the other player. On the other hand, the extensive form (also called a game tree) is a description of how a game is played over time. It is generally assumed that a single player can move based on observation of the prior choices of other players when the game is at a given stage. Generally speaking, games in extensive form deal with the situation where at least one player has partial information about other players' decision. There are two different scenarios for an extensive form game: a game of complete information is the strategic interaction when players are aware of each other's strategies or payoffs, i.e., all factors are common knowledge. In the game of incomplete information, at least one player is unaware of the payoffs or strategies of the other player.

In today's competitive communication market, it is impossible for service providers to divulge their payoffs or strategies to their rivals. In this chapter, all WSPs simultaneously and independently compute their quoted prices without the knowledge of their opponents' payoffs or strategies. Each WSP sets its own prices based on the users' response, but has no knowledge about other WSPs' prices and the users' response to other WSPs' prices in real time. Clearly, in this pricing game among multiple competing WSPs, the users' reaction to the WSPs' quoting prices and QoS is the determining factor. Therefore, this price competition game can be divided into two games:

- a game between the WSPs and users which can be expressed as a leader-follower game in strategic form with the users as the follower responding to the WSPs' prices and QoS; and
- a game among the WSPs which can be expressed as a simultaneous move game in strategic form.

A game in strategic form can be defined as $G = (i \in N, S_i, U_i)$, where N is the set of players, each of whom attempts to maximize his own particular utility. S_i represents the strategy space of player i, which is the set of all possible strategies of player i. A joint set of the strategy spaces of all players constitutes a strategy profile $s = \{s_1, s_2, \ldots, s_N\}$. $u_i(s)$ is payoff or utility that quantifies the outcome of game for player i given the strategy profile s. Fig. 1 illustrates a simplest example of two-player strategic form game with each player having two strategies. In our case the players are a set of WSPs whose strategy and payoff are price and profit, respectively, and a group of homogeneous users who need to decide to choose which WSP to submit their requests based on the combination of the WSPs' offered prices and corresponding QoS, which are factors of user's utility function. Note that we have assumed that the users are homogeneous in utility function in the introduction to this chapter. We further assume that the profit function are the same for all WSPs.

Player2 Strategy \ Player1 Strategy	s_{21}	s_{22}
s_{11}	$u_1(\{s_{11}, s_{21}\}), u_2(\{s_{11}, s_{21}\})$	$u_1(\{s_{11}, s_{22}\}), u_2(\{s_{11}, s_{22}\})$
s_{12}	$u_1(\{s_{12}, s_{21}\}), u_2(\{s_{12}, s_{21}\})$	$u_1(\{s_{12}, s_{22}\}), u_2(\{s_{12}, s_{22}\})$

Fig. 1. A two-player game in strategic form.

To solve the game, the concept of best response needs to be introduced first. The best response of player i to the profile of strategies of other players is a strategies $s_i^{'}$ such that

$$u_i(s_i^{'}, s_{-i}) > u_i(s_i, s_{-i}) \qquad \forall s_{-i} \in S_{-i}, \tag{1}$$

where subscript $-i$ represents all the players except player i himself. If all players' strategies are mutual best responses to each other, then no player would have a reason to deviate from the given strategy profile. The situation in which no players has incentive to unilaterally changing his current strategy is called a Nash equilibrium. Mathematically, a Nash equilibrium is a strategy profile $s^* = \{s_1^*, s_2^* \ldots, s_N^*\}$ such that for each player i

$$u_i(s_i^*, s_{-i^*}) \geq u_i(s_i, s_{-i}^*) \qquad \forall s_i \in S_i. \tag{2}$$

In a Nash equilibrium, none of the players can gain by unilateral deviation, which implies that no single player can leave this point without the cooperation of others in order to improve his own utility. In other words, a Nash equilibrium is a strategy profile comprised of mutual best responses of all the players[3].

2.2 Price and QoS competition in telecommunication networks

Pricing has been seen as a soft tool to control demand, to cope with congestion and to deal with heterogeneous applications with different QoS requirements. Therefore, there has been an increased research interest in telecommunication network pricing, which leads to many proposals for new pricing schemes motivated by different objectives, e.g. to allocate scarce network resources efficiently in order to maximize social welfare, i.e. (Kelly et al., 1998) (Low & Lapsley, 1999) (Yaïche et al., 2000) (Tassiulas et al., 2001) (La & Anantharam, 2002) (Shu & Varaiya, 2003) (Qiu & Marbach, 2003), to maximize service provider's revenue, i.e. (Basar & Srikant, 2002) (M. Bouhtou & Wynter, 2003), to guarantee fairness among users, i.e. (Kelly et al., 1998) (Kelly, 2000), to satisfy QoS requirements for differentiated network services (La & Anantharam, 1999) (Wang & Schulzrinne, 1999) (Mandjes, 2003). With the rapid growth of wireless data networks, e.g. wireless ad hoc networks and wireless mesh networks, recently many price-based resource allocation schemes also have been propose for wireless data networks, i.e. (Xue et al., 2003) (Das et al., 2004) (Xue et al., 2006) (Lüthi et al., 2006) (Kao & Huan, 2008). Pricing has also been used as an incentive mechanism to stimulate participation and collaboration of self-interested wireless node in wireless mesh networks, i.e. (Lam et al., 2006) (Lam et al., 2007).

A very large proportion of these proposed pricing schemes focus on the monopolistic case, where there is only one service provider dealing with a multitude of users and the the

[3]It should be noted that even though the Nash equilibrium indicates an equilibrium solution it may not be a solution that maximize the social welfare.

service provider is big enough to affect the entire market. However, as telecommunication networks have progressively switched from a monopolistic network to a oligopolistic one with competitive service providers, more attention has been given to price competition among service providers, see for example, (Gibbens et al., 2000) (Cao et al., 2002) (Sakurai et al., 2003) (Armony & Haviv, 2003) (Ros & Tuffin, 2004) (Khan, 2005) (Zhang et al., 2008).

In practice, markets are often partly regulated and partly competitive. In the rest of this chapter, we only discuss pricing game under perfect competition in a market without regulation, in which all service providers have certain market force and there is no provider so dominant that one of them can control the price. Therefore, no one is the leader and no one is the follower in such a price competition game. As a consequence, all service providers' prices are determined by the market in which users have the ability to switch from one service provider to another. The basic assumptions of the price competition game are that both service providers play the role of rational decision makers and each service provider knows that the opponents are also rational. A rational service provider always attempts to select the best response strategy.

In the rest of this section we will introduce some proposed pricing schemes reported in literature, which are related the pricing model presented in the next sections.

In (Gibbens et al., 2000), Gibbens *et al.* develop a framework to analyze competition between two ISPs, either or both of which may choose to offer multiple service classes. In their analytic framework, there are two ISPs: ISP1 and ISP2 charging prices p_1 and p_2 per unit time respectively. On joining ISPi, a user receives utility $U_i(\theta)$ per unit time. Utility $U_i(\theta)$ has three components: a positive benefit V which is independent of which ISP he/she joins; a dis-benefit which is a function of the degree of congestion on the network of the ISPi K_i and the user's preference for congestion θ; and a dis-benefit from having to pay a price p_i per unit time to ISPi for its service. To describe the range of preferences in the population of users in the simplest manner, assume that there is a continuum of users whose θ parameters form a population distribution which is uniformly distributed on the interval $[0,1]$. Thus the utility of a user with preference θ from joining ISPi is defined as

$$U_i(\theta) = V - \theta K_i - p_i \tag{3}$$

For analytical simplicity, congestion on a network is defined as the number of users, Q_i, divided by the capacity of the network, C_i: $K_i = \frac{Q_i}{C_i}$. Based on these assumptions, Gibbens *et al.* analyze the duopoly price competition for packet-based networks and show that the unique equilibrium outcome for both networks is to offer a single service class and charge the same price.

Sakurai *et al.* (Sakurai et al., 2003) propose an extended model based on Gibbens *et al.*'s game theoretic model for the case in which an opt-out strategy is introduced for users. In their model, the users have three strategy options: joining one of the ISPs and opting out of both of them. In (Sakurai et al., 2003) it is assumed that ISP1's price is higher than ISP2's price, $p_1 > p_2$, and both ISPs have the same fixed capacities $C_1 = C_2 = C$. A strategy for a user is a choice of ISP to join or opting out of both ISPs, given the prices quoted by the ISPs. If the user is indifferent between the two ISPs, his choice can be made randomly. Sakurai *et al.* suggest only the users who don't like congestion nor higher price opt out of ISP1, or mathematically only the users whose utility $U_1(\theta) < 0$ $(0 \leq \theta \leq 1)$ choose opting-out. Therefore, there are two types of marginal users as shown in Fig. 2: one is the users with congestion preference θ_{21}, who are indifferent between joining ISP2's lower priced network and joining ISP1's higher priced network; and the other one is the users with congestion preference θ_{10}, who are indifferent

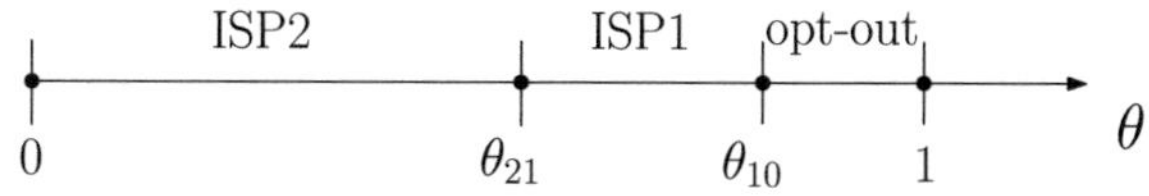

Fig. 2. Critical values of θ_{21} and θ_{10} for user preference (Sakurai et al., 2003)

between joining ISP1's higher priced network and opting out of ISP1. If there are N users in the market, the numbers of users who join ISP1 and ISP2 are given by $Q_1 = N(\theta_{10} - \theta_{21})$ and $Q_2 = N\theta_{21}$ respectively, provided that $0 < \theta_{21} < \theta_{10}$. Sakurai *et al.* conclude that Nash equilibrium for this non-cooperative game model greatly depends on three factors: the user's benefit from using the Internet, V, ISP's network capacity, C, and the number of users in market, N.

Notice that the fundamental of both models presented by Gibbens *et al.* and Sakurai *et al.* is to find the critical values of θ^* determined by the indifference relation $U_1(\theta^*) = U_2(\theta^*)$. The philosophy behind this is that when the user's utilities for joining either ISP are equal two ISPs can reach a Nash equilibrium, in which each ISP's pricing strategy is optimal in the sense that one ISP has no incentive to change its price strategy in response to the other ISP's strategy and vice versa. Because the users are indifferent between joining ISP1 or joining ISP2 when the utilities for joining either ISP are the same, no user of one ISP has an incentive to switch to the other ISP and all the users will stay where they are. This means that both ISP have no incentive to deviate from their current strategies. Indeed, if the users's compensated utility with ISP1 is lower than the one with ISP2 and both are positive, the users would switch to ISP2 until the compensated utility with ISP2 reaches the one with ISP1.

In fact, this philosophy is related to pricing in the presence of delay cost, which has received increasing interest in study of price competition among providers in communication network research literature. In (Ros & Tuffin, 2004), Ros *et al.* propose a mathematical model involving delay cost for a Paris Metro Pricing (PMP) network, where there are I classes and for the class i per packet price is p_i. In their analytical framework, a total cost function $p_i + \gamma d_i$ is associated to a class i, where d_i is the mean delay for a packet in the network and γ is a constant converting delay into money. A packet associated with a utility measure U, which is assumed to follow the same distribution for every packet, enters network i if

$$i = \min_{j \in I} p_j + \gamma d_j \qquad \text{and} \qquad U \geq p_i + \gamma d_i. \tag{4}$$

That means that the packet chooses the least expensive subnetwork in terms of total cost. If $U < \min_{j\in I} p_j + \gamma d_j$, the packet does not enter at all, meaning that the network is too expensive for it. In equilibrium, the distribution of packets among classes has to be stable, meaning that the total cost $p_j + \gamma d_j$ is the same for all classes j. If for a given class j the value $p_j + \gamma d_j$ were smaller than the total cost of the other classes, then new packets entering the network would choose class j until its total cost reaches that of other classes. This corresponds to a Wardrop equilibrium (Altman & Wynter, 2002) which can be described as: demand is distributed in such a way that all users choose one of the cheapest providers. Even though the aim of Ros *et al.*'s model is to analyze the so-called PMP scheme which separates the network into different and independent subnetworks, the analytical framework can be extended to analyze multi-providers competition, because each subnetwork in their model behaves equivalently and the customers (data packets) choose their subnetwork taking into account the prices and the QoS offered by different subnetwork, which are in common with the multi-providers competition.

Duopoly competition in the presence of a delay cost has also been studied by Armony *et al.* (Armony & Haviv, 2003). They analyze the price competition between two firms offering identical services under the assumption that all customers, belonging to one of two classes and differing by their waiting cost parameters, value the received service identically. Note that each type of customers, H-customers and L-customers, has its own waiting cost parameter, defined as the cost a customer incurs per unit of waiting time. Besides the choice between the two firms, the customers also have an option of balking, which is not included in Ros *et al.*'s work (Ros & Tuffin, 2004). The expected utility of a customer with a cost parameter C $(C = L, H)$ who joins firm i is $R - p_i - CW_i$, where R is customers' value for receiving service, p_i is the price charged by firm i and W_i is the expected waiting time (reponse time) determined using a M/M/1 queue. The corresponding utility associated with balking is assumed to be zero[4]. In their analysis, customers use mixed equilibrium strategies that specify the probability with which the customers choose each one of the firms given any pair of prices while firms use pure strategies in choosing what prices to charge per customer.
In (Zhang et al., 2008), a pricing competition model for packet-switching networks with a QoS guarantee in terms of an expected per-packet delay is studied. Zhang *et al.* propose a general framework in which service providers offering multi-class priority-based services compete to maximize their profits, while satisfying the expected delay guarantee in each class. The customer is assumed to have to choose a class of service from a service provider based on their preference for the guaranteed delay announced by the service providers. Zhang *et al.*'s work is also related to pricing in the presence of delay cost, which is assumed to be a linear function of the delay sensitivity h, γh, where γ is a constant and h is uniformly distributed between $[0,1]$. Then the expected net benefit to a user with (v, h) for sending a message in class i from provider j is $v - p_{i,j} - \gamma h d_i$, where, v is user's value assigned to the transmission of each packet in a message, and d_i is the expected per-packet delay guarantee, which is determined using $M^x/G/1/Pr$ queuing theory results. The benefit to the user for not choosing any service is zero. Two cases are studied in (Zhang et al., 2008): the case of fixed delay guarantee and the case that providers compete in both delay guarantee and price. For both cases, it is found that equilibrium outcome is symmetric $(p_1 = p_2)$.

3. A duopoly pricing model for wireless data networks under congestion-sensitive users

In this section, we present the basic model, which has been presented in (Zhu et al., 2009), for the pricing game under two WSPs competition based on the works of (Gibbens et al., 2000) and (Sakurai et al., 2003), which addresses the question of duopoly with demand-dependent quality. In this pricing game, the players are:

1. two WSPs: WSP_1 and WSP_2, who compete to maximize their individual profits in a market;
2. a group of homogeneous users who are price-as well as congestion-sensitive.

For analytical convenience, we assume that both WSPs' capacities are fixed and equal so that the only strategy for the WSPs is related to setting its price. We focus on the pricing strategies of the WSPs and analyze a Nash equilibrium for this two WSPs competition with regard to their pricing strategies. Given the prices and QoS offered by the WSPs, a strategy for a user is a choice of which WSP to join or opting out of both WSPs.

[4] This excludes the very real situation where the utility of balking is actually negative.

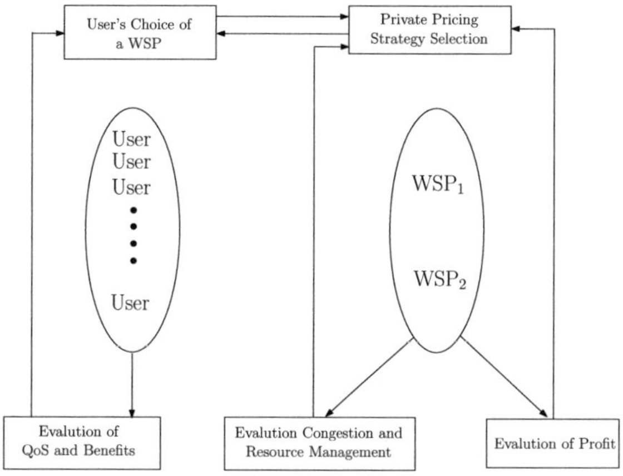

Fig. 3. Pricing model for two WSPs

3.1 Basic model

We model this oligopoly as a two-stage non-cooperative game: first, in stage 1, both WSPs set their prices to maximize their profits respectively. Then, in stage 2, given prices quoted by both WSPs and their QoS, the users decide whether purchase the service, and if so, from which WSP. Note that the two stages are solved sequentially. Given prices quoted by the WSPs and perceived QoS, the users decide in Stage 2 to choose which WSP. Based on the decisions of the users, in Stage 1 the WSPs adjust their optimal prices. This sequential decision-making process is illustrated in Fig. 3.

Suppose there are two WSPs: WSP$_1$ and WSP$_2$ in a market competing to maximize their individual profits. Assume WSP$_1$ and WSP$_2$ set prices p_1 and p_2 respectively per packet transmission, and the costs for providing per packet transmission are c_1 and c_2 respectively. Let the profit of WSP$_1$ and WSP$_2$ be denoted as Π_1 and Π_2 respectively. Obviously, Π_i is a function of the price charged per packet, p_i, and the WSP$_i$'s cost, c_i.

In this duopoly, both WSPs are able to change their prices based on their congestion status while the price- and congestion-sensitive users who are connecting to one WSP are able to switch to WSP anytime they want. In other words, the users' association with the WSPs would be on a per-service or per-session basis. However, each WSP only knows its own quoting prices, it own cost and the users response to its quoting price, and has no knowledge about its rival's price, cost and the users response to its rival's price in real time. It is a realistic assumption because it is not allowed for a WSP to divulge its private information to its rivals. Therefore, the users' reaction to the multiple WSPs' prices and QoS is the determining factor in this two WSPs' price competition game.

Assume the behavior of all users are identical and economically rational, so we can view all the requests as being from the same user and simply use the singular word "user". On joining WSP$_i$ ($i = 1,2$), a user receives gross utility U_i. Consider each WSP as a system that can only serve a finite population of potential users, meaning that the gross utility that the user obtains when subscribing to one WSP depends partly on the level of congestion or QoS of the WSP. Obviously, generally speaking, the larger the number of users subscribe to a WSP, the lower the gross utility the users can obtain because the level of congestion increases as users join. However, QoS may take many different forms, such as response time, bit-error rate, packet delay, and so forth. For the purpose of facilitating analysis, mean packet delay (or response time), ED, is used to determine the utility experienced by a user in this chapter. Clearly, the

mean packet delay is affected by the number of users who subscribe to the same WSP.
A strategy for the user has three options: subscribing to WSP_1; subscribing to WSP_2; opting out of both of them. To define the strategy of the user mathematically, we have to introduce the concept of compensated utility, which is gross utility minus price. As defined in (Mandjes, 2003) without loss of generality, the compensated utility curves can be defined as:

$$\mathbb{U}(ED) = U(ED) - p \qquad \text{with } U(ED) = ED^{-\theta}, \tag{5}$$

where $U(\cdot)$ is gross utility and $\theta > 0$. To simplify the calculation, we choose $\theta = 1$. Thus, the user's gross utility $U(ED) = 1/ED$ and the compensated utility $\mathbb{U}(\mathbb{ED}) = 1/ED - p$. Note that $U(ED)$ monotonically decreases with its argument ED.
The user wants to use the service as long as his compensated utility is positive. If the compensated utilities that the user receives from both WSPs are negative, the user will choose to submit neither of them. Then, if the compensated utilities that a user receives from one WSP or both of them are positive, the user will choose the WSP from which he receives the higher compensated utility. In other words, the user's strategy strongly depends on the price difference and the QoS performance difference between these two WSPs.
Suppose that mean packet delays for WSP_1 and WSP_2 are ED_1 and ED_2 respectively. Denote $U_1(ED_1)$ and $U_2(ED_2)$ as the user's gross utility for WSP_1 and WSP_2 respectively. Thus,

- if $U_1(ED_1) - p_1 < 0$ and $U_2(ED_2) - p_2 < 0$, the user will opt out of both WSPs;
- if $U_1(ED_1) - p_1 > U_2(ED_2) - p_2 > 0$, the user will subscribe to WSP_1;
- if $U_2(ED_2) - p_2 > U_1(ED_1) - p_1 > 0$, the user will subscribe to WSP_2;
- only if $U_1(ED_1) - p_1 = U_2(ED_2) - p_2 > 0$, the user is indifferent between WSP_1 and WSP_2.

In the situation in which the indifference relation

$$U_1(ED_1) - p_1^* = U_2(ED_2) - p_2^* \tag{6}$$

holds, a newly arriving user randomly selects WSP_i with probability 50%, and there is no incentive for a user who has already joined one WSP to unilaterally change his current strategy because user derives no benefit from switching to another WSP. Therefore, the pair of prices (p_1^*, p_2^*), which also maximize Π_1 and Π_2 simultaneously, is a Nash equilibrium. In equilibrium, the distribution of users between the two APs is stable beacuse the compensated utility $\mathbb{U}_i(ED_i)$ with both WSPs are the same. Indeed, if for a newly arriving user the compensated utility with the WSP_1, $\mathbb{U}_1(ED_1)$, is greater than the compensated utility with the WSP_2, $\mathbb{U}_2(ED_2)$, the newly arriving user would choose WSP_1 and the existing users with WSP_2 would switch to WSP_1 until the compensated utility with WSP_1 reaches that with WSP_2. Addtionally, in equilibrium, both WSPs also have no unilateral incentive to change their current optimal prices, because changing price could lead to an increase or decrease in the user' compensated utility and create an incentive for the user to change his strategy.

3.2 Duopoly queuing model

We first assume that there are N independent users in this duopoly. All the users generate information packets that they feed into the system after they submit to a WSP. The users who submit to the same WSP share a First-In-First-Serve (FIFS) based queuing and scheduling system. To simplify the analysis, we assume the information packets arrival process and the service time distributions, respectively, are Poisson and Exponential, which is called M/M/1 model (Hock, 1996). In this M/M/1/FCFS system, packets generated by the N users arrive according to a Poisson process with mean rate λN (this rate includes those who select to opt out of both WSPs) and the service times of individual packet for both WSPs are i.i.d. exponentially distributed with mean μ^{-1} given the assumption that both APs have the same bandwidth. Thus, the mean packet delay for WSP_i is:

$$ED = \frac{1}{\mu - \lambda N E_i(p_i)}, \tag{7}$$

where $E(p_i)$ is the expectation of the acceptance for the price p_i. Note that Equation 7 only holds provided that $\mu > \lambda N E(p_i)$. We will show later that even though these system parameters are important for our analysis the equilibrium prices in our pricing model are independent of any system parameter. Since the user's compensated utility is a function of the response time, ED, the only information that a user needs to know is the response times of the packets generated by him when making his choice. We believe that it is practically possible for a WSP to inform each user of the response time of the packets generated by him. In (Jagannatha et al., 2002) Jagannathan *et al.* suggest a parameterized customer behavior model for customer's willingness-to-pay to a given price using a Pareto distribution of customer capacity to pay. Every customer has the capacity to pay based on a Pareto distribution with scale b and shape α, where all customers have capacities at least as large as b and α determines how the capacities are distributed. The greater the value of α, the fewer the users who can pay more than b. When $\alpha \to \infty$, all users have the same capacity b. However, for a normal service, the shape α would be expected to be a very large but finite number. It is reasonable to assume that users' willingness-to-pay is associated with their capacities to pay. Therefore, the expectation of acceptance for a given price p_i is:

$$E_i(p_i) = \begin{cases} 1 - \frac{\alpha_i}{\alpha_i+\delta_i}\left(\frac{p_i}{b_i}\right)^{\delta_i} & 0 \le p_i \le b_i; \\ \frac{\delta_i}{\alpha_i+\delta_i}\left(\frac{b_i}{p_i}\right)^{\alpha_i} & p_i > b_i, \end{cases} \tag{8}$$

where shape α_i, scale b_i and user-willingness elasticity δ_i are determined by WSP_i based on its own observation. Different WSPs should have different values of these parameters. Since a WSP provider can observe the users' acceptance to the quoted price online, these parameters can be learned using an adaptive algorithm suggested by Jagannathan *et al.* in (Jagannatha et al., 2002) from the observed acceptance rate for a given price. In fact, the process of learning these parameters is a dynamic process with an aim to adjust the quoted price in line with the change of the user's compensated utility. The objective of dynamically learning these parameters is to capture the time-varying feature of customer behavior.

Then the expression for the user's compensated utility U_i with WSP_i can be written as:

$$\mathbb{U}_i = U_i(ED_i) - p_i = 1/ED_i - p_i = \mu - \lambda N E_i(p_i) - p_i. \tag{9}$$

According to the analysis of Nash equilibrium in previous section, a pair of prices (p_1^*, p_2^*) is in Nash equilibrium if and only if the following condition is satisfied:

$$\begin{aligned} & E_2(p_2^*) - E_1(p_1^*) = \frac{p_1^* - p_2^*}{\lambda N}, \\ \text{subject to} \quad & \mu - \lambda N E_1(p_1^*) - p_1^* > 0 \;\text{ or }\; \mu - \lambda N E_2(p_2^*) - p_2^* > 0. \end{aligned} \tag{10}$$

In the next section we will study the problem of identifying the Nash equilibrium prices (p_1^*, p_2^*) between the two WSPs.

3.3 The price selection problem

The profit of WSP_i is defined as the expected number of packets transmitted by the users who subscribe to WSP_i per unit time, multiplied by the difference between the price per packet, p_i, and the cost per packet, c_i. Thus the profit function for APi per unit time, Π_i, for a given price p_i is given by:

$$\Pi_i = \lambda N E_i(p_i)(p_i - c_i) \tag{11}$$

Since an AP provider's prime concern is cost recovery, it is reasonable to assume that an WSP will set its price greater than or at least equal to its cost c_i,
The objective of each WSP is to select a price that will maximize its profit. Therefore, a strategic equilibrium $(p_1^*,\ p_2^*)$ for the two WSPs has to satisfies the following relations first:

$$\begin{aligned} \forall p_1 : \Pi_1(p_1^*) \geq \Pi_1(p_1) \\ \forall p_2 : \Pi_1(p_2^*) \geq \Pi_2(p_2) \end{aligned} \tag{12}$$

Mathematically, the profit of APi is maximized for the first order condition $\frac{\partial \Pi_i}{\partial p_i} = 0$. There are two cases need to be discussed:

- **CASE 1**: When $p_i > b_i$, $E_i(p_i) = \frac{\delta_i}{\alpha_i+\delta_i}(\frac{b_i}{p_i})^{\alpha_i}$, with which

$$\Pi_i(p_i) = \lambda N \frac{\delta_i}{\alpha_i + \delta_i}(\frac{b_i}{p_i})^{\alpha_i}(p_i - c_i); \tag{13}$$

- **CASE 2**: When $0 \leq p_i \leq b_i$, $E_i(p_i) = 1 - \frac{\alpha_i}{\alpha_i+\delta_i}(\frac{p_i}{b_i})^{\delta_i}$, with which

$$\Pi_i(p_i) = \lambda N[1 - \frac{\alpha_i}{\alpha_i + \delta_i}(\frac{p_i}{b_i})^{\delta_i}](p_i - c_i). \tag{14}$$

It can be proved that there is at least one maximization point in the range of $c_i \leq p_i \leq b_i$ and $p_i > b_i$ respectively. Thus, solving the following maximization problem gives the optimal price at which the WSP_i maximizes its profit:

$$\max \Pi_i(p_i) = \begin{cases} \lambda N[1 - \frac{\alpha_i}{\alpha_i+\delta_i}(\frac{p_i}{b_i})^{\delta_i}](p_i - c_i) & c_i \leq p_i \leq b_i; \\ \lambda N \frac{\delta_i}{\alpha_i+\delta_i}(\frac{b_i}{p_i})^{\alpha_i}(p_i - c_i) & p_i > b_i. \end{cases} \tag{15}$$

Since a user has three options: subscribing to WSP_1, subscribing to WSP_2 or opting out of both of them, $\sum_{i=1}^{2} E_i(p_i)$ must be smaller than or equal to 1. Note that here $E_i(p_i)$ is the acceptance rate at which WSP_i's expected profit is maximized. In other words, only if WSP_i ensures the acceptance rate $E_i(p_i)$, can the best payoff be achieved by choosing the optimal price $p_{i_{opt}}$. Combining the constraint condition $\sum_{i=1}^{2} E_i(p_i) \leq 1$ with Equation 10 and Equation 15, the optimal price $p_{i_{opt}}$ that maximizes WSP_i's excepted profits must satisfy:

$$\max_{i \in \{1,2\}} \Pi_i(p_i) = \begin{cases} \lambda N[1 - \frac{\alpha_i}{\alpha_i+\delta_i}(\frac{p_i}{b_i})^{\delta_i}](p_i - c_i) & c_i \leq p_i \leq b_i \\ \lambda N \frac{\delta_i}{\alpha_i+\delta_i}(\frac{b_i}{p_i})^{\alpha_i}(p_i - c_i) & p_i > b_i, \end{cases} \tag{16}$$
$$\text{subject to } \mathbb{U}_1 = \mathbb{U}_2 > 0 \text{ and } E_1(p_1) + E_2(p_2) \leq 1.$$

However, Equation 16 is difficult to solve mathematically. To investigate the Nash equilibrium for this model, we provide numerical examples in the following section.

3.4 Numerical Examples

Since there is at least one maximization point for Π_i defined in Equation 16 within the ranges of $c_i \le p_i \le b_i$ and $p_i > b_i$ respectively, we have to study the expression for Π_i further to determine which one is the maximization point for all $p_i > c_i$. Indeed, there is only one maximum for the profit function defined in Equation 15 within the ranges of $p_i \ge c_i$ as shown in Fig. 4(a). Fig. 4(b) further plots the expected profit per unit time for different assumed values of willingness elasticity parameter, δ_i, with $\alpha_i = 4$, $b_i = 8$ and $c_i = 5$. As can be observed, the willingness elasticity parameter, δ_i, has no affect on the value of the optimal price, $p_{i_{opt}}$. Without loss of generality and so as to simplify the analysis, we assume $\delta_i = 2$ for both WSPs. Fig. 4(c) plots the expected profit per unit time for different assumed values of shape, α_i, with $b_i = 8$, $\delta_i = 2$ and $cost_i = 5$. It can be seen that the value of the optimal price, $p_{i_{opt}}$, is sightly affected by the value of shape, α_i. For instance, when $\alpha_i = 3$ the corresponding $p_{i_{opt}}$ is 7.9 while when $\alpha_i = 13$ the corresponding $p_{i_{opt}}$ is 6.9. Note that the shape α is supposed to be a very large number for a normal service. For ease of illustration, we assume $\alpha_i = 10$ and $\delta_i = 2$. Then the maximization problem in Equation 16 becomes:

$$\max_{i \in \{1,2\}} \Pi_i(p_i) = \begin{cases} \lambda N[1 - \frac{5}{6}(\frac{p_i}{b_i})^2](p_i - c_i) & c_i \le p_i \le b_i; \\ \frac{1}{6}\lambda N(\frac{b_i}{p_i})^{10}(p_i - c_i) & p_i > b_i, \end{cases} \tag{17}$$
$$\text{subject to } \mathbb{U}_1(p_{1_{opt}}) = \mathbb{U}_2(p_{2_{opt}}) > 0 \text{ and } E_1(p_{1_{opt}}) + E_2(p_{2_{opt}}) \le 1.$$

Thus, we obtain the optimal price as

$$p_{i_{opt}} = \begin{cases} \frac{1}{3}(c_i + \sqrt{c_i^2 + \frac{18}{5}b_i^2}) & c_i \le p_i \le b_i; \\ \frac{10}{9}c_i & p_i > b_i. \end{cases} \tag{18}$$

Clearly, when $b_i \ge \frac{10}{9}c_i$, $p_i = \frac{1}{3}(c_i + \sqrt{c_i^2 + \frac{18}{5}b_i^2})$ is the maximum of the profit function defined in Equation 17, while when $b_i < \frac{10}{9}c_i$, $p_i = \frac{10}{9}c_i$ is the maximum. Since dynamic pricing is our main concern, in the rest of the chapter we will confine attention to the case of $b_i \ge \frac{10}{9}c_i$ and study the Nash equilibrium prices for this case.

Summarily, under the assumption $\alpha_i = 10$ and $\delta_i = 2$, a pair of strategic equilibrium prices $(p_1^*, \ p_2^*)$ for the two WSPs has to satisfies the following maximization problem:

$$\max_{i \in \{1,2\}} \Pi_i(p_i) = \lambda N[1 - \frac{5}{6}(\frac{p_i}{b_i})^2](p_i - c_i) \tag{19}$$
$$\text{subject to } \mathbb{U}_1(p_{1_{opt}}) = \mathbb{U}_2(p_{2_{opt}}) > 0 \text{ and } E_1(p_{1_{opt}}) + E_2(p_{2_{opt}}) \le 1,$$

where $b_i \ge \frac{10}{9}c_i$ and $E_i(p_i) = 1 - \frac{\alpha_i}{\alpha_i + \delta_i}(\frac{p_i}{b_i})^{\delta_i}$. Note that the corresponding optimal price $p_{i_{opt}} = \frac{1}{3}[c_i + \sqrt{c_i^2 + \frac{18}{5}b_i^2}]$.

In order to investigate the Nash equilibrium in this model, we performed the following simulations in which two WSPs provide wireless services in a market with $N = 30$ users. Each user generates packets according to a Poisson process with rate $\lambda = 10$packet/sec, and the service times of individual packet are i.i.d. exponentially distributed with mean 300^{-1}sec. The costs of WSP$_1$ and WSP$_2$ per packet are $c_1 = 7$ and $c_2 = 5$ units respectively. Assume that both WSPs choose $\alpha = 10$ and $\delta = 2$. Then WSP$_i$'s optimal price, $p_{i_{opt}}$, is $\frac{1}{3}[c_i + \sqrt{c_i^2 + \frac{18}{5}b_i^2}]$.

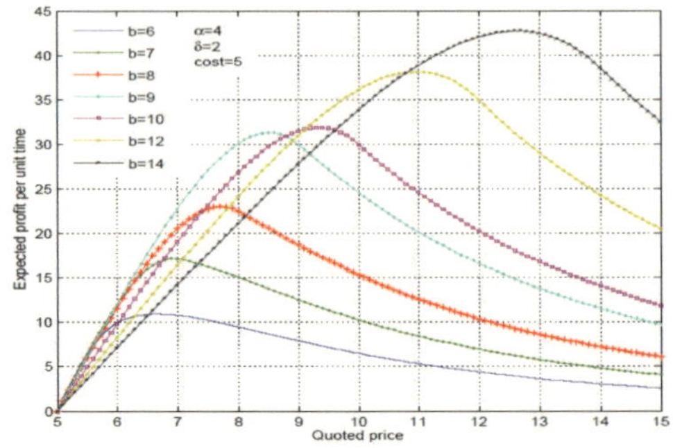

(a) Expected profit per unit time versus quoted price (varying scale parameter b)

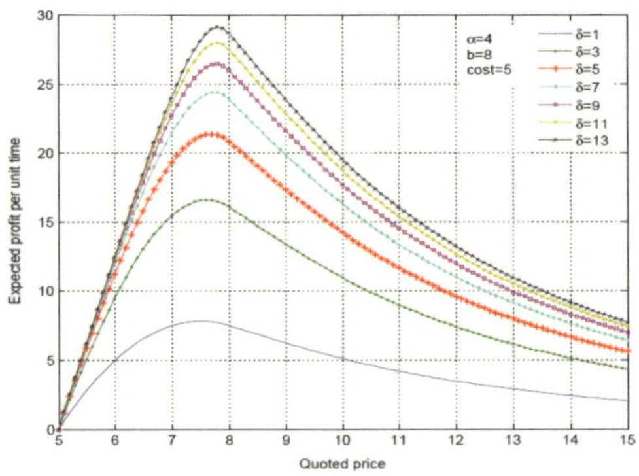

(b) Expected profit per unit time versus quoted price (varying willingness elasticity parameter δ)

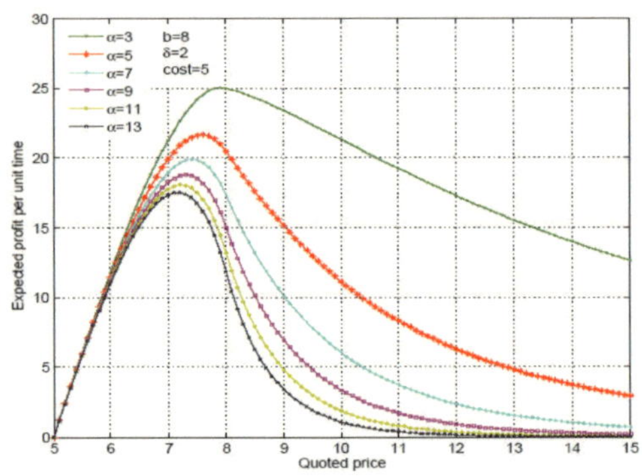

(c) Expected profit per unit time versus quoted price (varying willingness elasticity parameter α)

Fig. 4. Expected profit per unit time versus quoted price under different parameters

For the sake of ease of simulation, WSP$_i$'s prices fall into the range of $[c_i, 2c_i]$, which implies that $\frac{10}{9}c_i \leq b_i \leq \sqrt{\frac{20}{3}}c_i$ and $0.167 \leq E_i(p_{i_{opt}}) \leq 0.5$.
Since a Nash equilibrium would exist when Equation 10 holds, the following expression can be derived:

$$p^*_{1_{opt}} - p^*_{2_{opt}} = \frac{5}{54}\lambda N[(\xi_1 + \sqrt{\xi_1^2 + \frac{18}{5}})^2 - (\xi_2 + \sqrt{\xi_2^2 + \frac{18}{5}})^2], \tag{20}$$

where $\xi_1 = \frac{c_1}{b_1}$ and $\xi_2 = \frac{c_2}{b_2}$. Combining the constraint condition $\sum_{i=1}^{2} E_i(p_i) \leq 1$, Nash equilibrium prices $(p^*_{1_{opt}}, p^*_{2_{opt}})$ can be obtained. Evidently, Nash equilibrium prices $(p^*_{1_{opt}}, p^*_{2_{opt}})$ are dependent on the ratio of the cost, c_i, and the parameter, b_i, of the probabilistic model for user's willingness-to-pay.
In the simulation, a two-dimensioned numerical search procedure is employed to obtain the equilibrium prices. The search procedure is described as follows. Firstly, WSP$_i$ $(i = 1,2)$ determines b_i according to algorithm 1 and calculates corresponding $p_{i_{opt}}$, respectively. Note that here variable b_i is used to represent the users' different responses to the WSPs' quoted prices. For convenience, in the rest of section, $p_{i_{opt}}$ and p_i are interchangeable. Let $\{p_1^1, p_1^2, \cdots, p_1^n\}$ and $\{p_2^1, p_2^2, \cdots, p_2^n\}$ be optimal price strategy forms of WSP$_1$ and WSP$_2$ respectively. We start by finding the expected user's compensated utilities with WSP$_1$ and WSP$_2$ respectively, given that WSP$_0$ and WSP$_1$ set their prices according to their optimal price strategy forms respectively. We first keep WSP$_1$'s price fixed at p_1^1, while WSP$_2$'s price continuously changes according to its price strategy form $\{p_2^1, p_2^2, \cdots, p_2^n\}$. The continuous price change allows us to identify the equilibrium compensated utilities, namely $\mathbb{U}_0 = \mathbb{U}_1$. Thus, the equilibrium prices, (p_1^{1*}, p_2^{1*}), that correspond to the equilibrium compensated utilities are identified. We then proceed to the next price in $\{p_1^1, p_1^2, \cdots, p_1^n\}$.

Algorithm 1
begin
$b_1^1 \leftarrow \frac{10}{9}c_1$
for $b_1 = \frac{10}{9}c_1$ ***to*** $\sqrt{\frac{20}{3}}c_1$ ***do***
$b_1^i = b_1^{i-1} + 0.005$
$b_2^1 \leftarrow \frac{10}{9}c_2$
for $b_2 = \frac{10}{9}c_2$ ***to*** $\sqrt{\frac{20}{3}}c_2$ ***do***
$b_2^i = b_2^{i-1} + 0.005$
end
end
end

Fig. 5 illustrates the value difference between $\mathbb{U}_2(p_2)$ and $\mathbb{U}_1(p_1)$, $\mathbb{U}_2(p_2) - \mathbb{U}_1(p_1)$, under different pairs of prices (p_1, p_2). The pairs of prices that make $\mathbb{U}_2(p_2) = \mathbb{U}_1(p_1)$ are Nash equilibrium prices we are looking for. Note that using the b_i obtained according to the above algorithm one could only find out pairs of prices (p_1, p_2) which make $\mathbb{U}_2(p_2) - \mathbb{U}_1(p_2) \approx 0$. Here, we actually refer to the pairs of prices (p_1, p_2) which make $\mathbb{U}_2(p_2) - \mathbb{U}_1(p_1) \approx 0$ as Nash equilibrium prices, which is represented as (p_1^*, p_2^*)
Table 1 lists some pairs Nash equilibrium prices and the corresponding acceptance rates. As is observed, both WSPs have to set their price within a certain range so that the constraint

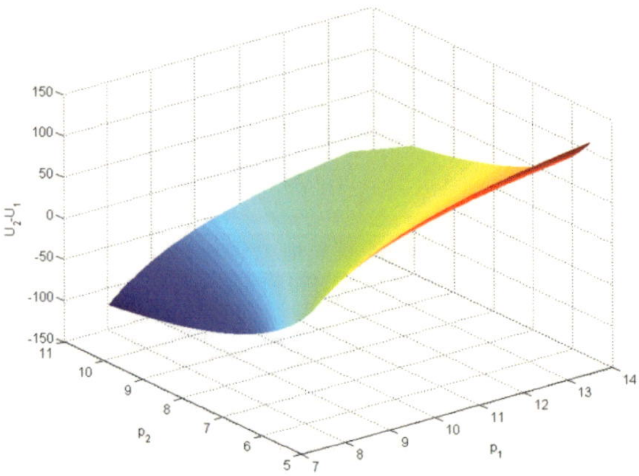

Fig. 5. $\mathbb{U}_2(p_2) - \mathbb{U}_1(p_1)$

p_1	9.8033	10.77	11.71	12.78	13.28	13.688
$E_1(p_1)$	0.3638	0.4118	0.4458	0.4749	0.4861	0.4942
p_2	7.114	7.8227	8.6116	9.4802	9.9847	10.236
$E_2 p_2)$	0.3727	0.4192	0.4562	0.4859	0.4973	0.5057

Table 1. Nash equilibrium price versus Nash equilibrium acceptance rate

$E_1(p_1^*) + E_2(p_2^*) \leq 1$ can be satisfied. For instance, as shown in Table 1, when the WSP$_2$ with lower price increases its price higher than twice its cost, $E_2(p_2^*) > 0.5$, as a response, WSP$_1$ with higher price also increases its price, which in turn lead to $E_1(p_1^*) + E_2(p_2^*) > 1$. It implies that these two WSPs cannot increase their price as high as they want without cooperation.

Table 2 lists the user's compensated utilities with the two WSPs under different prices. As can be observed, for example, if WSP$_1$ initially sets its price at 12.906 while WSP$_2$ initially sets its price as 5.635, $\mathbb{U}_1(p_1) < \mathbb{U}_2(p_2)$ and the users will choose to subscribe or switch to WSP$_2$. Then WSP$_1$ will have to decrease its price to attract more users, while WSP$_2$ would increase its price considering its congestion situation or just leave its price unchanged. This price adjusting process will be repeated until p_1 and p_2 converge to the point at which the compensated utility experienced by the users with WSP$_1$, $\mathbb{U}_1$, is equal to the compensated utility experienced by the users with WSP$_2$, $\mathbb{U}_2$. For instance, WSP$_1$ decreases its price to 11.084 while WSP$_2$ increases its price to 8.112. At this point, $\mathbb{U}_1(p_1^*) = \mathbb{U}_2(p_2^*)$, no WSP has any incentive to deviate from its price without the cooperation of the other, which may not happen, until existing users voluntarily disconnect or new users join in.

When the number of users in this market N changes, $\mathbb{U}_1(p_1)$ and $\mathbb{U}_2(p_2)$ change accordingly, which in turn leads to the Nash equilibrium moving. This results in another round of price adjustment among the two WSPs and the users. As shown in Table 3, if we suppose in the first round the number of users $N = 30$ and the two WSPs end up with a Nash equilibrium with prices $p_1 = 11.818$ and $p_2 = 8.6976$ respectively. When the number of users N increases or decreases, even if WSP$_1$ keeps its price $p_1 = 11.818$, WSP$_2$ has to change its price such that both WSPs and the users can reach a new equilibrium. Additionally, when some users with one WSP disconnect or some new users join in one WSP, the WSP could adjust its price considering its resource utility or congestion situation. This also could lead to a new round of adjustment.

p_1	8.246	8.995	9.791	10.237	11.084	11.818	12.478	12.906	13.685
$\mathbb{U}_1$	222.13	198.83	181.27	173.54	161.64	153.44	147.26	143.73	138.06
p_2	5.635	6.602	7.552	7.774	8.112	8.474	9.187	9.452	10
$\mathbb{U}_2$	239.16	195.38	171.46	167.29	161.64	156.37	147.75	145.02	140

Table 2. Users' compensated utilities

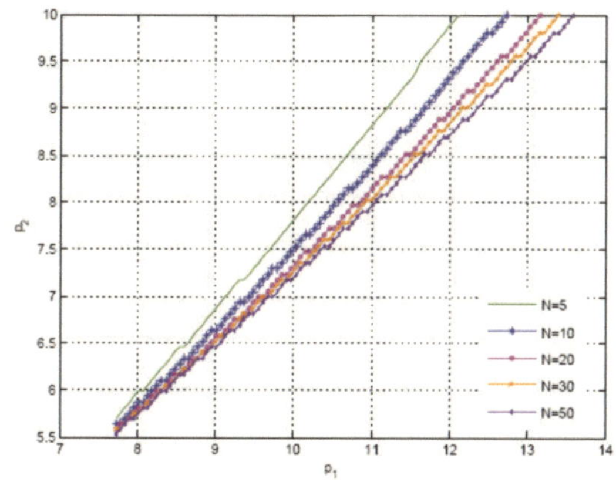

Fig. 6. Equilibrium prices for different number of users N

Fig. 6 plots the equilibrium points (at which $\mathbb{U}_1(p_1^*) - \mathbb{U}_2(p_2^*) \approx 0$) over p_1 and p_2 for various N. As can observed, for a given optimal price p_1 of WSP$_1$, there is a unique optimal price p_2 of WSP$_2$ so that WSP$_1$ and WSP$_2$ reach Nash equilibrium, and there is a quasi-linear relationship between p_1^* and p_2^*. It is worth noting that this property is useful for a dynamic pricing competition. Since each optimal price $p_{i_{opt}}$ corresponds to a certain value of b_i, which is determined by the traffic load of WSP$_i$, if we simply assume, for WSP$_i$, b_i is varied with the number of users arriving at WSP$_i$, sequentially, the optimal price $p_{i_{opt}}$ also varies in response to the change of the number of the arriving users. Suppose WSP$_1$ and WSP$_2$ set their optimal prices p_1 and p_2 based on the numbers of their arriving users respectively, but p_1 and p_2 are not in equilibrium. The users who receive a lower compensated utility will switch to another WSP, which in turn results in an increase of users at one WSP and a decrease of users at the alternative. Accordingly, p_1 and p_2 vary gradually until they converge to a Nash equilibrium point (p_1^*, p_2^*).

4. An extended pricing model for wireless oligopolies

In this section, we extend the two-stage noncooperative game model described in Section 3 to a multi-provider setting as shown in Fig. 7.

4.1 Model description

Assume that there is a set $\mathbb{I} = \{0,1,2,\ldots,I-1\}$ of WSPs in a certain area to provide wireless services to N potential users. Denote p_i and c_i as WSP$_i$'s price and cost per packet transmission respectively. Each user generates packets according to a Poisson process with mean rate λ. Then the potentially total mean arrival rate of packets in the whole network is given by λN. Note that λN can be seen as an arrival rate when all potential users send out

$N=5$	p_1	8.246	8.9975	9.7912	10.237	11.084	11.818	12.112	—	—
	$\mathbb{U}_1$	280.15	275.64	272.05	270.39	267.7	265.73	265	—	—
	Π_1	14.46	30.64	50.68	62.71	86.63	108.19	117	—	—
	p_2	6.1613	6.8438	7.6099	8.0513	8.9159	9.6874	10	—	—
	$\mathbb{U}_2$	280.15	275.65	272.05	270.39	267.7	265.72	265	—	—
	Π_2	15.90	32.28	53.09	65.78	91.56	115.26	125	—	—
$N=10$	p_1	8.246	8.9945	9.7912	10.237	11.084	11.818	12.478	12.7	—
	$\mathbb{U}_1$	268.55	260.28	253.9	251.02	246.49	243.27	240.77	240	—
	Π_1	28.92	61.28	101.36	125.41	173.27	216.37	256.11	269.61	—
	p_2	6.0265	6.6407	7.3191	7.7071	8.4705	9.1531	9.7964	10	—
	$\mathbb{U}_2$	268.56	260.29	253.89	251.03	246.49	243.27	240.77	240	—
	Π_2	26.09	54.26	89.96	111.7	156.3	197.58	236.68	250	—
$N=20$	p_1	8.246	8.9945	9.7912	10.237	11.084	11.821	12.481	12.906	13.187
	$\mathbb{U}_1$	245.34	229.56	217.59	212.28	204.06	198.33	194	191.52	189.99
	Π_1	57.83	122.56	202.71	250.82	346.54	433.11	512.59	584.49	599.03
	p_2	5.9594	6.5336	7.1592	7.5159	8.2071	8.8205	9.3845	9.7544	10
	$\mathbb{U}_2$	245.33	229.57	217.59	212.28	204.05	198.34	194.01	191.5	190
	Π_2	47.74	97.99	162.48	201.78	281.39	354.68	423.58	469.5	500
$N=30$	p_1	8.246	8.9945	9.7912	10.237	11.084	11.818*	12.478	12.906	13.685
	$\mathbb{U}_1$	222.13	198.83	181.24	173.54	161.64	153.44	147.26	143.73	138.06
	Π_1	86.75	183.84	304.07	376.23	519.8	649.12	768.32	846.73	991.04
	p_2	5.9361	6.498	7.105	7.4492	8.1124	8.6976*	9.2333	9.5853	10.236
	$\mathbb{U}_2$	222.13	198.83	181.27	173.54	161.64	153.44	147.26	143.73	138.06
	Π_2	67.34	141.81	234.96	291.49	405.39	509.75	607.48	672.59	794.28
$N=35$	p_1	8.246	8.9975	9.7912	10.237	11.084	11.821	12.478	12.906	13.474
	$\mathbb{U}_1$	210.53	183.38	163.12	154.17	140.42	130.95	123.89	119.84	115
	Π_1	101.21	214.97	354.74	438.94	606.44	757.95	896.38	987.85	1110.5
	p_2	5.9303	6.4891	7.09	7.431	8.8049	8.6638	9.1901	9.5358	10
	$\mathbb{U}_2$	210.48	183.41	161.1	154.14	140.43	130.96	123.88	119.83	115
	Π_2	77.76	163.91	271.3	336.53	467.33	587.59	699.46	773.97	875
$N=40$	p_1	8.246	8.9975	9.7912	10.237	11.084	11.818*	12.478	12.909	13.53
	$\mathbb{U}_1$	198.92	168.11	144.96	134.8	119.21	108.53	100.51	95.91	90
	Π_1	115.67	245.12	405.42	501.64	693.07	865.5	1024.4	1129.7	1282.9
	p_2	5.9245	6.4801	7.0779	7.4159	8.0635	8.6362*	9.1562	9.4988	10
	$\mathbb{U}_2$	198.93	168.09	144.95	134.78	119.23	108.51	100.51	95.92	90
	Π_2	87.96	185.66	307.48	381.24	529.1	664.89	791.08	875.37	1000
$N=50$	p_1	8.246	8.9945	9.7912	10.237	11.084	11.818*	12.481	12.906	13.614
	$\mathbb{U}_1$	175.72	137.39	108.65	96.06	76.78	63.62	53.72	48.16	39.99
	Π_1	144.58	306.4	506.78	627.06	806.34	1081.9	1281.5	1411.2	1629.6
	p_2	5.9187	6.4683	7.0599	7.3947	8.0361	8.5962*	9.1099	9.4432	10
	$\mathbb{U}_2$	175.64	137.41	108.69	96.06	76.77	63.63	53.73	48.15	40
	Π_2	209.95	288.67	517.49	629.44	707.32	842.3	935.83	1087.6	1181.6

Table 3. Users' compensated utilities for different number of users N

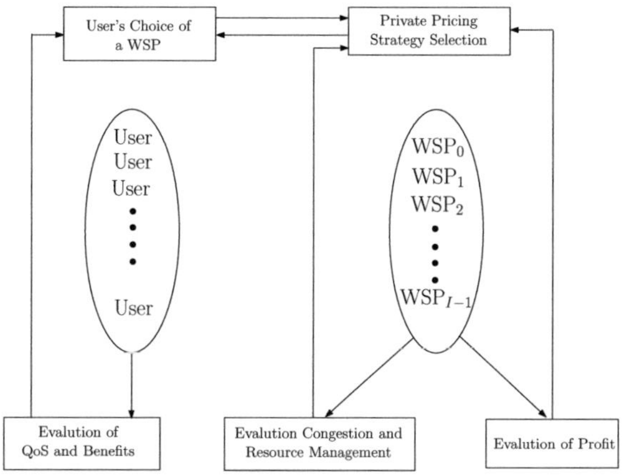

Fig. 7. Pricing model for multiple WSPs

all their requests without considering price or QoS. In addition, we stick to the mathematical definition of user's utility and compensated utility in Section 3. Then the user's compensated utility $\mathbb{U}_i$ with WSP_i can be expressed as:

$$\mathbb{U}_i = U_i(ED_i) - p_i = 1/ED_i - p_i = \mu - \lambda N E_i(p_i) - p_i, \tag{21}$$

where $i = 0,1,2,\ldots,I-1$.
Similar to the duopoly case, from a user's perspective,

- if $\mathbb{U}_i < 0$ for all $i \in \mathbb{I}$, the user will opt out of all the WSPs;
- if $\mathbb{U}_i > 0$ and $\mathbb{U}_i > \mathbb{U}_{-i}$, where the subscript $-i$ represents all the WSPs belonging to $\mathbb{I}$ except i itself, the user will subscribe to WSP_i;
- only if $\mathbb{U}_0 = \cdots = \mathbb{U}_i = \cdots = \mathbb{U}_{I-1} > 0$, the user is indifferent among these WSPs.

In the situation in which the indifference relation

$$U_0(ED_0) - p_0^* = \cdots = U_0(ED_0) - p_i^* = \cdots = U_{I-1}(ED_{I-1}) - p_{I-1}^* > 0 \tag{22}$$

holds, all the WSPs and users reach a Nash equilibrium and the set of prices $\{p_0^*, p_1^*, p_2^*, \cdots, p_{I-1}^*\}$ is the Nash equilibrium prices.
Similarly, the profit function for WSP_i per unit time, Π_i, is given by $\Pi_i = \lambda N E_i(p_i)(p_i - c_i)$. Thus, solving the following maximization problem gives WSP_i's optimal price at which WSP_i maximizes its profit:

$$\max \Pi_i(p_i) = \begin{cases} \lambda N[1 - \frac{\alpha_i}{\alpha_i+\delta_i}(\frac{p_i}{b_i})^{\delta_i}](p_i - c_i) & c_i \le p_i \le b_i; \\ \lambda N \frac{\delta_i}{\alpha_i+\delta_i}(\frac{b_i}{p_i})^{\alpha_i}(p_i - c_i) & p_i > b_i, \end{cases} \tag{23}$$

where $i = 0,1,\ldots,I-1$.
According to the previous analysis in Section 3, the maximization problem Equation 23 can be reduced to the following maximization problem:

$$\begin{aligned} &\max_{i\in 0,1,\ldots,I-1} \Pi_i(p_i) = \lambda N[1 - \frac{5}{6}(\frac{p_i}{b_i})^2](p_i - c_i) c_i \le p_i \le b_i \\ &\text{subject to } \mathbb{U}_0(p_{0_{opt}}) = \mathbb{U}_1(p_{1_{opt}}) = \ldots = \mathbb{U}_{I-1}(p_{I-1_{opt}}) > 0 \\ &\text{and } \sum_{i=0}^{I-1} E_i(p_{i_{opt}}) \le 1, \end{aligned} \tag{24}$$

where $b_i \geq \frac{10}{9}c_i$ and $E_i(p_i) = 1 - \frac{\alpha_i}{\alpha_i+\delta_i}(\frac{p_i}{b_i})^{\delta_i}$. Note that the corresponding optimal price $p_{i_{opt}} = \frac{1}{3}[c_i + \sqrt{c_i^2 + \frac{18}{5}b_i^2}]$.
In the next section, we will show the existence and uniqueness of Nash equilibrium in this oligopoly model.

4.2 Equilibrium: existence and uniqueness

In equilibrium, the distribution of users among the WSPs has to be stable from a macro perspective, meaning that a new arrival user will randomly subscribe to a particular WSP_i $(i = 0,1,2,\cdots,I-1)$ and the users who already subscribed to a WSP have no incentive to switch to a different WSP. This indicates that the compensated utilities $\mathbb{U}_i$ $(i = 0,1,2,\cdots,I-1)$ should be non-negative and equal to each other for all WSP_i $(i = 0,1,2,\cdots,I-1)$. Without loss of generality, we assume that $c_0 < c_1 < c_2 < \cdots < c_{I-1}$. We then take WSP_0's optimal price $p_{0_{opt}}$ and cost c_0 as references such that the equilibrium price of WSP_i $(i = 1,2,\cdots,I-1)$ can be expressed as a function of $p_{0_{opt}}$ and c_0. The condition that all the compensated utilities $\mathbb{U}_i$ $(i = 0,1,2,\cdots,I-1)$ are equal in equilibrium can be expressed as:

$$\begin{aligned} &\mu - \lambda N E(p_{0_{opt}}) - p_{0_{opt}} = \mu - \lambda N E(p_{i_{opt}}) - p_{i_{opt}}, \quad (25\text{-}1) \\ \text{where} \quad &E_i(p_i) = 1 - \frac{5}{6}(\frac{p_i}{b_i})^2; \quad (25\text{-}2) \\ \text{and} \quad &p_{i_{opt}} = \frac{1}{3}(c_i + \sqrt{c_i^2 + \frac{18}{5}b_i^2}), \quad (25\text{-}3) \end{aligned} \tag{25}$$

for all $i = 0,1,2,\cdots,I-1$.
From Equation 25-3, we obtain:

$$6b_i^2 = 15p_{i_{opt}}^2 - 10p_{i_{opt}}c_i. \tag{26}$$

Substituting b_i into Equation 25-2, we get:

$$E_i(p_{i_{opt}}) = 1 - \frac{p_{i_{opt}}}{3p_{i_{opt}} - 2c_i}. \tag{27}$$

Substituting $E_i(p_{i_{opt}})$ into Equation 25-1 results in:

$$p_{i_{opt}} - p_{0_{opt}} = \lambda N(\frac{p_{i_{opt}}}{3p_{i_{opt}} - 2c_i} - \frac{p_{0_{opt}}}{3p_{0_{opt}} - 2c_0}). \tag{28}$$

Dividing $p_{0_{opt}}$ on both sides of Equation 28 and denoting $\frac{p_{i_{opt}}}{p_{0_{opt}}}$ by ξ_i, we have

$$\xi_i - 1 = \lambda N(\frac{\xi_i}{3p_{i_{opt}} - 2c_i} - \frac{1}{3p_{0_{opt}} - 2c_0}). \tag{29}$$

Substituting $\frac{6b_i^2}{5p_{i_{opt}}}$ for $3p_{i_{opt}} - 2c_i$, Equation 29 can be rewritten as:

$$\xi_i(\frac{5p_{i_{opt}}\lambda N}{6b_i^2} - 1) = \frac{5p_{0_{opt}}\lambda N}{6b_0^2} - 1. \tag{30}$$

Dividing $p_{0_{opt}}$ on both sides of Equation 30 again, we can obtian

$$\xi_i(\frac{5\xi_i\lambda N}{6b_i^2} - \frac{1}{p_{0_{opt}}}) = \frac{5\lambda N}{6b_0^2} - \frac{1}{p_{0_{opt}}}. \quad (31)$$

Here Equation 31 is a quadratic equation in ξ_i, which can be rewritten as:

$$a_1\xi_i^2 + a_2\xi_i + a_3 = 0. \quad (32)$$

where

$$a_1 = \frac{5\lambda N}{6b_i^2}, \quad a_2 = -\frac{1}{p_{0_{opt}}}, \quad a_3 = \frac{1}{p_{0_{opt}}} - \frac{5\lambda N}{6b_0^2}.$$

The solutions to Equation 32 are $\frac{-a_2 \pm \sqrt{a_2^2 - 4a_1a_3}}{2a_1}$, in which

$$a_2^2 - 4a_1a_3 = \frac{1}{p_{0_{opt}}^2} + \frac{25\lambda^2N^2}{9b_0^2b_i^2} - \frac{10\lambda N}{3p_{0_{opt}}b_i^2}.$$

It is straightforward to prove that $\frac{25\lambda^2N^2}{9b_0^2b_i^2} > \frac{10\lambda N}{3p_{0_{opt}}b_i^2}$ given that $5p_{0_{opt}}\lambda N > 6b_0^2$, which can be rewritten as $\lambda N > 3p_{0_{opt}} - 2c_0$. Later we will show that $p_{0_{opt}} \leq 2c_0$. Thus $\lambda N > 3p_{0_{opt}} - 2c_0 \geq 4c_0$. Note that λN represents the mean rate that all N users generate packets per second while c_0 is WSP$_0$'s per packet transmission. Therefore c_0 cannot be directly compared to λN. However, it is reasonable to assume that the cost per packet can be converted into a small enough unit such that $c_0 << \lambda N$, which means $5p_{0_{opt}}\lambda N >> 6b_0^2$ always holds.

Since $\frac{25\lambda^2N^2}{9b_0^2b_i^2} > \frac{10\lambda N}{3p_{0_{opt}}b_i^2}$, it can be proved that $a_2^2 - 4a_1a_3 > \frac{1}{p_{0_{opt}}^2} > 0$ and $\sqrt{a_2^2 - 4a_1a_3} > -a_2$, meaning that the solutions to Equation 32 are real numbers and only one of them is positive. Note that $p_{i_{opt}}$ for all $i = 0, 1, 2, \cdots, I-1$ are positive. Therefore Equation 32 has a unique solution, which is a positive real number ($\frac{-a_2 + \sqrt{a_2^2 - 4a_1a_3}}{2a_1}$). Now it can be concluded that for all $i = 1, 2, \cdots, I-1$, $p_{i_{opt}}$ is linear with $p_{0_{opt}}$ with a coefficient $\xi_i = \frac{-a_2 + \sqrt{a_2^2 - 4a_1a_3}}{2a_1}$. In other words, under the constraint $\sum_{i=0}^{I-1} E_i(p_{i_{opt}}) \leq 1$, for an arbitrary $p_{0_{opt}}$, there exists a unique $p_{i_{opt}}$ $(i = 1, 2, \cdots, I-1)$ in equilibrium with it.

Furthermore, by investigating the unique solution to Equation 32, we find that $\xi_i \approx \frac{\frac{1}{p_{0_{opt}}} + \sqrt{\frac{25\lambda^2N^2}{9b_0^2b_i^2}}}{\frac{5\lambda N}{3b_i^2}} > \frac{b_i}{b_0}$. Let $\frac{b_i}{b_0}$ be k_i. Thus $\xi_i = \gamma_i k_i$, where $\gamma_i \cong \frac{\frac{1}{p_{0_{opt}}}}{\frac{5\lambda N}{3b_i^2}} + 1$. It can be proved that γ_i approximates to 1 but is strictly greater than 1. According to Equation 26,

$$k_i^2 = \frac{b_i^2}{b_0^2} = \frac{3p_{i_{opt}}^2 - 2p_{i_{opt}}c_i}{3p_{0_{opt}}^2 - 2p_{0_{opt}}c_i} = \frac{3\gamma_i^2k_i^2p_{0_{opt}}^2 - 2\gamma_ik_ip_{0_{opt}}c_i}{3p_{0_{opt}}^2 - 2p_{0_{opt}}c_i}. \quad (33)$$

Solving Equation 33, k_i is given by

$$k_i = \frac{2\gamma_i c_i}{3\gamma_i^2 p_{0_{opt}} - 3p_{0_{opt}} + 2c_0}. \quad (34)$$

Since $\gamma_i \cong 1$, $k_i \cong \gamma_i \frac{c_i}{c_0}$.

In summary, $\frac{p_{i_{opt}}}{p_{0_{opt}}} \cong \gamma_i \frac{b_i}{b_0} \cong \gamma_i^2 \frac{c_i}{c_0}$, where $\gamma_i \cong 1$, and under the constraint $\sum_{i=0}^{I-1} E_i(p_{i_{opt}}) \leq 1$, for an arbitrary $p_{0_{opt}}$, the higher the cost c_i the greater γ_i, meaning for WSP$_i$ the higher the cost c_i the higher the corresponding equilibrium price, $p_{i_{opt}}$.
The constraint $\sum_{i=0}^{I-1} E_i(p_{i_{opt}}) \leq 1$ can be expressed as:

$$\begin{aligned} & I - \frac{5}{6}\left(\frac{p_{0_{opt}}^2}{b_0^2} + \frac{p_{1_{opt}}^2}{b_1^2} + \cdots + \frac{p_{I-1_{opt}}^2}{b_{I-1}^2}\right) \\ =& I - \frac{5}{6}\frac{p_{0_{opt}}^2}{b_0^2}(1 + \gamma_1^2 + \cdots + \cdots + \gamma_{I-1}^2) \\ \approx& I - I\frac{5}{6}\frac{p_{0_{opt}}^2}{b_0^2} \\ =& IE_0(p_{0_{opt}}) \leq 1. \end{aligned} \tag{35}$$

Here Equation 35 indicates that the maximum value of $E_0(p_{0_{opt}})$, which is denoted by $E_{0_{max}}$, approximates to $\frac{1}{I}$. The corresponding optimal price is denoted as the maximum optimal price $p_{0_{max}}$. Then, it is straightforward to prove that the greater the number of providers I the lower the maximum optimal price $p_{0_{max}}$ that WSP$_0$ could reach. This in turn means that, in equilibrium, all other WSPs' maximum optimal prices, $p_{i_{max}}$ for all $i = 1, 2, \cdots, I-1$, are lower accordingly. Thus, when more WSPs enter into the market, in equilibrium, the maximum prices they can charge decreases.

4.3 Numerical examples

In order to verify the analytical results obtained in the previous section, we performed the following simulations where several WSPs provide wireless services in a market with $N = 40$ users. Each user generates packets according to a Poisson process with rate $\lambda = 10$packet/sec, and the service time of individual packet are i.i.d. exponentially distributed with mean 300^{-1}sec.
To study the impact of an entry of a new WSP on the existing WSPs, we first conducted a simulation where there are two WSPs: WSP$_0$ and WSP$_1$ competing in the market. The costs of WSP$_0$ and WSP$_1$ per packet are $c_0 = 5$ and $c_1 = 7$ units respectively. For convenience, in the rest of section, $p_{i_{opt}}$ and p_i are interchangeable. Let $\{p_0^1, p_0^2, \cdots, p_0^n\}$ and $\{p_1^1, p_1^2, \cdots, p_1^n\}$ be optimal price strategy forms of WSP$_0$ and WSP$_1$ respectively. Using the same approach that we used to identify the equilibrium in Section 3, we find the set of the equilibrium prices $\{(p_0^{1*}, p_1^{1*}), (p_0^{2*}, p_1^{2*}), \ldots, (p_0^{n*}, p_1^{n*})\}$, which is represented as (p_0^*, p_1^*) in the rest of the section. Fig. 8(a) plots (p_0^*, p_1^*). As can be seen, in equilibrium, for an arbitrary p_0^*, there is a unique p_1^* corresponding to it. In particular, p_0^* is quasi-linear with p_1^*. Under the same assumptions, $p_{0_{max}}^*$ and $p_{1_{max}}^*$ are 9.997 and 13.526 respectively.
We then simulated another scenario where the third WSP, WSP$_2$, joins WSP$_0$ and WSP$_1$. WSP$_2$'s cost per packet $c_2 = 9$ units and its optimal price strategy form is $\{p_2^1, p_2^2, \cdots, p_2^n\}$. Using the same approach, we obtained the equilibrium prices $(p_0^{'*}, p_1^{'*}, p_2^{'*})$, which are plotted in Fig. 8(b) to compare with those obtained from the two-WSP scenario. As can be observed, in equilibrium, for an arbitrary $p_0^{'*}$, there are unique $p_1^{'*}$ and $p_2^{'*}$ corresponding to it respectively. Again, $p_0^{'*}$ is linear with $p_1^{'*}$ and $p_2^{'*}$ respectively. In addition, the range of values for the equilibrium prices (p_0^*, p_1^*) in two-WSP case shown in Fig. 8(a) is larger than that in three-WSP

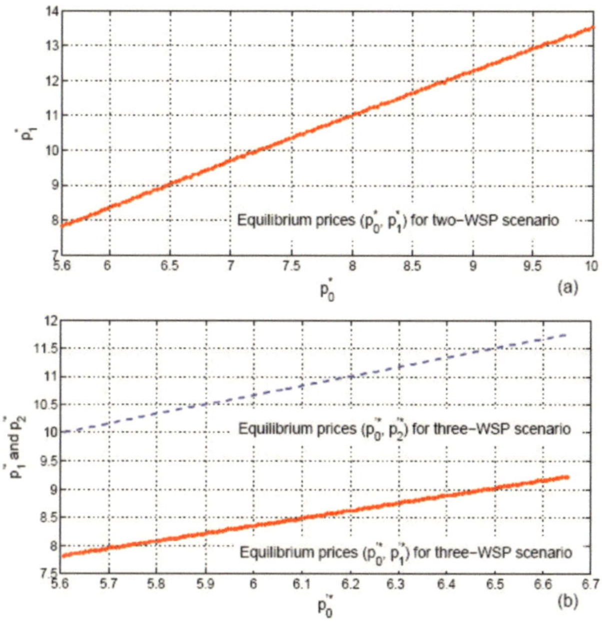

Fig. 8. Equilibrium prices: two-WSP scenario vs three-WSP scenario

case shown in Fig. 8(b). This is because of the equilibrium condition $\sum_{i=0}^{I-1} E_i(p_{i_{opt}}) \leq 1$ $\forall i = 0,1,2,\cdots,I-1$, which restricts the allowed values for p_i^* as analyzed in Equation 35. When the number of WSPs increases to 3 from 2, under the same assumptions, the maximum equilibrium prices $p'^*_{0_{max}}$ and $p'^*_{1_{max}}$ are 6.623 and 9.186 respectively. Evidently, $p'^*_{0_{max}}$ and $p'^*_{1_{max}}$ are significantly lower than those obtained in two-WSP scenario.

Fig. 9 illustrates the expected profits of all WSPs associated with the equilibrium prices plotted in Fig. 8. Similarly, the range of the values for the expected profit in two-WSP case is larger than that of three-WSP case. As we can see, in three-WSP scenario, the expected profits of WSP_0 and WSP_1 are lower than those of two-WSP scenario, because WSP_2 takes some market share.

When the number of potential users, N, in this market changes, the expected compensated utility $\mathbb{U}_i$ changes, which in turn changes the current Nash equilibrium. This leads to another round of price adjustment among the WSPs and the users. Table 3 lists some equilibrium prices, the corresponding expected compensated utilities and the corresponding expected acceptance rate with various number of potential users, N. As shown in Table 3, if we suppose in the first round $N = 30$ and the three WSPs end up with a Nash equilibrium with $p_0 = 6.2970$, $p_1 = 8.7274$ and $p_2 = 11.114$. When the number N changes, even if WSP_0 keeps its price $p_0 = 6.2970$ unchanged, WSP_1 and WSP_2 have to change their price such that the three WSPs and the users can reach a new equilibrium.

It has been shown that, with the proposed pricing scheme, when a new WSP enters into a market with two or more WSPs already existing, the maximum equilibrium prices that the existing WSPs can reach will decrease. For a WSP, besides its traffic load status, cost is another factor determining its optimal prices, which in turn affects the WSPs' equilibrium prices. Thus, a follow-up question would be how the cost of the new WSP affects the equilibrium prices of the existing WSPs. Then another three-WSP scenario simulation, in which costs

	p_0	5.7624	5.7769	5.9918	6.1528	6.2970*	6.3827	6.4864	6.5726	6.9011
	$\mathbb{U}_0$	273.31	273.03	269.09	266.59	264.53	263.39	262.09	261.06	257.58
	E_0	0.2092	0.2120	0.2487	0.2726	0.2918	0.3023	0.3143	0.3236	0.3552
	p_1	7.9153	7.9325	8.2041	8.4020	8.5776*	8.6833	8.807	8.9133	9.3052
$N = 10$	$\mathbb{U}_1$	273.30	273.03	269.10	266.58	266.53	263.38	262.10	261.05	257.56
	E_1	0.1878	0.1904	0.2262	0.2502	0.2689	0.2794	0.2912	0.3004	0.3313
	p_2	10.003	10.023	10.336	10.565	10.766*	10.885	11.026	11.146	11.583
	$\mathbb{U}_2$	273.29	273.03	269.09	266.58	264.54	263.39	262.10	261.06	259.57
	E_2	0.1670	0.1695	0.2051	0.2285	0.2470	0.2573	0.2690	0.2780	0.3084
	p_0	5.6587	5.7769	5.9918	6.1528	6.2970*	6.3827	6.4864	6.5726	6.7395
	$\mathbb{U}_0$	256.57	251.83	244.26	239.33	235.35	233.16	230.66	228.70	225.17
	E_0	0.1888	0.2120	0.2487	0.2607	0.2918	0.3023	0.3143	0.3240	0.3405
	p_1	7.8493	8.0045	8.2883	8.4984	8.6862*	8.7982	8.9340	9.0436	9.2575
$N = 20$	$\mathbb{U}_1$	256.57	251.87	244.28	239.36	235.38	233.17	230.64	228.70	225.18
	E_1	0.1779	0.2010	0.2372	0.2726	0.2797	0.2902	0.3021	0.3113	0.3278
	p_2	10.003	10.195	10.539	10.792	11.017*	11.152	11.314	11.444	11.699
	$\mathbb{U}_2$	256.59	251.83	244.26	239.38	235.15	233.15	230.63	228.70	225.16
	E_2	0.1670	0.1899	0.2260	0.2493	0.2680	0.2785	0.2903	0.2993	0.3157
	p_0	5.6243	5.7769	5.9918	6.1528	6.2970*	6.3827	6.4864	6.5726	6.6828
	$\mathbb{U}_0$	239.88	230.64	219.39	212.07	206.17	202.93	199.23	196.33	192.84
	E_1	0.1817	0.2120	0.2487	0.2726	0.2918	0.3023	0.3143	0.3236	0.3349
	p_1	7.8264	8.0305	8.3203	8.5365	8.7274*	8.8424	8.9784	9.0940	9.2397
$N = 30$	$\mathbb{U}_1$	239.87	230.70	219.40	212.05	206.19	202.91	199.25	196.31	192.81
	E_1	0.1744	0.2042	0.2409	0.2647	0.2836	0.2941	0.3059	0.3153	0.3265
	p_2	10.003	10.258	10.614	10.879	11.114*	11.255	11.420	11.562	11.737
	$\mathbb{U}_2$	239.89	230.65	219.42	212.09	206.21	202.92	199.27	196.31	192.84
	E_2	0.1670	0.1970	0.2332	0.2568	0.2756	0.2861	0.2977	0.3071	0.3181
	p_0	5.6071	5.7769	5.9918	6.1528	6.2970*	6.3827	6.4864	6.5726	6.6559
	$\mathbb{U}_0$	223.20	209.44	194.52	184.81	177	172.70	167.80	163.97	160.44
	E_0	0.1780	0.2120	0.2487	0.2726	0.2918	0.3023	0.3143	0.3233	0.3323
	p_1	7.8149	8.0449	8.3378	8.5541	8.7481*	8.8660	9.0021	9.1208	9.2307
$N = 40$	$\mathbb{U}_1$	223.18	209.48	194.49	184.84	177.04	172.64	167.85	164.01	160.43
	E_1	0.1726	0.2062	0.2429	0.2665	0.2855	0.2962	0.3079	0.3172	0.3258
	p_2	10.003	10.29	10.655	10.926	11.167*	11.311	11.479	11.622	11.761
	$\mathbb{U}_2$	223.19	209.53	194.54	184.81	177	172.65	167.85	164.02	160.44
	E_2	0.1670	0.2005	0.2370	0.2607	0.2796	0.2901	0.3017	0.3109	0.3195
	p_0	5.5956	5.7769	5.9918	6.1528	6.2970*	6.3827	6.4864	6.5726	6.6381
	$\mathbb{U}_0$	206.65	188.25	169.65	157.56	147.82	142.47	136.37	131.60	128.14
	E_0	0.1775	0.2141	0.2510	0.2750	0.2944	0.3048	0.3166	0.3262	0.3304
	p_1	7.8064	8.0507	8.3466	8.56	8.7657*	8.8779	9.0199	9.1356	9.2218
$N = 50$	$\mathbb{U}_1$	206.58	188.46	169.68	157.57	147.79	142.48	136.31	131.59	128.19
	E_1	0.1712	0.2082	0.2453	0.2689	0.2880	0.2986	0.3103	0.3197	0.3252
	p_2	10	10.31	10.681	10.955	11.199*	11.343	11.518	11.663	11.773
	$\mathbb{U}_2$	206.67	188.37	169.62	157.51	147.81	142.47	136.38	131.57	128.12
	E_2	0.1667	0.2026	0.2394	0.2633	0.2820	0.2924	0.3042	0.3135	0.3202

Table 4. Equilibrium prices, expected compensated utilities and expected acceptance rates for different number of users N

		$p'^{*}_{0_{max}}$	$p'^{*}_{1_{max}}$
$c0 = 5$ $c1 = 7$	$c2 = 1$	6.5428	9.0792
	$c2 = 3$	6.6559	9.2307
	$c2 = 9$	6.6232	9.1861
	$c2 = 13$	6.5964	9.1534
	$c2 = 20$	6.5815	9.1297

Table 5. Maximum equilibrium prices of WSP_0 and WSP_1 $(p_0'^*, p_1'^*)$ for three-WSP scenario with various cost of WSP_2.

of WSP_0 and WSP_1 are kept unchanged while WSP_2's cost varies, was conducted. Fig. 10 presents the equilibrium prices $(p_0'^*, p_1'^*)$ with WSP_2 taking different costs. For ease of illustration, the maximum equilibrium prices corresponding to Fig. 10 (a), (b) and (c) are listed in Table 5. It can be observed that when WSP_2's cost is lower than costs of WSP_0 and WSP_1, in equilibrium, the maximum prices that WSP_0 and WSP_1 could reach are lower compared to the case where WSP_2'cost is higher than the costs of WSP_0 and WSP_1. For the latter case, the entry of a new WSP with a higher cost results in slight lower maximum equilibrium prices for both WSP_0 and WSP_1.

5. Advanced thoughts

The presented material can be usefully extended in a number of ways. In this section only the extensions will be identified and potential game theory modeling modes indicated.
Firstly an assumption in the section on the basic wireless duopoly was that all users show the same basic behavior. Relaxing this, a competition between the WSPs and a set of N types of users can be described. The N types of users could describe economic, social or regulatory groupings that have differing QoS and price utility definitions. This would allow for the development of a scaled preference analysis that could be used to gauge more accurately

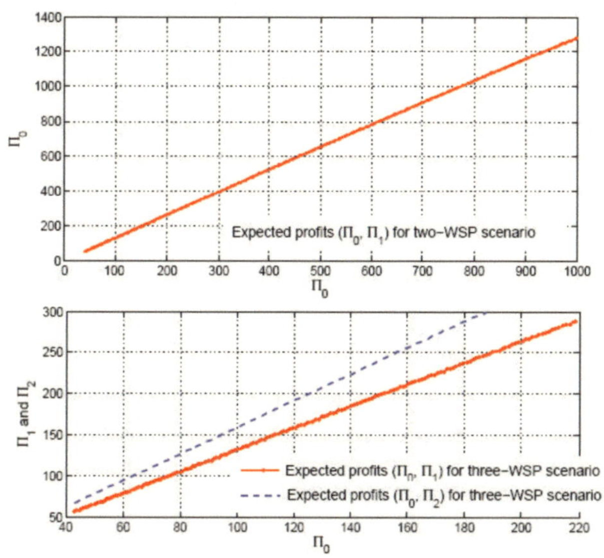

Fig. 9. Expected profits associated with equilibrium prices in Fig. 8

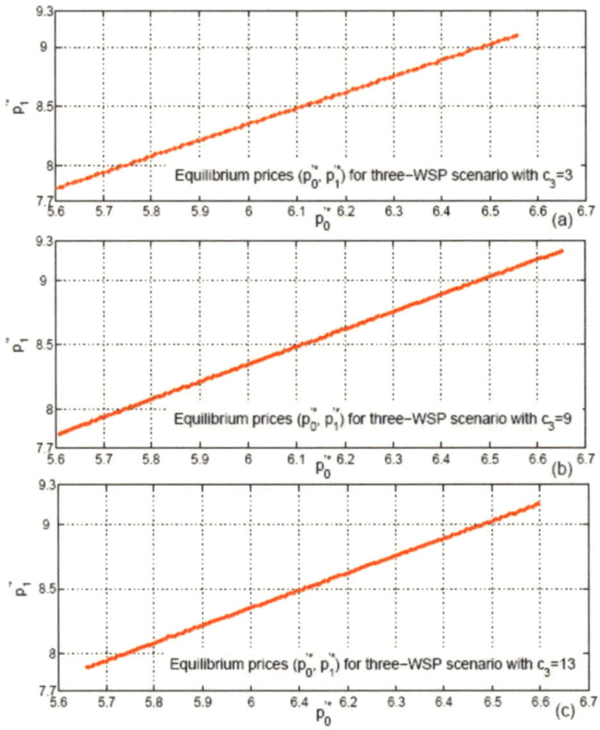

Fig. 10. Equilibrium prices of WSP_0 and WSP_1 $(p_0'^*, p_1'^*)$ for three-WSP scenario with various cost of WSP_2.

the social benefits of regulated access to the wireless bandwidth. Secondly the analysis of oligopoly based pricing depends on the assumption of a mature market where entrances and exits by WSPs are not relevant. In actual fact this is quite unrealistic and can profitably be expanded to take into account changes in the number of WSPs during a period and the impact on both relative profit and market share. Finally by changing the analysis basis queue to a more dynamic queue with memory effect models of user churn and brand loyalty can be developed that could show the benefit of branding compaigns.

6. References

Altman, E., Barman, D. & Azouzi, R. E. (2006). Pricing Differentiated Services: A Game-Theoretic Approach, *Computer Networks* 50: 982–1002.

Altman, E. & Basar, T. (1998). Multiuser Rate-Based Flow Control, *IEEE Transactions on Communication* 46: 940–949.

Altman, E. & Wynter, L. (2002). Equilibrium, Games and Pricing in Transportation and Telecommunication Networks, *Technical Report 4632, IRISA* .

Armony, M. & Haviv, M. (2003). Price and Delay competition Between Two Service Providers, *European Journal of Operational Research* 147: 32–50.

Basar, T. & Srikant, R. (2002). Revenue-maximizing Pricing and Capacity Expansion in a Many-user Regime, *Proceedings of the 21st Annual Joint Conference of the IEEE Computer and Communications Societies (IEEE INFOCOM 2002)*, New York, USA.

Cao, X., Shen, H., Milito, R. & Wirth, P. (2002). Internet Pricing with a Game Theoretical Approach: Concepts and Examples, *IEEE/ACM Transactions on Networking* 10: 208–216.

Das, S. K., Chatterjee, M. & Lin, H. (2004). An Econometric Model for Resource Management in Competitive Wireless Data Networks, *IEEE Network Magazine* 18(6): 20–26.

Dziong, Z. & Mason, L. (1996). Fair-efficient Call Admission Control Policies for Broadband Network - A Game Theoretical Approach, *IEEE/ACM Transactions on Networking* 4: 123–136.

Gibbens, R., Mason, R. & Steinberg, R. (2000). Internet Service Classes under Competition, *IEEE Journal on Selected Areas in Communications* 18(7): 2490–2498.

Hock, N. C. (1996). *Queuing Modeling Fundamentals*, John Wiley and Sons Ltd.

Jagannatha, S., Nayak, J., Almeroth, K. & Hofmann, M. (2002). A Model for Discovering Customer Value for E-content, *Proceedings of the Eighth ACM SIGKDD International Conference on Knowledge Discovery and Data Mining*, Edmonton, Alberta, Canada, pp. 23–26.

Kao, Y. & Huan, J. (2008). Price-based Resource Allocation for Wireless Ad Hoc Networks with Multi-rate Capability and Energy Constraints , *Computer Communications* 31: 3613–3624.

Kelly, F. P. (2000). Models for A Self-Managed Internet, *Philosophical Transactions of the Royal Society* A358: 2335–2348.

Kelly, F. P., Mauloo, A. K. & Tan, D. K. H. (1998). Rate Control in Communication Networks: Shadow Prices, Proportional Fairness and Stability, *Journal of the Operational Research Society* 49: 237–252.

Khan, S. Q. (2005). Optimizing Providers' Profit in Per Networks Applying Automatic Pricing and Game Theory, *PhD thesis, The University of Kansas* .

La, R. J. & Anantharam, V. (1999). Network Pricing using Game Theoretic Approach, *Proceedings of 38th IEEE Conference on Decision and Control*, Vol. 4, Phoenix, AZ Piscataway, NJ, pp. 4008–4013.

La, R. J. & Anantharam, V. (2002). Utility-based Rate Control in the Internet for Elastic Traffic, *IEEE/ACM Transactions on Networking* 10(2): 272–286.

Lam, R. K., Chiu, D. & Lui, J. C. S. (2007). On the Access Pricing and Network Scaling Issues of Wireless Mesh Networks, *IEEE transactions on Computers* 56: 1456–1469.

Lam, R. K., Lui, J. C. S. & Chiu, D. (2006). On the Access Pricing Issues of Wireless Mesh Networks, *Proceedings of the 26th IEEE International Conference on Distributed Computing Systems (ICDCS 2006)*, Lisboa, Portugal.

Low, S. H. & Lapsley, D. E. (1999). Optimization Flow Control: Basic Algorithm and Convergence, *IEEE/ACM Transactions on Networking* 7(6): 861–874.

Lüthi, M., Nadjm-Tehrani, S. & Curescu, C. (2006). Comparative Study of Price-based Resource Allocation Algorithms for Ad Hoc Networks, *Proceedings of the 20th International Parallel and Distributed Processing Symposium (IPDPS 2006)*, Rhodes Island, Greece.

M. Bouhtou, M. D. & Wynter, L. (2003). Capacitated Network Revenue Management through Shadow Pricing, *Networked Group Communication* 2816: 342–351.

Mandjes, M. (2003). Pricing Strategies under Heterogeneous Service Requirements, *Computer Networks* 42(2): 231–249.

Musacchio, J. & Walrand, J. (2006). WiFi Access Point Pricing as a Dynamic Game, *IEEE/ACM Transactions on Networking* 14: 289–301.

Parsons, S., Gmytrasiewicz, P. J. & Wooldridge, M. J. (2002). *Game Theory and Decision Theory in Agent-Based Systems*, Kluwer Academic Publishers.

Qiu, Y. & Marbach, P. (2003). Bandwith Allocation in Ad-Hoc Networks: A Price-Based Approach, *Proceedings of the 22nd Annual Joint Conference of the IEEE Computer and Communications Societies (IEEE INFOCOM 2003)*, Vol. 2, San Francisco, California, USA, pp. 797–807.

Ros, D. & Tuffin, B. (2004). A Mathematical Model of the Paris Metro Pricing Scheme for Charging Packet Network, *Computer Networks* 46: 73–85.

Sakurai, H., Kasahara, S. & Adachi, N. (2003). Internet Pricing and User Opt-out Strategy under two ISPs Competition, *IEICE Technical Report (Institute of Electronics, Information and Communication Engineers)* 102(694(IN2002 233-305)): 153–156.

Shu, J. & Varaiya, P. (2003). Pricing Network Services, *Proceedings of the 22nd Annual Joint Conference of the IEEE Computer and Communications Societies (IEEE INFOCOM 2003)*, San Franciso, CA, USA.

Tassiulas, L., Kar, K. & Sarkar, S. (2001). A Simple Rate Control Algorithm for Maximizing Total User Utility, *Proceedings of the 20th Annual Joint Conference of the IEEE Computer and Communications Societies (IEEE INFOCOM 2001)*, Vol. 1, Anchorage, Alaska, USA, pp. 133–141.

Wang, X. & Schulzrinne, H. (1999). RNAP: A Resource Negotiation and Pricing Protocol, *Proceedings of International Workshop on Network and Operating System Support for Digital Audio and Video (NOSSDAV'99)*, Basking Ridge, New Jersey, pp. 77–93.

Xue, Y., Li, B. & Nahrstedt, K. (2003). Price-based Resource Allocation in Wireless Ad Hoc Networks, *Proceedings of the 11th International Conference on Quality of Service (IWQoS 2003)*, Berkeley, CA, USA.

Xue, Y., Li, B. & Nahrstedt, K. (2006). Optimal Resource Allocation in Wireless Ad Hoc Networks: A Price-Based Approach, *IEEE transactions on mobile computing* 5: 347–364.

Yaïche, H., Mazumdar, R. R. & Rosenberg, C. (2000). A Game Theoretic Framework for Bandwidth Allocation and Pricing in Broadband Networks, *IEEE/ACM Transactions on Networking* 8: 667–678.

Zemlianov, A. & de Veciana, G. (2005). Cooperation and Decision-Making in a Wireless Multi-Provider Setting, *Proceedings of 24th Annual Joint Conference of the IEEE Computer and Communications Societies (IEEE INFOCOM 2005)*, Miami, FL, USA.

Zhang, Z., Dey, D. & Tan, Y. (2008). Price and QoS Competition in Data Communication Services, *European Journal of Operational Research* 187: 871–886.

Zhu, H., Nel, A. & Clarke, W. (2009). A Duopoly Pricing Model for Wireless Mesh Networks under Congestion-sensitive Users, *South African Institute of Electrical Engineers Africa Research Journal* 100: 48–58.